Sound Capture and Processing

Sound Capture and Processing
Practical Approaches

Ivan J. Tashev

Microsoft Research, USA

A John Wiley and Sons, Ltd., Publication

This edition first published 2009
© 2009 John Wiley & Sons Ltd.,

Registered office
John Wiley & Sons Ltd, The Atrium, Southern Gate, Chichester, West Sussex, PO19 8SQ, United Kingdom

For details of our global editorial offices, for customer services and for information about how to apply for permission to reuse the copyright material in this book please see our website at www.wiley.com.

Library of Congress Cataloging-in-Publication Data

Tashev, Ivan J. (Ivan Jelev)
  Sound capture and processing : practical approaches / Ivan J. Tashev.
      p. cm.
  Includes index.
  ISBN 978-0-470-31983-3 (cloth)
 1. Speech processing systems. 2. Sound–Recording and reproducing–Digital
techniques. 3. Signal processing–Digital techniques. I. Title.
  TK7882.S65T37 2009
  621.382'8–dc22
                                                        2009011987

A catalogue record for this book is available from the British Library.

ISBN 9780470319833 (H/B)

Typeset in 11/13pt Times by Thomson Digital, Noida, India.

To my family: the time to write this book was taken from them

Contents

About the Author

Dr Ivan Tashev took both his Engineering Diploma in Electronics and PhD in Computer Science degrees at the Technical University of Sofia, Bulgaria, in 1984 and 1990 respectively. After his graduation he worked as R&D engineer and researcher in the R&D Department of the same university. Dr Tashev became assistant professor in 1989. He created and taught two courses, "Data and signal processing" and "Programming of real-time systems" to the students of fourth and fifth year in the Department of Electronics.

Dr Tashev joined Microsoft in 1998 and held positions in various product teams until 2001 when he moved to Microsoft Research. Here he was involved in projects such as RingCam (now a Microsoft product – Round Table Device), microphone array (currently part of Windows Vista), and many others related to sound capturing devices and audio signal processing. Currently he is a member of the Speech Technology Group in Microsoft Research lab at the Microsoft headquarters in Redmond, Washington.

Dr Ivan Tashev is senior member of IEEE and IEEE Signal Processing Society, member of Audio Engineering Society and its Pacific Northwest Committee. He is reviewer for most of the audio and signal processing journals and conferences. Dr Tashev has published three books, more than fifty scientific papers and is listed as inventor of five granted U.S. patents and seventeen U.S. patent applications.

The research interests of Dr Tashev include sound capturing devices, signal processing for arrays of transducers, speech enhancement algorithms, and signal processing of audio, speech and biological signals.

Foreword

Just a couple of decades ago we would think of "sound capture and processing" as the problems of designing microphones for converting sounds from the real world into electrical signals, as well as amplifying, editing, recording, and transmitting such signals, mostly using analog hardware technologies. That's because our intended applications were mostly analog telephony, broadcasting, and voice and music recording. We have come a long way: small digital audio players have replaced bulky portable cassette tape players, and people make voice calls mostly via digital mobile phones and voice communication software in their computers. Thanks to the evolution of digital signal processing technologies, we now focus mostly on processing sounds not as analog electrical signals, but rather as digital files or data streams in a computer or digital device. We can do a lot more with digital sound processing, such as transcribe speech into text, identify persons speaking, recognize music from humming, remove noises much more efficiently, add special effects, and so much more. Thus, today we think of sound capture as the problem of digitally processing the signals captured by microphones so as to improve their quality for best performance in digital communications, broadcasting, recording, recognition, classification, and other applications.

This book by Ivan Tashev provides a comprehensive yet concise overview of the fundamental problems and core signal processing algorithms for digital sound capture, including ambient noise reduction, acoustic echo cancellation, and reduction of reverberation. After introducing the necessary basic aspects of digital audio signal processing, the book presents basic physical properties of sound and propagation of sound waves, as well as a review of microphone technologies, providing the reader with a strong understanding of key aspects of digitized sounds. The book discusses the fundamental problems of noise reduction, which are usually solved via techniques based on statistical models of the signals of interest (typically voice) and of interfering signals. An important discussion of properties of the human auditory system is also presented; auditory models can play a very important role in algorithms for enhancing audio signals in communication and recording/playback applications, where the final destination is the human ear.

Microphone arrays have become increasingly important in the past decade or so. Thanks to the rapid evolution and reduction in cost of analog and digital electronics in recent years, it is inexpensive to capture sound through several channels, using an array of microphones. That opens new opportunities for improving sound capture, such as detecting the direction of incoming sounds and applying spatial filtering techniques. The book includes two excellent chapters whose coverage goes from the basics of microphone array configurations and delay-and-sum beamforming, to modern sophisticated algorithms for high-performance multichannel signal enhancement.

Acoustic echoes and reverberation are the two most important kinds of signal degradations in many sound capture scenarios. If you're a professional singer, you probably don't mind holding a microphone or wearing a headset with a microphone close to your mouth, but most of us prefer microphones to be invisible, far away from our mouths. That means microphone will capture not only our own voices, but also reverberation components because of sound reflections from nearby walls, as well as echoes of signals that are being played back from loudspeakers. Removing such undesirable artifacts presents significant technical challenges, which are well addressed in the final two chapters, which present modern algorithms for tackling them.

A key quality of this book is that it presents not only fundamental theoretical analyses, models, and algorithms, but it also considers many practical aspects that are very important for the design of real-world engineering solutions to sound capture problems. Thus, this book should be of great appeal to both students and engineers.

I have had the pleasure of working with Ivan on research and development of sound capture systems and algorithms. His enthusiasm, deep engineering and mathematical knowledge, and pragmatic approaches were all contagious. His work has had significant practical impact, for example the introduction of multichannel sound capture and processing modules in the Microsoft Windows operating system. I have learned a considerable amount about sound capturing and processing from my interactions with Ivan, and I am sure you will, as well, by reading this book. Enjoy!

Henrique Malvar
Managing Director
Microsoft Research
Redmond Laboratory

Preface

Capturing and processing sounds is critical in mobile and handheld devices, communication systems, and computers using automatic speech recognition. Devices and technologies for proper conversion of sounds to electric signals and removing unwanted parts, such as noise and reverberation, have been used since the first telephones. They evolved, becoming more and more complex. In many cases the existing algorithms exceed the abilities of typical processors in these devices and computers to provide real-time processing of the captured signal.

This book will discuss the basic principles for building an audio processing stack, sound capturing devices, single-channel speech-enhancement algorithms, and microphone arrays for sound capture and sound source localization. Further, algorithms will be described for acoustic echo cancellation and de-reverberation – building blocks of a sound capture and processing stack for telecommunication and speech recognition. Wherever possible the various algorithms are discussed in the order of their development and publication. In all cases the aim is to try to give the larger picture – where the technology came from, what worked and what had to be adapted for the needs of audio processing. This gives a better perspective for further development of new audio signal processing algorithms.

Even the best equations and signal processing algorithms are not worth anything before being implemented and verified by processing of real data. That is why, in this book, stress is placed on experimenting with recorded sounds and implementation of the algorithms. In practice, frequently a simpler model with fewer parameters to estimate works better than a more precise but more complex model with a larger number of parameters. With the latter one has either to sacrifice estimation precision or to increase the estimation time. This balance of simplicity, precision, and reaction time is critical for real-time systems, where on top of everything we have to watch out for parameters such as latency, consumed memory, and CPU time.

Most of the algorithms and approaches described in this book are based on statistical models. In mathematics, a single example cannot prove but can disprove a theorem. In statistical signal processing, a single example is . . . just a sample. What matters is careful evaluation of the algorithms with a good corpus of speech or audio signals, distributed in their signal-to-noise ratios, type of noise, and other parameters – as close as possible to the real problem we are trying to solve.

The solution of practically any signal processing problem can be improved by tuning the parameters of the algorithm, provided we have a proper criterion for optimality. There are always adaptation time constants, thresholds, which cannot be estimated and their values have to be adjusted experimentally. The mathematical models and solutions we use are usually optimal in one or another way. If they reflect properly the nature of the process they model, then we have a good solution and the results are satisfactory. In all cases it is important to remember that we do not want a "minimum mean-square error solution," or a "maximum-likelihood solution," or even a "log minimum mean-square error solution." We do not want to improve the signal-to-noise ratio. What we want is for listeners to perceive the sound quality of the processed signal as better – improved – compared to the input signal. From this perspective, the final judge of how good is an algorithm is the human ear, so use it to verify the solution. Hearing is an important sense for humans and animals. In many places in this book are provided examples of how humans and animals hear and localize sounds – this explains better some signal processing approaches and brings biology-inspired designs for sound capture and processing systems.

In many cases the signal processing chain consists of several algorithms for sound capture and speech enhancement. The practice shows us that a sequence of separately optimized algorithms usually provides suboptimal results. Tuning and optimization of the designed sound capturing system end-to-end is a must if we want to achieve best results.

For further information please visit http://www.wiley.com/go/tashev_sound

Ivan Tashev
Redmond, WA
USA

Acknowledgements

I want to thank the Book Program in MathWorks and especially Dee Savageau, Naomi Fernandes, and Meg Vulliez for the help and responsiveness. The MATLAB® scripts, part of this book, were tested with MATLAB® R2007a, provided as part of this program.

I am grateful to my colleagues from Microsoft Research Alex Acero, Amitav Das, Li Deng, Dinei Florencio, Cormac Herley, Zicheng Liu, Mike Seltzer, and Cha Zhang. They read the chapters of this book and provided valuable feedback.

And last, but not least, I want to express my great pleasure working with the nice and helpful people from John Wiley & Sons, Ltd. During the long process from proposal, through writing, copyediting, and finalizing the book with all the details, they were always professional, understanding, and ready to suggest the right solution. I was lucky enough to work with Tiina Ruonamaa, Sarah Hinton, Sarah Tilley, and Catlin Flint – thank you all for everything you did during the process of writing this book!

1

Introduction

1.1 The Need for, and Consumers of, Sound Capture and Audio Processing Algorithms

The need for capturing sound and converting it to electric signals came with the first telephones. This is why the first microphones were designed. For a long time tele-communications was the only user of captured sound. The radio broadcasting and music recording industries increased the demand for high-quality microphones, good amplifiers, and systems for sound reproduction.

Sound capture and audio processing in general stayed in the analog signal processing domain until after World War II. At that time the first programmable digital computers were designed and researchers started to work on digital signal processing algorithms. Initially communications were the major consumer of signal processing algorithms, such as echo cancellation and digital speech compression. In the meantime, digital computers become more powerful, with more memory and faster processors. They invaded offices and homes and far exceeded their initial role as a tool for increased productivity for information workers. Modern computers are communication and entertainment centers, many of them having attached or integrated loudspeakers, microphone, and web camera. They are used for storing and playing music and videos. Programs for audio and video chat are widely used. Sound capture and audio processing algorithms today are an integral part of every personal computer. Mobile phones for the first time took the phone out of quiet rooms and exposed the microphones to substantially higher noise levels. This increased the demand for real-time implementations of noise suppression and speech enhancement algorithms running on inexpensive processors.

Automatic speech recognition had an initial task of speech dictation as the primary scenario in an office or home environment. The microphone was placed in its best

Sound Capture and Processing   Ivan J. Tashev
© 2009 John Wiley & Sons, Ltd

position, close to the mouth, and provided good-quality sound. With the advancement of the underlying technology, speech recognition gradually became an integral part of the human–machine interface. Speech-enabled dialog systems are deployed in mobile phones and cars and they even greet us when we dial the phone lines of many companies. They are used in a wide range of tasks: from song selection to booking hotels and plane tickets. Speech recognition emerges as the other large consumer of sound capture and audio processing algorithms.

Humans do not like to wear headsets and close-talk microphones, which drives the demand for hands-free sound capture and processing algorithms. Acoustic echo cancellation and microphone array processing are engaged to provide comfortable communication. Some emerging scenarios are speech-enabled dialog systems and voice-controlled devices – from mobile phones to the multimedia equipment in the family room.

Increasing the speed and the memory of computing devices and reducing the power consumption leads to the creation of personal devices with a small size and rich functionality. Features include voice and text communications, media player, navigation, and information access via the Internet. Owing to their small size and light weight, people carry these devices with them everywhere. Small size means a small number of keys or any other means for a rich user interface. This increases the role of speech recognition in interaction with these devices because it can provide a more convenient and natural interface. Increasing the bandwidth will allow most of the phones to have video communication features. This is the moment when the microphone is removed from its best position close to the mouth and placed an arm's length away. Such devices will be used in practically every place, which means an increased demand for better microphones, devices, and algorithms for sound capture and processing for the needs of real-time communications and speech recognition.

1.2 Typical Sound Capture System

Capturing sounds starts with converting the acoustic wave in the air to an electrical signal by one or more microphones. These microphones can have very different characteristics and – if properly designed, selected, and positioned – can provide a substantially better sound for the next processing steps. The microphone signals are amplified, filtered, and pre-processed. An analog-to-digital convertor performs discretization and quantization and converts the sound into a stream of numbers. This is where the digital signal processing starts.

One of the first stages of the sound capturing system is the *acoustic echo reduction system*. It removes from the captured signal the sound coming from the loudspeakers – the voice from the other side in a telecommunication session, or the sound track from a CD or DVD. What is left at the end of this type of processing is the local sound – the voice we want to capture, plus some noise and reverberation.

If we have multiple microphones arranged in a device called a *microphone array*, we can combine the signals from the microphones such that the microphone array will listen in the direction of the desired sound source, suppressing the sounds coming from other directions. This process is referred to as *beamforming*. The microphone array is electronically capable of changing the listening direction and following the movements of a human speaker. To do this it employs algorithms for *beamsteering* and *sound source localization*.

The microphone array output still contains some residual noise and reverberation. The audio quality is further improved using *speech enhancement techniques* such as *noise suppression* and *de-reverberation*.

At the end of the sound capture system and audio processing chain we should have a speech signal with good enough quality for the final consumer – telecommunication or speech recognition.

1.3 The Goal of this Book and its Target Audience

The book provides a reference for most of the audio signal processing algorithms in the areas of speech enhancement, microphone array processing, sound source localization, acoustic echo reduction systems, and dereverberation. It contains information about various types of sound capturing devices. The exercises provide sample MATLAB® scripts and audio files to play with the algorithms, improve them, and create new ones. Most of the presented algorithms have been implemented and evaluated by the author. They are illustrated with real-life audio recordings, in scenarios close to the major application of these algorithms.

The material in the book allows for a quick building of a complete end-to-end sound capture and audio processing system with relatively good quality. This can be a good baseline for further work on improvement of some of them. Software engineers and designers working on sound capture and processing systems can benefit from this audio processing toolbox to build the initial prototype of the system with off-the-shelf algorithms, to evaluate end to end, and to focus on the processing blocks that need improvement.

The target audience includes graduate and undergraduate students. If this book can spark interest in digital signal processing and influence the decision to further study audio and signal processing it will have achieved one of its goals. It can help a grad school student quickly acquire the necessary information for the available algorithms and what they can achieve. The book provides comparisons of the existing approaches, evaluation parameters, and methodology on how to measure them.

University students, researchers, or academic staff, working on projects where design of audio processing algorithms is not the primary goal, can benefit from the book as well. They can use the presented algorithms for projects like 'we just need a noise suppression algorithm to clean up the sound we capture during our user studies'. Many of the algorithms described in this book are directly applicable or can become a good starting point for modification by researches working in neighboring areas – processing of biological signals (EEG, ECG, EMG), for example.

A separate goal of this book is to shorten the distance between the audio research community and the industry. On the one hand the industry needs education and information about the state of the art in sound processing algorithms. The information should be from their own perspective: what works well in real practice and adds value to the designed product and what is a very cool research algorithm, which may open the door for further research and the creation of better algorithms, but at this point it brings little value or requires an unattainable amount of resources such as memory or CPU time. The research community needs feedback about the issues the industry faces, which will stimulate the creation of robust and practically applicable algorithms.

We want to close the gap between 'algorithms for sound capture and processing' and 'algorithms for manufacturable systems for sound capture and processing'. Under the conditions of increased demand of such algorithms this will happen sooner or later, but this book will help to speed up this process.

1.4 Prerequisites

While the book starts with the basics of audio processing, an understanding of the foundations of digital signal processing (sampling theory, conversion to the frequency domain, filtering, and adaptive filters) is necessary to fully benefit from the book. Knowledge of MATLAB is required for the execution, modification, or porting of the provided sample code to another programming language. A general mathematical background (integrals, differentials, matrix operations, probability, and statistics) is needed to understand the mathematical equations and follow the algorithm evaluations. Some basic knowledge about sound as a mechanical wave in the air and how it propagates will be handy to understand the sound capturing devices part.

1.5 Book Structure

The book covers digital signal processing algorithms and devices for capturing sounds, mostly human speech, and enhancing these signals to achieve a sufficient quality for the needs of real-time communication and speech recognition. It starts with the digital sound processing basics in Chapter 2, which includes defining the noise and speech properties, definition of the basic terminology, the overlap-add processing chain for processing in the frequency domain, and some aspects of sound quality evaluation. Chapter 3 covers the sound capturing sensors: microphones of various types. The parameters of these microphones are described and models for differential and unidirectional microphones are provided. The chapter gives some ideas about what can be achieved for capturing better sound using acoustical means and decreasing the noise and reverberation at the entrance of the system. Chapter 4 is dedicated to single-channel speech enhancement. Various algorithms for gain-based noise suppression are described and the overall architecture of a noise suppressor provided. Other speech enhancement techniques and approaches are discussed. Chapter 5 covers the microphone arrays as sound capture devices and the corresponding processing algorithms, while Chapter 6 is dedicated to using microphone arrays for sound source localization. Chapters 7 and 8 cover acoustic echo reduction systems and de-reverberation algorithms – important parts of the sound capturing system. Each chapter ends with a list of reference materials for further reading. Most of them are commented on briefly in the chapter.

Out of the scope of the book are sound processing algorithms and approaches that are computationally expensive, iterative, and in general not suitable for implementation in real-time working sound processing systems.

1.6 Exercises

Each chapter contains several exercises, which consist of writing or modifying MATLAB implementations of the discussed algorithms. A computer with MATLAB installed on it is required to execute the sample code, modify it, and evaluate the presented techniques. We chose MATLAB as one of the most common programming systems used in the signal processing community for modeling and experimenting with digital signal processing algorithms. It has a good set of mathematical functions and well-developed graphical output, which makes for ease of implementation, debugging, and illustration of waveforms and signals.

Every signal processing algorithm is an abstraction before it is implemented and evaluated with real audio signals. Implementing and experimenting with processing algorithms is the best way to learn more and understand better. We understand that for practical usage these algorithms have to be ported to a compliable programming language, such as C or C++, which is specific to each particular project.

All sample sound files are provided in a non-compressed WAV format. It is common and MATLAB reads and writes it well. The absence of compression does not introduce or mask distortions in the signals. Good headsets or high-quality loudspeakers are recommended for listening. Readers are encouraged to record and process their own audio files. To do this an additional microphone should be connected to the personal computer.

2

Basics

This chapter is dedicated to the sound and audio signal processing basics. We will discuss the properties of noise and speech signals, which will allow us to distinguish them later. The mathematical model and main properties of conversion to the frequency domain are described, which will let us build the overlap-and-add audio processing chain – typical architecture for any type of audio processing. At the end we will discuss ways for evaluation of speech and sound quality: subjective and objective quality measurements of the audio signal.

2.1 Noise: Definition, Modeling, Properties

Captured sound is a mixture of wanted and unwanted signals. Wanted signal is the speech we want to capture; unwanted signals include other speech signals, reverberation, and what in general we call noise. Algorithms for reducing and suppressing various types of noise will be described in the course of this book. To fight successfully you have to know your "enemy." Defining and studying the noise properties is an important part of the knowledge necessary for distinguishing, reducing, and suppressing it.

2.1.1 Statistical Properties

A good statistical model of a noise signal in the time domain is a zero-mean Gaussian process $\mathbb{N}(\mu_N, \sigma_N^2)$ with a given variation σ_N^2, mean $\mu_N = 0$, and normal probability density function (PDF) given by

$$p(x|\mu_N, \sigma_N) = \frac{1}{\sigma_N \sqrt{2\pi}} \exp\left(-\frac{(x-\mu_N)^2}{2\sigma_N^2}\right) = \frac{1}{\sigma_N \sqrt{2\pi}} \exp\left(-\frac{x^2}{2\sigma_N^2}\right). \qquad (2.1)$$

Sound Capture and Processing   Ivan J. Tashev
© 2009 John Wiley & Sons, Ltd

This simple model with just one parameter actually describes the noise signals quite well. In the discrete time domain this means that every sample has a random value with a Gaussian PDF. This statistical model is called *Gaussian noise*. Figure 2.1 shows the PDF of noise recorded in the cabin of a passenger airliner, and the shape of a Gaussian PDF. The PDF of the noise file matches the Gaussian distribution very well.

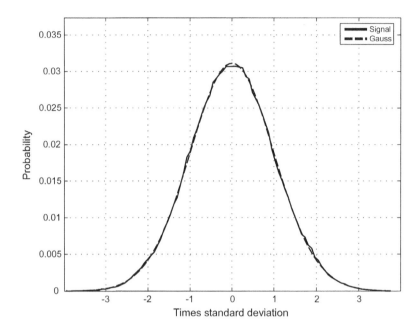

Figure 2.1   Probability density of noise signal compared with Gaussian PDF

The statistical model of the noise as a random signal does not cover all unwanted signals. Another group of such signals are predictable and can be represented as a combination of a small number of sinusoidal signals. This is the power hum, inducted in the audio cables with a main frequency of 60 Hz in North America or 50 Hz in Europe and Asia. For some mobile power circuitry in trains or planes the main frequency can be different, 400 Hz for example. Another group of signals are the frame or line frequencies from TV and computer monitors. The frame frequency can vary between 50 and 80 Hz, depending on the monitor, and has multiple harmonics all over the audible spectrum. The line frequency can vary from 12 to 70 kHz. Usually just the first harmonic is in the audible part of the spectrum. Another way of acquiring these signals is acoustically, by placing the microphones near transformers, TV monitors, or cameras. These signals do not meet the statistical model above, have a different PDF, and should be treated differently.

EXERCISE

To verify the Gaussian PDF of the noise signal experimentally, use the provided MATLAB® script *PlotWavePDF.m*. This script needs two arguments: a WAV file name and the type of expected PDF, Gaussian is by default. To plot the PDF of this noise, run from MATLAB:

```
>> PlotWavePDF('NoiseAirplane.WAV')
```

The result of the execution of this script is similar to Figure 2.1. Try plotting the PDFs of other provided noise files. Record your own noise files and plot their PDFs.

2.1.2 Spectral Properties

While noise is defined as a random signal, it has very specific spectral properties, such as spectral density (the power of the distribution of the frequency spectrum).

The spectral representation of white noise has an equal average magnitude in each frequency and random phase. This is why it is called "white noise" – an analogy with light, which is white when all colors (frequencies) are present. While white noise is used widely for modeling processes in sound capture systems, it does not reflect the spectrum of real noises very well. The magnitudes in their spectrum usually decline towards the high frequencies. The first attempt to better model real noises is the so-called "pink noise." It has a 6 dB/octave decline in magnitude towards the high frequencies. Pink noise can be modeled as white noise passed through a first-order low-pass filter (integrator). The name pink comes again from the analogy with light – red color has a lower frequency and increasing the amount of lower frequencies leads to a more reddish color.

The term "blue noise" can be found in literature as well. This occurs when noise magnitudes increase 6 dB/octave towards the high frequencies. The light analogy is the same – blue color has the highest frequency in the visible spectrum. Noises with spectral characteristics similar to blue noise are relatively rare in the normal environment.

Based on the same analogy, the literature often refers to "colored noise," which is a noise signal with a given magnitude spectrum (mimicking a given color) and random phases. This type of noise can be used for replacing existing background signals. Human ears easily detect even traces of tones and organized signals. This makes residuals audible after some suppression or cancellation procedures. On the other hand, replacing the background residuals with white or pink noise makes the transition moments audible. When the background residual is replaced with noise with the same magnitude spectrum it sounds better to human ears and the transitions are not audible.

So-called "Hoth noise" is used to model indoor ambient noise when evaluating communication systems such as telephones. The first study was done by D.F. Hoth [1]. The official definition of Hoth noise is in the IEEE Standard [2]. The spectral density of

Hoth noise is given in Table 2.1. Hoth noise is frequently mentioned in ITU standards for evaluation of codecs and testing measurement algorithms. Figure 2.2 shows the magnitude spectra of white, pink, and Hoth noises.

Table 2.1   Hoth noise: spectrum density adjusted in level to produce reading of 50 dBA

Frequency (Hz)	Spectrum density (dB SPL/Hz)	Bandwidth 10.log(f) (dB)	Total power in each 1/3 octave band (dB SPL)	Tolerance (dB)
100	32.4	13.5	45.9	±3
125	30.9	14.7	45.5	±3
160	29.1	15.7	44.9	±3
200	27.6	16.5	44.1	±3
250	26.0	17.6	43.6	±3
315	24.4	18.7	43.1	±3
400	22.7	19.7	42.3	±3
500	21.1	20.6	41.7	±3
630	19.5	21.7	41.2	±3
800	17.8	22.7	40.4	±3
1000	16.2	23.5	39.7	±3
1250	14.6	24.7	39.3	±3
1600	12.9	25.7	38.7	±3
2000	11.3	26.5	37.8	±3
2500	9.6	27.6	37.2	±3
3150	7.8	28.7	36.5	±3
4000	5.4	29.7	34.8	±3
5000	2.6	30.6	33.2	±3
6300	−1.3	31.7	30.4	±3
8000	−6.6	32.7	26.0	±3

EXERCISE

Listen to the provided noise files *NoiseWhite.WAV*, *NoisePink.WAV*, and *NoiseHoth.WAV* and compare how they sound. Note how "sharp" the white noise sounds and how natural is the sound of the Hoth noise.

Listen to the naturally recorded noises *NoiseAirplane.WAV* and *NoiseOffice.WAV*. Plot and compare the spectra of these noises by running *PlotWaveSpectr.m* from MATLAB. The script parameters are one or two WAV file names:

```
>> PlotWaveSpectr('NoiseWhite.WAV', 'NoisePink.WAV')
```

Write a simple MATLAB script to generate white noise and store it into a WAV file. Modify the script to generate pink noise and Hoth noise (use interpolation for the spectral densities between the frequencies in Table 2.1). Listen to the generated noises and verify their spectra with *PlotWaveSpectr.m*.

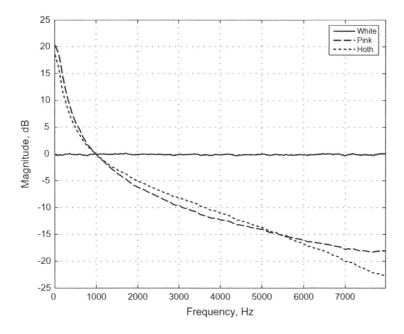

Figure 2.2   Spectra of white, pink, and Hoth noises

2.1.3 Temporal Properties

One of the widely used temporal properties of a noise signal is the assumption that the noise is a stationary signal. This means that its statistical properties, noise variation, and PDF in the time domain, and magnitudes, mean, and variation in the frequency domain, do not change with time. Frequently this property is referred to as "quasi-stationary." This means that the noise signal properties change much more slowly than the other signal properties it is compared with. In most of the cases in this book this is a human speech signal.

A separate group includes the so-called "non-stationary noises." Examples are door opening/closing, keyboard key clatter, passing cars, and so on. They are actually unwanted signals and require different approaches for recognition and removal.

2.1.4 Spatial Characteristics

Spatial characteristics of a noise signal are given by the positions of the noise sources. In most cases there is no strongly defined position of the noise source, or there are many noise sources. In real conditions the reverberation smears in addition to the distinct positions of the noise sources. As a result, in most of the cases we are dealing with ambient noise; that is, there are no noise sources with strongly defined positions. In modeling we can assume that the noise energy comes with equal probability from every direction, which is called "isotropic ambient noise."

Under special conditions, point noise sources are used for evaluation and testing of some sound capture algorithms. For example, we can put a point white noise source at a certain position to verify how well it is suppressed. In some scientific papers such unwanted signal sources are referred to as "jammers."

Usually the ambient noise assumption is a better model to follow, rather than the assumption of a finite number of noise sources and their positions.

2.2 Signal: Definition, Modeling, Properties

When referring to a signal we will mean the wanted part of the mixture of sounds captured from the microphone. Unless stated explicitly, by signal we will mean human speech. From the signal processing perspective, human speech is a complex signal – a sequence of voiced, unvoiced, plosive, and silence segments. Some segments may have characteristics of more than one of these categories.

- *Voiced segments* are characterized by their fundamental frequency, called "pitch," and its harmonics. The pitch frequencies for male voices vary from 50 to 250 Hz, while for female voices they vary from 120 to 500 Hz. The magnitude of the pitch harmonics varies with the frequency and the envelope usually has at least two well-defined maxima. They are the so-called "formant frequencies." Male voices usually have up to three formant frequencies, while female voices can have up to five. The first two formant frequencies determine the vowel, while the third and above are different for each person and allow people to recognize who is talking.
- *Unvoiced segments* are noise-like and characterized mainly by the shape of their spectral envelopes. They may occur simultaneously with voiced segments.
- *Plosive segments* contain highly dynamic transient changes in the signal spectrum. They are usually preceded by a silence segment and are sudden, burst-like sounds. In the frequency domain they are signals with a wide, fast-changing spectrum.
- *Silence segments* are used to separate words and phonemes and are an integral part of human speech. There is no signal during silence segments.

It should be noted that the main recipient of human speech signals is a human with ears and brain. How humans perceive sounds should be accounted for and used to create better sound capturing systems and audio processing algorithms. We will return to this topic later in the book, but a simple example here is the telephone bandwidth from 300 to 3400 Hz. This means that when we speak on the phone in most cases we do not hear the main pitch frequency. The human brain is capable of restoring it and we can understand each other over the phone without any problems. In nature, most of the noise energy is concentrated in the lower part of the frequency band. This means that the ratio between the voice signal energy and the noise energy there is low and the main pitch frequency can drown out the noise. This is why humans can understand speech with a missing (or masked) main pitch frequency. If the bandwidth is reduced further down to 2000 Hz, people can still understand well, but experience difficulties in recognizing the speaker owing to the absence of the upper formant frequencies.

Something different is observed in the voices of trained singers. They can put more energy in the high formants (frequencies of 3000 Hz and above, the so-called "singer's formant") which makes their voices clearly audible even when performing with a large orchestra.

Details of speech production from humans, the physiology of hearing, segments classification, and speech recognition are beyond the scope of this book. Refer to Chapter 2 *Models of Speech Production and Hearing* in [3] for more detail.

2.2.1 Statistical Properties

As mentioned above, the speech signal is a complex mixture and sequence of segments with quite different statistical properties. Therefore in most cases the speech signal is assumed to have a Gaussian PDF; that is, the captured sound is viewed as a mixture of two Gaussian signals, noise and speech. This model works quite well in practice. The reason for this is that it is generic and equally well (or badly) approximates the various types of speech segments.

More precise approximation of long-term speech signals provides a Laplace distribution with a probability density function for variation σ^2 and mean μ given by

$$p(x|\mu,\sigma) = \frac{1}{\sigma\sqrt{2}}\exp\left(-\frac{|x-\mu|}{\sigma\sqrt{2}}\right). \tag{2.2}$$

For the speech case, the mean is $\mu = 0$ and the PDF has the form

$$p(x|\sigma) = \frac{1}{\sigma\sqrt{2}}\exp\left(-\frac{|x|}{\sigma\sqrt{2}}\right). \tag{2.3}$$

Another approximation used is the symmetric gamma distribution. The gamma distribution PDF is given by

$$f(x|\mu,\gamma,\beta) = \frac{1}{\beta\Gamma(\gamma)}\left(\frac{x-\mu}{\beta}\right)^{\gamma-1}\exp\left(-\frac{x-\mu}{\beta}\right) \tag{2.4}$$

where $x \geq \mu; \gamma, \beta > 0$ and $\Gamma(\cdot)$ is the gamma function. Here μ is the location parameter, γ is the shape parameter, and β is the scale parameter. For the signal above, $\mu = 0$, $\beta = 1$, and $\gamma = 0.5$ we have:

$$p(x|\sigma) = \frac{\sqrt[4]{3}}{2\sqrt{\pi\sigma}\sqrt[4]{2}}|x|^{-\frac{1}{2}}\exp\left(-\frac{\sqrt{3}|x|}{\sqrt{2}\sigma}\right). \tag{2.5}$$

Note that this symmetric gamma distribution is just an approximation of the real process and does not reflect or model any physical processes in the generation of human speech. Figure 2.3 shows the computed PDF of a speech signal compared with normal, Laplace, and gamma distributions. It is clear that the gamma distribution better approximates the statistical distribution of the speech signal. For more details see [19] and [4].

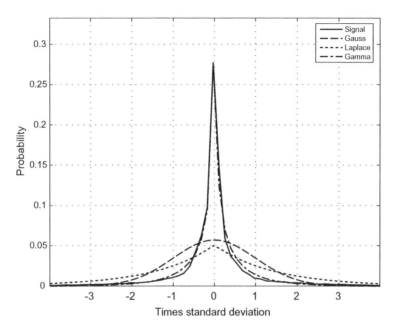

Figure 2.3   Probability density of speech signal compared with various PDFs

EXERCISE

Record with a microphone 2- to 3-second voiced segments ("a-a-a-a-a-a" or "u-u-u-u-u") and unvoiced segments ("s-s-s-s-s-s"). Plot and compare their PDFs using *PlotWavePDF.m.*

Do the same with short recordings of plosive segments ("t," "d," "p," etc.). Plot and compare their PDFs.

Record short text (10–20 seconds) or use the provided *Speech.WAV* file and plot its PDF with the same MATLAB script. Compare it with different types of PDFs:

```
>> PlotWavePDF('Speech.WAV','G')
```

You can use "G" for Gaussian, "L" for Laplace, and "M" for Gamma probability density functions.

Look at the MATLAB code and add other PDFs; Poisson and chi-squared for example. Compare the PDF of the speech signal with them.

(a)

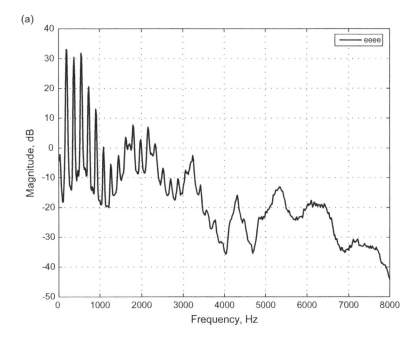

(b)

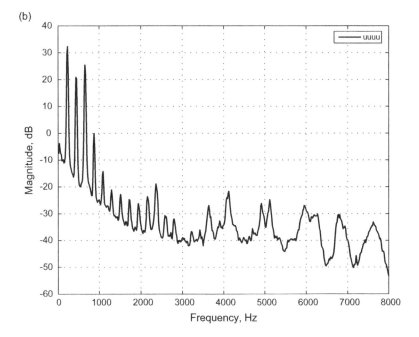

Figure 2.4   Magnitude spectra of two vowels:p (a) e, and (b) u

2.2.2 Spectral Properties

In general we can say that human speech is a signal with frequency content mostly concentrated between 100 and 7000 Hz. With some reduction of the quality the bandwidth of the speech signal can be reduced to 300–3400 Hz, so-called "telephone quality" because this is the bandwidth used in communications. The spectrum of the speech signal depends on the type of segment.

Voiced segments have a well-defined harmonic structure: a pitch and its harmonics. The spectral envelope forms the formant frequencies. Figure 2.4 shows the magnitude spectra of the English sounds "e" and "u" with approximately the same pitch frequency. The formant frequencies are very clear around 500 and 2300 Hz for "e" and at 320 and 800 Hz for "u." Table 2.2 shows the centers of the vowel formants for some vowels in the English language.

Unvoiced segments resemble more a random noise signal. They have an almost flat magnitude spectrum and can be distinguished from the ordinary noise by its duration. The first plot in Figure 2.5 shows the time–frequency content of the sound "sh-sh-sh-sh."

Plosive segments have a wide spectral content for a short period of time. The second plot in Figure 2.5 shows the time-frequency content of the plosive sound "p."

Table 2.2 Centers of vowel formants

Vowel	Formant f1	Formant f2
u	320 Hz	800 Hz
o	500 Hz	1000 Hz
ø	700 Hz	1150 Hz
a	1000 Hz	1400 Hz
ø	500 Hz	1500 Hz
y	320 Hz	1650 Hz
æ	700 Hz	1800 Hz
e	500 Hz	2300 Hz
i	320 Hz	3200 Hz

EXERCISE

Plot the spectra of the recorded voice segments in the previous paragraph using *PlotWaveSpectr.m*.

Do the same with recorded short text or use the provided *Speech.WAV*. Estimate the speech signal approximate bandwidth.

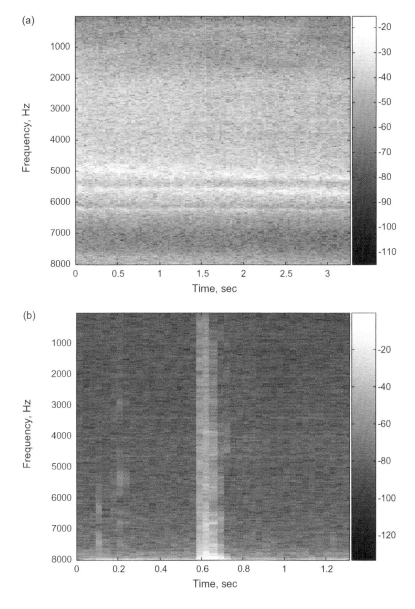

Figure 2.5   Time–frequency content of (a) unvoiced sound sh, and (b) plosive sound p

2.2.3 Temporal Properties

The temporal properties of the speech signal are important and for some segments this is the only way to distinguish them from noise. It is assumed that each speech segment has a duration between 0.1 and 0.5 seconds. The speech signal cannot change faster since it is a result of moving the muscles around the vocal tract. If there are segments where the signal changes faster, they do not belong to the speech signal. In normal speech the vowels are usually no longer than 0.5 seconds (excluding singing and humming). If there is a signal with spectral content changing slower than this, it is most probably not part of the speech signal either.

2.2.4 Spatial Characteristics

The speech signal is considered a point sound source. This is the human mouth, which for distances of one to two meters has an angular size of 2–3 degrees. We should say here that with increasing the distance the proportion between the direct signal and the reverberation decreases and the speech signal gradually loses its point sound source properties.

Together with the head, the speech signal source forms a specific directivity pattern. Figure 2.6 shows the directivity pattern of the human head in the horizontal plane on the level of the mouth, 0 degrees is pointing forward. The sound emitted from the mouth diffracts around the head differently for different frequencies. For the lower part of the spectrum there is practically no directivity; that is, the sound is emitted evenly in all directions. While increasing the frequency the directivity also increases. The human torso plays a substantial role in forming the directivity pattern as well. This directivity

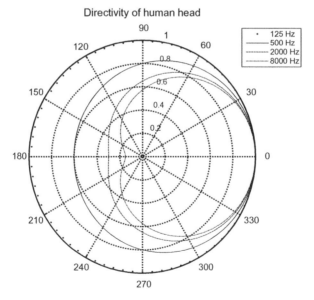

Figure 2.6   Directivity of a human head for various frequencies

pattern affects the reverberation and this is why specialized "mouth simulators" are used to model the human speech source. These devices are designed in a way such that their directivity pattern is as close as possible to the human head. In some cases modeling with such simulators is not enough and then full-scale torso simulators are used. These devices are equipped with mouth simulators and microphones in the ears, which makes them suitable for experiments with both sound production and human hearing.

The human head and torso directivity pattern affects the spectral content in different places around the body, where microphones can be positioned. This is important for the design of headsets and hearing-aid devices. For example, a miniature headset with a short boom requires frequency correction. The microphone is positioned on the cheek, close to the ear. Owing to the human head directivity, the upper part of the frequency band is lowered around 8 dB. Similar frequency distortions are observed when designing devices with microphones positioned around the neck, on top of the head, or on the ear. For more detail see [5].

2.3 Classification: Suppression, Cancellation, Enhancement

Once the sound is converted from minute changes in atmospheric pressure to an electrical signal, sampled with a given sampling rate and quantized with an analog-to-digital converter (ADC), it is nothing but a stream of numbers. Various algorithms for improving the quality of the captured sound exist. They fall into several major groups.

2.3.1 Noise Suppression

Noise suppression algorithms are the most popular method for removing broadband noise. The process involves application of a time-varying filter to the frequency-domain transformation of the noisy signal. Let $x_n = x(nT)$ represent values from a finite-duration analog signal sampled at a regular interval T, in which case a corrupted sequence may be represented by the additive observation model $y_n = x_n + d_n$, where y_n represents the observed signal at time index n, x_n is the original signal, and d_n is additive random noise, uncorrelated with the original signal. The goal of noise suppression is then to form an estimate $\hat{x}_n$ of the underlying signal x_n based on only the observed signal y_n. In many implementations where efficient online performance is required, the set of observations $\{y_n\}$ is filtered using the overlap–add method of short-time Fourier analysis and synthesis, in a manner known as "short-time spectral attenuation." We will discuss the process in detail later in this chapter. Taking the discrete Fourier transformation on windowed intervals of length N yields K frequency bins per interval: $Y_k = X_k + D_k$ ($k = 0, 1, \ldots, k-1$) where these quantities are complex. Noise suppression in this manner may be viewed as the application of a nonnegative real-valued gain H_k, to each bin k of the observed signal spectrum Y_k, in order to form an estimate $\hat{X}_k$ of the original signal spectrum $\hat{X}_k = H_k Y_k$. This spectral estimate is then converted back to the time domain to obtain the signal reconstruction. The key here is estimation of the nonnegative real-valued gain filter H_k, frequently called the "suppression rule." This is the most common case – we have captured the sound and know only the captured signal. Note that the noise suppression procedure adds distortions and artifacts to the estimated signal. Algorithms for noise suppression will be discussed in Chapter 4 of this book.

2.3.2 Noise Cancellation

While noise suppression algorithms are based only on the knowledge of the corrupted signal y_n, noise cancellation algorithms use knowledge of the corrupting signal d_n, eventually obtained through the channel with some impulse response g_n. The noise cancellation system has two input signals: the corrupted signal $y_n = x_n + d_n$ and the corrupting signal convolved through an unknown filter $g_n * d_n$. In the frequency domain these signals are $Y_k = X_k + D_k$ and $G_k \cdot D_k$ respectively. The goal is to estimate the original signal spectrum by estimating and subtracting the corrupting noise $\hat{X}_k = Y_k - \hat{G}_k^{-1}(G_k D_k)$. The inverted filter $\hat{G}_k^{-1}$ is estimated iteratively using adaptive algorithms like LMS, NLMS, or RLS. Note that the estimated filter is a complex vector, and if G_k has no zeros and is stationary, a perfect reconstruction of the source signal is possible without introducing distortions. The price for this is the additional hardware for acquiring one more channel and placing additional microphones where the noise signal can be captured. One potential application is noise reduction in cars, planes, or other vehicles. A hands-free microphone, placed on the dashboard or in the rear-view mirror in a car, captures a lot of engine noise. Additional microphones in the engine compartment can be used to substantially reduce the noise in the captured signal.

2.3.3 Active Noise Cancellation

Active noise cancellation is a process and a system used to reduce the noise in certain areas by emitting sound with the opposite phase from loudspeakers so that they cancel out with the ambient noise. A typical system consists of one or more microphones to capture the disturbing sound, a set of loudspeakers, and eventually a microphone or microphones in the quiet area for feedback on how efficient the noise cancellation is. Note that reducing the noise by adding noise energy means that somewhere the sound emitted from the loudspeakers will be in phase with the noise and will cause an increase in the noise level. Another limitation is that the quiet zone is limited by the sound wavelength; that is, the noise cancellation is either limited to the lower part of the frequency band for a larger area or the quiet area is very small. The last situation is completely acceptable when used in noise-canceling headphones, when the goal is to make the quiet area just around the entrance of the ear canal. Another application is in industrial systems for air conditioning, where this technique can be used to reduce the noise coming from the pipes delivering the air. From the algorithmic perspective, the algorithms used are quite similar to noise cancellation and therefore the specifics of the active noise cancellation algorithms and systems are not going to be discussed further in this book.

2.3.4 De-reverberation

Reverberation is the reflection of sound waves from walls and other objects. This means that the microphone captures not only the direct signal from the sound source, but also multiple delayed copies of the signal with reduced magnitude and changed spectral content. Reverberation affects the consumers of captured speech signals: humans and automatic speech-recognition (ASR) systems. When in a room humans use their two ears and the brain to focus on the speaker and ignore and reduce the reverberation and ambient noise. When the speech signal is captured by a mono microphone and translated to another room via a communication channel, or just recorded and listened to later, the listener tries to do the same. The problem is that the spatial cues are lost and the brain cannot do the de-reverberation any more. This is not only annoying, but also the listeners get tired and lose their attention faster. Increasing the reverberation in the captured signal leads to severe degradation of the recognition rate in automatic speech recognition systems.

The process of reverberation can be described by the room impulse response $h(t)$. The microphone captures the source signal $s(t)$, convolved with the room impulse response $x(t) = h(t) * s(t)$. The length of the room impulse response varies from 60 ms to about one second for highly reverberant spaces. This impulse response changes with movements in the room or changing the positions of the microphones. The frequency representation of the room impulse response can have nulls. At some point the reflections' arrival rate exceeds even the sampling rate. All these factors make estimation of the inverse room impulse response a difficult task. Algorithm details and implementation are discussed in Chapter 8 of this book.

2.3.5 Speech Enhancement

Speech enhancement is a generic description of all algorithms and techniques leading to improvement in the measured or perceived quality of the captured speech signal, contaminated with noise, reverberation, and other factors. From this perspective all the algorithms above fall in this category.

2.3.6 Acoustic Echo Reduction

This technique is an important part of every communication system. Acoustic echo reduction is the process of removing the signal coming from the local loudspeaker from the captured signal. The voice of the correspondent during an audio communication session, called the "far signal," is played from the local loudspeaker. It is captured from the microphone together with the voice of the local speaker, called the "near signal." If we send this mixture through the communication line, it will be played by the far-end loudspeaker, captured by the far-end microphone and returned back. This will lead to audible and annoying echoes, and under some circumstances to feedback; that is, generation of high amplitude signals on certain frequencies. Therefore, the echo reduction is part of every communication system using a speakerphone type of audio communication. Echo reduction is not necessary if headphones are used at least in one of the ends. From the algorithmic perspective the problem is similar to noise cancellation. We have the signal sent to the loudspeakers and the problem is to estimate the transfer function speaker-microphone. What makes this group of algorithms more complex is that in this case we deal with nonstationary signals in both cases. This means that proper measures should be taken to adapt filters when we have only the far-end signal. The block that detects this is called a "double talk detector" and is one of the most critical parts in systems for acoustic echo control.

There are two major approaches for solving the echo reduction problem: acoustic echo cancellation and acoustic echo suppression. Algorithms for acoustic echo cancellation (AEC) try to estimate the complex transfer function of the channel loudspeaker-microphone. They use similar techniques to those of noise cancellation algorithms. The complication here is caused by the reverberation. The direct path and the first-order of reflections from the walls usually can be estimated well. The reverberation tail is already a stochastic process and this reduces the efficiency of the echo cancellation algorithms. Another serious problem is that these algorithms are sensitive to the sampling rates drift. If the ADCs of the microphones and the digital-to-analog-converters (DACs) of the loudspeakers are synchronized – that is, they have the same clock generator – there is no sampling rate drift. In personal computers, however, the microphones can be connected to one sound card or become external USB devices and the loudspeakers can be connected to another sound card or become external USB loudspeakers as well. In this case we have different clock generators for the capturing and rendering parts and there is a sampling rate drift that has to be accounted for.

Acoustic echo suppression (AES) uses techniques similar to those of noise suppression. Under the assumption that the speech is a sparse signal – that is, many of the frequency bins do not carry information – we can say that even when we have signals from both far and near ends, their frequency representations do not overlap much. Then we can use the far-end speech to estimate the energy transfer function of the channel speaker-microphone for each frequency and to use it for suppression of the echo signal. This group of algorithms deals with the reverberation tail better and in general is more stable and robust to clock drifts. The price for this is distortions and artifacts introduced in a similar manner to noise suppression algorithms.

A well-designed system for acoustic echo control usually has both approaches combined. Details of these algorithms and their implementation are discussed in Chapter 7.

2.4 Sampling and Quantization

2.4.1 Sampling Process and Sampling Theorem

In the surrounding world the signals are continuous functions of time and have continuous magnitudes. Theoretically they spread in an infinite time interval. In real sound capture and processing systems we cannot have all the values of a signal with infinite duration. These systems capture the magnitude of the time domain signal in certain moments of time. If these moments are equidistant, we call the distance between them the "sampling period" T and define the sampling frequency as $F_s = 1/T$. The process itself is called "sampling" or "discretization." Under perfect conditions it can be described as multiplication of the time domain signal and the sampling function:

$$x(kT) = x(t).\text{Ш}_T(t), \tag{2.6}$$

where $\text{Ш}_T(t)$ (pronounced as Sha) is the sampling function, which is an infinite sum of Dirac pulses:

$$\text{Ш}_T(t) = \sum_{k=-\infty}^{+\infty} \delta(t-kT). \tag{2.7}$$

Here $\delta(t)$ is defined as

$$\delta(t) = \begin{vmatrix} 1 & t = 0 \\ 0 & \forall t \neq 0 \end{vmatrix}. \tag{2.8}$$

The graphic representation of the sampling function is shown in Figure 2.7. It is known as a *Dirac comb* and resembles the letter Ш in the Cyrillic alphabet (pronounced in English as Sh), hence the denotation. Figure 2.8 is a picture of the back window of a car taken during the winter in Moscow. The triangle sign with this letter is an abbreviation of the Russian word шипы (spikes) and warns drivers that this car has spiked tires and a shorter braking distance on icy winter roads. It closely resembles the sampling function.

Representing the time domain processes in the frequency domain is quite common and useful. We will discuss the properties of this representation further in this chapter. The time and frequency domain representations describe the same process and are identical in information content. Further in this book we will denote this relationship as

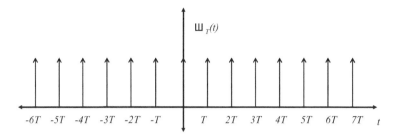

Figure 2.7   Sampling function $\text{III}_T(t)$

Figure 2.8   Sign "Attention spike tires"

$X(f) \Leftrightarrow x(t)$. It is common to denote the time domain signals with lower-case letters and the corresponding frequency domain representation with capital letters.

The spectrum of the sampling function $\text{III}_T(t)$ is the same series of Dirac pulses, but in the frequency domain and at a distance of F_S:

$$\text{III}_T(t) \Leftrightarrow F_S \text{III}_{F_S}(f). \tag{2.9}$$

Then the discrete signal can be described as a convolution of the continuous signal spectrum and sampling function spectrum:

$$x(kT) \Leftrightarrow X(f) * [F_S . \text{III}_{F_S}(f)] \tag{2.10}$$

which is shown graphically in Figure 2.9. The continuous signal spectrum is repeated periodically with period F_S. The intuitive explanation of the sampling theorem is that to prevent loss of information the sampling frequency F_S should be at least two times higher than the highest frequency in the spectrum of the sampled signal $x(t)$; that is, $F_S > 2F_{max}$. Otherwise, after sampling, the repeated spectrum will overlap with its copies, causing information uncertainty. This theorem is known as the Nyquist, or Shannon, or Kotelnikov theorem and in general states that if we have a bandwidth-limited signal with given bandwidth B then the sampling frequency should be $F_S > 2B$ to prevent information loss.

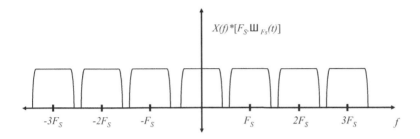

Figure 2.9   Signal spectrum multiplication after discretization

In real systems the sampling process is performed by so-called "track-and-hold" circuits in front of the ADC.

2.4.2 Quantization

A quantized signal $x_n(t)$ is a signal that has a finite set of amplitudes $h_1, h_2, \ldots, h_n$. Especially interesting to us are quantized discrete signals $x_n(kT)$, where $k = 0, 1, 2, \ldots$ and $x_n \in \{h_1, h_2, \ldots, h_n\}$. During the quantization the sample $x(kT)$ is assigned the closest discrete value h_l; that is, when

$$h_l - \frac{h_l - h_{l-1}}{2} \leq x(kT) < h_l + \frac{h_{l+1} - h_l}{2}. \qquad (2.11)$$

In the most common case the levels are numbered consecutively. In some sensors it is more convenient that the relationship between the level and number representing it is encoded in a different way. A classic example here is the Grey code in which the difference between the binary representations of neighbor levels can be only one bit as shown in Table 2.3 for three-bit binary numbers. An example of an optical sensor for reading an angular position is shown in Figure 2.10. It consists of optical couples (light-emitting diode and photosensor) and a disk with darkened areas. The optical sensors

Table 2.3 Grey code for three-bit binary number

Number	0	1	2	3	4	5	6	7
Binary code	000	001	011	010	110	111	101	100

give the angular position in 2^N discrete levels, where N is the number of the optical couples, three in our case. If the levels were numbered using consequent binary numbers, then the sensor would be error prone. In a position right between 000_2 and 111_2 the smallest vibrations can cause one or more of the optical sensors to switch between the levels and as result we can get practically any three-bit number. When using Grey encoding, as shown in this figure, the maximum error can be one bit. Of course after reading the sensor indication it can be encoded to consecutively numbered levels.

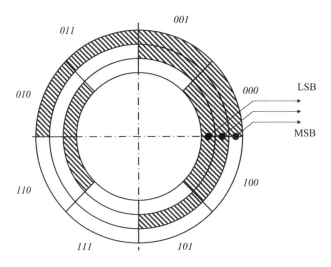

Figure 2.10 Optical sensor for angular position with Grey code

Nonuniform quantization has its application in audio signals compression. Certain compression algorithms use banks of pre-computed quantization levels – more coarse for higher signal levels, more dense for lower signal levels. A typical example here is μ- or A-law encoding in communications [6]. This type of encoding allows reduction of the representation of one sample from 16 bits to 8 bits, which at a sampling rate of 8 kHz reduces the bit rate from 128 to 64 kbps. Audio compression algorithms are out of the scope of this book. Therefore, in the future we will assume consequent numbering of the equidistant levels; that is, $h_l = lq$, where q is the quantization step.

The process of quantization causes quantization of the probability density function of the signal in the following way:

$$p(x' = nq) = \int_{nq-\frac{q}{2}}^{nq+\frac{q}{2}} p(x)dx = \int_{-\infty}^{+\infty} \Pi_q(x-nq)p(x)dx \qquad (2.12)$$

where

$$\Pi_q(x) = \begin{vmatrix} 1 & -\frac{q}{2} \leq x < \frac{q}{2} \\ 0 & \text{otherwise} \end{vmatrix}. \qquad (2.13)$$

If $P(\alpha)$ is the Fourier image of $p(x)$, and there exists an α_{max} for which $P(\alpha) \equiv 0$ for all $\alpha > \alpha_{max}$, then we can apply the sampling theorem and use quantization step $q < 1/(2\alpha_{max})$ to preserve the statistical parameters of $x(t)$ after the quantization process. Assuming a Gaussian probability density distribution, then

$$P(\alpha) = e^{-\frac{4\pi^2\alpha^2\sigma^2}{2}}. \qquad (2.14)$$

This function declines rapidly, and at $\alpha = 4/(2\pi\sigma)$ it has a value of just 0.0034. This means that we can use quantization step $q = \pi\sigma/4 = 0.8\sigma$. Assuming that the magnitudes are in the interval of $\pm 3\sigma$, which holds for more than 99% of the signal values, we can use number of quantization levels $n = 6\sigma/0.8\sigma \approx 8$, which is just three-bit numbers! In real practice for voice signals we use at least 8 bits, and in most cases the number of bits is 16 and even 24. Besides covering the larger dynamic range of the input signals, this resolution allows us to ignore the effects of quantization. In most of the cases we assume knowledge of the exact magnitude. For more precise modeling quantization noise $\mathbb{Q}$ can be added with a uniform distribution between $[-q/2, +q/2)$, frequently approximated as Gaussian with normal distribution:

$$x_n(kT) = x(kT) + \mathbb{Q}\left(-\frac{q}{2}, +\frac{q}{2}\right) \approx x(kT) + \mathbb{N}\left(0, \frac{q^2}{12}\right). \qquad (2.15)$$

The process of quantization is performed by the analog-to-digital converters (ADC). After sampling and quantization, the continuous signal $x(t)$ is converted to a sequence of numbers, ready for further processing in a digital computer.

2.4.3 Signal Reconstruction

Assume a limited-bandwidth signal $x(t)$ sampled to $x(kT)$ under the conditions of the Shannon theorem. This means that there is no information loss and we should be able to restore the signal magnitudes between the sampling moments using only the discretized signal. As the representations in the time and frequency domains are equivalent, the signal restoration is equivalent to restoring the spectrum of the signal. Looking at Figure 2.9, the signal spectrum restoration is just a multiplication of the discretized signal spectrum with the rectangular window function defined as

$$\Pi_{F_S}(f) = \begin{vmatrix} 1 & \text{if} & |f| \leq \dfrac{F_s}{2} \\ 0 & & \text{otherwise} \end{vmatrix}. \tag{2.16}$$

This function has a time domain representation

$$\Pi_{F_S}(f) \Leftrightarrow F_S \frac{\sin(\pi F_s t)}{\pi F_s t} \tag{2.17}$$

and the signal spectrum restoration will be

$$\left[X(f) {*} F_S \sum_{n=-\infty}^{+\infty} \delta(f - nF_s) \right] \Pi_{F_S}(f) = F_S X(f); \tag{2.18}$$

that is, we have spectrum restoration. The process is shown graphically in Figure 2.11. Multiplication in the frequency domain is equivalent to convolution in the time domain, as will be discussed later in this chapter. This is known as *Plansherel's theorem* and the signal restoration in the time domain is nothing but a convolution of the discretized signal with the image of the rectangular function in the time domain:

$$\hat{x}(t) = \sum_{k=-\infty}^{+\infty} x(kT) \frac{\sin(\pi F_S(t - kT))}{\pi F_S(t - kT)}. \tag{2.19}$$

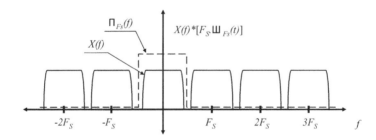

Figure 2.11   Signal reconstruction by multiplication with $\Pi_{F_S}(f)$

This equation is known as the *Shannon approximation formula*. While the sampling theorem just states the condition under which we will not have information loss, the

Shannon approximation tells us how to restore the signal values between the sampling moments.

In real systems this process is performed by the digital-to-analog converters (DAC). They are actually interpolators of the zero-th order; that is, they hold the value of the output signal constant between the fetching moments kT. This is shown in Figure 2.12. The smoothing filter after the DAC performs the convolution with the spectrum restoration function. Its frequency response is an approximation of the ideal frequency response $\Pi_{F_s}(f)$. Note that in certain cases the requirements for this filter may not be so strong. For example, if there is a loudspeaker connected after the DAC, then this device cannot reproduce the higher copies of the discretized signal spectrum, playing the role of the filter.

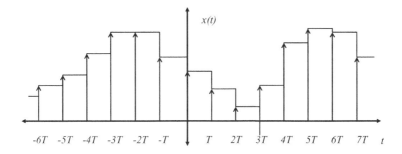

Figure 2.12   Digital-to-analog convertor as zeroth-order interpolator

2.4.4 Errors During Real Discretization

The process of discretization, described so far, is performed under ideal conditions: infinitely thin, equally spaced, sampling pulses and assuming the signal is known for an infinitely long time interval. In real systems we do not have the luxury of an infinite frequency band and time. This fact reflects the precision of sampling, processing, and restoration. We will consider the influence of a non-ideal sampler, averaging ADCs and a finite length time interval.

2.4.4.1 Discretization with a Non-ideal Sampling Function

Instead of using the sequence of perfect Dirac pulses we can use a set of pulses $h(t)$, as shown in Figure 2.13, to discretize the signal $x(t)$:

$$x_h(kT) = \int_{-\infty}^{+\infty} x(t)h(t-kT)dt = x(t)*h(t)|_{t=kT}. \tag{2.20}$$

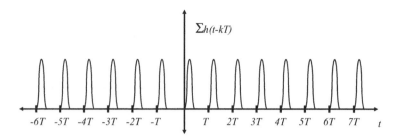

Figure 2.13 Non-ideal sampling function

As we use only the values in moments kT, we can present the equation above as

$$x_h(t) = [x(t)*h(t)].\sum_{k=-\infty}^{+\infty}\delta(t-kT). \qquad (2.21)$$

This simply means that the process of discretization with a non-ideal sampling function $h(t)$ can be presented as an ideal discretization of $x(t) * h(t)$. This is equivalent to placing a filter with an impulse response $h(t)$ before discretizing the signal $x(t)$. The spectrum of the discretized signal will be $X(f)\cdot H(f)$.

These equations model a non-ideal sampler; that is, the finite frequency response of the track-and-hold circuits in real ADCs.

2.4.4.2 Sampling with Averaging

Certain ADCs cannot perform instant sampling of the magnitude of the input signal. Instead they measure the average of the signals in a given sampling time interval θ. In this case the process of discretization can be described as

$$x_a(kT) = \frac{1}{\theta}\int_{kT}^{kT+\theta} x(t)dt = \frac{1}{\theta}x(t)*\Pi_\theta(t)|_{t=kT}. \qquad (2.22)$$

As we fetch only the signal values at moments kT, then after the discretization we have

$$x_a(kT) = \frac{1}{\theta}[x(t)*\Pi_\theta(t)]\sum_{k=-\infty}^{+\infty}\delta(t-kT). \qquad (2.23)$$

This is equivalent to an ideal discretization of the signal $x(t) * \Pi_\theta(t)$, which has spectrum

$$X_a(f) = X(f) \frac{\sin(\pi f \theta)}{\pi f \theta} e^{-j2\pi f \theta}. \tag{2.24}$$

We now have the ideal signal spectrum, passed through a filter with impulse response $\Pi_\theta(t)$. The second term of the equation reflects the magnitude distortions, the third is phase rotation.

Assume that we have a signal with a spectrum of up to 8 kHz. We want to discretize the signal using an averaging ADC with frequency distortions of less than 0.1 dB. This means that at 8 kHz the second term of the equation should have a value higher than 0.99855. Then the averaging interval θ should be less than 3.5 µs. This requirement makes discretization with averaging not very applicable for the capture of sound signals. However, there are other applications. The frequency distortions are not always bad. Note that after discretization with averaging we have nulls for the frequency $1/\theta$ and its harmonics. This is used in pocket voltmeters, where the integration interval is set to be as long as the period of the power net. The device's indication is stable even when measuring low input direct voltages because there are nulls for the frequencies of the major disturbing factor.

2.4.4.3 Sampling Signals with Finite Duration

The signal $x_\Gamma(t)$ with finite duration Γ can be presented as the multiplication of the infinite signal $x(t)$ with a rectangular window $\Pi_\Gamma(t)$:

$$x_\Gamma(t) = x(t)\Pi_\Gamma(t) = \begin{vmatrix} x(t) & -\dfrac{\Gamma}{2} \le t < \dfrac{\Gamma}{2} \\ 0 & t \notin \left[-\dfrac{\Gamma}{2}, \dfrac{\Gamma}{2} \right) \end{vmatrix}. \tag{2.25}$$

Let the spectrum $X(f)$ of the infinite-duration signal $x(t)$ be finite and $X(f) \equiv 0$ for each $|f| > B$. Then

$$X_\Gamma(f) = X(f) * \left[\Gamma \frac{\sin(\pi f \Gamma)}{\pi f \Gamma} \right]. \tag{2.26}$$

Because the function $\{\sin(\pi f \Gamma)/(\pi f \Gamma)\}$ has an infinite spectrum, $X_\Gamma(F)$ will have an infinite spectrum as well. This means that theoretically there is no sampling frequency high enough to meet the requirements of the sampling theorem and precise signal restoration is not possible when sampling a finite interval of the signal. It can be proven, however, that outside of the frequency interval $[-B, +B]$ the signal spectrum has magnitudes below a certain boundary:

$$|X_\Gamma(f)| < \frac{1}{B\Gamma}, \quad \forall |f| > B. \tag{2.27}$$

With this we can compute the momentary values of $x_\Gamma(t)$ from the discretized sequence $x_\Gamma(kT)$ using the Shannon approximation formula:

$$x_\Gamma(t) = \left[\sum_{k=-\frac{N}{2}}^{\frac{N}{2}} x_\Gamma(kT) \frac{\sin[\pi(t-kT)F_s]}{\pi(t-kT)F_s} \right] \Pi_\Gamma(t), \tag{2.28}$$

with a quadratic error of $\varepsilon < 1/(B\Gamma)^{1/2}$. In the approximation formula above, $N = (\Gamma/T)$ is the total number of samples. For 10 seconds of audio signal with bandwidth of 8 kHz the quadratic error is below -50 dB. This is still far above the normal resolution of a 16-bit ADC, which is -96 dB, but below the usual signal-to-noise ratio of the captured signal.

2.5 Audio Processing in the Frequency Domain

2.5.1 Processing in the Frequency Domain

The majority of audio processing algorithms today work in the frequency domain. There are two major reasons for this: easier mathematical representation and computational effectiveness.

Most processing algorithms are described more easily in the frequency domain. The assumption of statistical independence of the frequency bins is a good approximation for most cases, which makes the algorithms more easily described per frequency bin via vector and matrix operations. In the time domain representation of audio signals this is not true, and consequent samples of the speech signal are correlated and predictable to a certain degree.

Audio processing algorithms frequently use digital filtering, applying time-invariant or adaptive filter banks to the processed signal. A simple finite impulse response filter of order N in the time domain is given by

$$y(kT) = \sum_{i=0}^{N-1} b_i x((k-i)T) \tag{2.29}$$

which involves N multiplications and $(N-1)$ additions. Applying a filter bank of N filters will require N^2 multiplications and $N^2 - N$ additions, and in the time domain this is roughly an $O(N^2)$ algorithm. Increasing the filter order requires a substantial amount of computing power. In the frequency domain, the digital filter bank is just vector multiplication, which makes it an $O(N)$ algorithm. Conversion to the frequency domain usually uses a fast Fourier transformation (FFT) algorithm, which is an $O(N\log_2(N))$ algorithm. This simply means that, with increasing filter length N, soon or later a point will be reached where the frequency domain processing is faster. For modern processors with integrated floating-point instructions this already happens for the commonly used filter lengths. In many cases it is not necessary even to convert the signal back to the time domain if the next block works in the frequency domain – a speech encoder or speech recognizer, for example.

Processing in the frequency domain should not be considered as the only type that is acceptable. With processors that have low computational power and only integer instructions, some simple algorithms can run faster and be more efficient in the time domain. Such processing units are used in hearing aids, wireless headsets, battery powered microphones, and so on. There are models of specialized digital signal processors that perform digital filtering in the time domain as part of their natural instruction set. Most of the audio processing algorithms described for the frequency domain can be converted to the time domain, but this process depends heavily on the particular design. That is why in this book we will describe and explain all algorithms in the frequency domain.

2.5.2 Properties of the Frequency Domain Representation

Generic conversion of a given time domain signal $x(t)$ to the frequency domain represents it as a sum of two orthogonal basic functions with continuous or discrete varying frequency. While any two orthogonal functions can be used, the most common today are $\cos(2\pi ft)$ and $\sin(2\pi ft)$. The reason is that a sinusoidal signal does not change shape when passed through a linear filter. This representation is called a "spectrum" – another word borrowed from optics where it means light as a sum of different colors – that is, signals with varying wavelength. The most common representation of the signal as a sum of sinusoidal signals is the Fourier transformation, which in the continuous variant has the form

$$X(f) = \int_{-\infty}^{+\infty} x(t)e^{-j2\pi ft}dt. \tag{2.30}$$

Recall that the Gaussian formula is $e^{-\varphi} = \cos(\varphi) + j\sin(\varphi)$, which clearly makes the representation of $x(t)$ as a sum $X(f)$ of sinusoidal signals with given frequency f with magnitude

$$|X(f)| = \sqrt{\operatorname{Re}(X(f))^2 + \operatorname{Im}(X(f))^2}$$

and phase

$$\varphi(f) = -\arctan\left(\frac{\text{Im}(X(f))}{\text{Re}(X(f))}\right),$$

where $\text{Re}(\cdot)$ and $\text{Im}(\cdot)$ denote the real and imaginary part of the signal spectrum. Note, therefore, that the function $X(f)$ is complex. The inverse Fourier transformation allows one to restore the time domain representation of the signal from its spectrum:

$$x(t) = \frac{1}{2\pi} \int_{-\pi}^{+\pi} X(f)e^{j2\pi ft}df. \tag{2.31}$$

Note that in the common case $x(t)$ can be complex as well.

Conversion to the frequency domain and back has the following useful properties:

- **Linearity**: if $x(t) = a \cdot y(t) + b \cdot z(t)$, then $X(f) = a \cdot Y(f) + b \cdot Z(f)$.
- **Conjugation**: if $x^*(t)$, then $X^*(-f)$.
- **Parseval's theorem**: if $x(t) \cdot y^*(t)$ then $X(f) \cdot Y^*(f)$.
- **Convolution**: if $x(t) * y(t)$ then $X(f) \cdot Y(f)$; and if $X(f) * Y(f)$ then $x(t) \cdot y(t)$. This is known as Plansherel's theorem.
- **Energy preservation**: $\int_{-\infty}^{+\infty} x(t)^2 dT = \int_{-\infty}^{+\infty} X(f)^2 df$, which simply means that the energies of the signal and its spectrum are equal.
- **Time shift**: if $x(t - \tau)$ then $e^{-j2\pi f\tau} \cdot X(f)$.
- **Scaling**: if $x(at)$ $(a>0)$ then $\frac{1}{a}X\left(\frac{f}{a}\right)$.
- **Derivatives**: if $\frac{d^n x(t)}{dt^n}$ then $(j2\pi f)^n X(f)$. Note that the derivatives have the signal spectrum multiplied with increasing magnitudes towards the high frequencies. With real signals we assume that the spectrum is equal to zero after a given frequency when the magnitudes are below some acceptable error. Then we can apply Shannon's theorem and use a sampling frequency at least two times higher. If we want to preserve the derivatives, however, then we should apply Shannon's theorem to the spectrum of the derivative we want to preserve. This usually increases the required sampling frequency.
- **Correlation function**: $C_{xy}(\tau) = \int_{-\infty}^{+\infty} x(t)y(t-\tau)dt \Leftrightarrow X(f) \cdot Y^*(f)$. This is known as Wiener–Khinchin's theorem and provides a way to compute correlation and autocorrelation functions from the signal's spectra.

An important property of the frequency domain representation is its *symmetry*. As we are going to use symmetry frequently, it deserves a more detailed explanation. If $x(t)$ is a real function (the most common case, all natural signals), then $X(f) = X^*(-f)$. This means that the spectrum is conjugate-symmetric around zero. If we process real signals and expect to have a real signal as output, then we can process only one half of the spectrum and restore the second half just before converting back to the time domain. This reduces the processing time to one half in the frequency domain. Other symmetric properties of the representation in the frequency domain are:

- if $x(t)$ is real and even, then $X(f)$ has an imaginary part equal to zero;
- if $x(t)$ is real and odd, then $X(f)$ has a real part equal to zero.

2.5.3 Discrete Fourier Transformation

After discretizing, the input signal is just a sequence of real numbers in the memory of the computer system. Finding the signal spectrum requires an implementation of the Fourier transformation as a numerical algorithm. Converting the Fourier integral for numerical computation is relatively easy. Computing M points of the spectrum from K discrete points of the signal $x(kT)$ is given by

$$X(m \cdot v) = \sum_{k=0}^{K-1} x(kT)e^{-j2\pi\frac{km}{M}} \quad \text{where} \quad m \in [0, M), \quad k \in [0, K), \quad v = \frac{F_s}{M}.$$

(2.32)

Note that the frequency spectrum is discrete by itself and is computed with a given frequency resolution v. The transformation above is known as a discrete Fourier transformation (DFT).

Inverse transformation is trivial as well:

$$x(kT) = \frac{1}{M} \sum_{m=0}^{M-1} X(m \cdot v)e^{j2\pi\frac{km}{K}} \quad \text{where} \quad m \in [0, M), \quad k \in [0, K), \quad v = \frac{F_s}{M}.$$

(2.33)

Computation of the signal spectrum using DFT is possible, but this is an O(KM) algorithm and computation effort increases faster with an increasing number of points. The most common algorithm for computing Fourier transformations is the fast Fourier transformation (FFT) and was created by Cooley and Tukey [7]. In their initial implementation they assumed $K = M = 2^l$; that is, the same number of input and output points which is an integer number, power of two. This algorithm is covered very well in the literature and so will not be discussed in detail here. It is an O($K\log_2 K$) algorithm, which for 512 points makes it $K^2/(K\log_2 K) \approx 56.8$ times faster than the direct DFT, assuming the same computational time necessary for a one-point transformation. For actual implementations the ratio of the execution times is not so large, as FFT needs some additional time to support the algorithm; but it is still very impressive. Practically, this algorithm made possible the processing of audio signals in real time.

The FFT algorithm has been continuously improved and now there are implementations for practically any number of input points. The assumption $M = K$ usually stays.

Note that the FFT algorithm is based on representing K as a multiplication of prime numbers. The algorithm's efficiency increases when these numbers are small. This is why Cooley and Tukey used the smallest prime number, two. The efficiency of FFT algorithms for any number of points decreases with an increasing value of prime multiplicands and, if K is a prime number itself, the algorithm efficiency degrades back to $O(K^2)$. In the future we will assume $M = K$ and will denote K as the number of frequency bins, and k as the current frequency bin.

Later in this book the direct and inverse Fourier transformations will be just blocks in the described algorithms. In software implementation this is a function with an array of K numbers as the argument, which returns an array of M numbers. For practically all applications today $K = M$, and the FFT input is an array of real numbers (real signals) and the output is an array of complex numbers (complex spectrum) with the same length. For the inverse transformation, denoted as iFFT, the input is a complex array and the output is a real array. If we have 512 samples of an audio signal, sampled at $F_S = 16\,000$ Hz, then the resolution in the frequency domain will be $\nu = F_s/K = 31.5$ Hz. Each frequency point is called a frequency bin and we have 512 frequency bins with central frequency $k \cdot \nu$. The spectrum is known at the points 0, 31.5, 63, ..., 15 968.75 Hz. Owing to spectrum multiplication after the discretization, the point at $F_S = 16\,000$ Hz is the same as the point at 0 Hz.

Under the assumption of real signals and using the symmetry of the spectrum, some fast FFT implementations return only the first half of the spectrum because the second half is a symmetrically complex conjugate. The corresponding iFFT implementation adds the missing part and then performs the inverse transformation. This helps to speed up the processing in the frequency domain. From this perspective, the signal spectrum is an array of complex numbers, representing the spectrum values at 0, 31.5, 63, ..., 7968.75 Hz. Note that the point at $F_S/2$, which is the number 257 in our case, is not processed and will be assumed to be zero at the inverse transformation time.

2.5.4 Short-time Transformation, and Weighting

Processing of audio and sound signals is not performed on the entire signal. The reason for this is real-time applications. When processing the audio signal as part of a communication audio stack, the delay (latency) is critical and should be kept as low as possible. Humans can detect a latency larger than 50 milliseconds during normal conversation; at 100 milliseconds it is already annoying, and above 250 milliseconds conversation through this communication line is difficult. The logical solution is to slice the audio signal into short pieces and to process them sequentially. This guarantees low latency and real-time processing. These short pieces are called "audio frames" and practically all audio signal processing systems process the signals frame after frame. The naive approach is to separate the time intervals, which is equivalent to convolving with consequent rectangular windows:

$$x^{(n)}(kT) = x((k+nN)T).\Pi_{NT}(t-nNT) \qquad (2.34)$$

where $x^{(n)}(kT)$ is the n-th frame, containing N samples; that is, $k \in [0,N)$. Working with a short epoque with duration $\Gamma = NT$ leads to a decrease in precision and frequency distortions, as was described in Section 2.4.4.3. The sharp slopes of the rectangular window cause "ringing" (i.e., additional frequencies) in the entire spectrum owing to its wide spectral content.

Generalizing the frame extraction function as $w_L(t)$ with finite duration L, we can write

$$x^{(n)}(kT) = x(kT+nL) \cdot w_L(t-nL) \qquad (2.35)$$

and try to find a weighting function $w_L(t)$ with better characteristics. Many weighing functions are described in the literature, usually named after the researcher who first published them. The most commonly used weighing functions and their spectra are given in Table 2.4. The table provides the weighting function in the time domain and its spectrum. The magnitude and power coefficient should be applied if we want to preserve the magnitude or power, respectively, for measurement purposes. The natural weighting function represents just a selection of samples within given duration T. It has a $\mathrm{sinc}(x) = (\sin x)/x$ shape of the spectrum which leads to smearing of the spectral lines after the Fourier transformation. A better weighting window is the triangle (Bartlett). It has lower side spikes since its frequency domain representation is $\mathrm{sinc}^2(x)$. The Parzen weight window goes further in this direction with a frequency domain representation of $\mathrm{sinc}^4(x)$. The most commonly used weighting window was named after Julius von Hann. It has a lesser smearing of the spectral lines and is symmetric – an important property for further restoration. The rest of the weighting functions in the table are optimal in one way or another; they are mostly used for spectral representations and measurements in the frequency domain. They are rarely used for real-time audio processing. For more detail see Chapter 14 of [8].

Figure 2.14 shows the spectrum of a 1024-point sinusoidal signal with a frequency of 1000 Hz sampled with 16 000 Hz and weighted with rectangular and Hann windows. The second spectrum is less spread out and looks closer to the theoretical – a single spectral line at the main frequency.

2.5.5 Overlap–Add Process

The process of frame extraction with a weighting function, described in the previous sub-section, works well provided we do not have to reconstruct the time domain waveform of the signal. If we try to process the frames, extracted in this way, and to convert them back to the time domain, the "ends" of the neighboring frames may not connect well together – that is, to have the same or close values. As a result of these discontinuities, annoying sounds will be heard. To ensure good stitching of the audio

Table 2.4 Weighting functions for frame duration T and their spectra

Weight window	Function	Magnitude	Power	Spectrum
Natural	$w_0(t) = \begin{vmatrix} 1 & |t| < T \\ 0 & |t| \geq T \end{vmatrix}$	1	1	$W_0(f) = 2T\dfrac{\sin 2\pi fT}{2\pi fT}$
Bartlett	$w_1(t) = \begin{vmatrix} 1 - \dfrac{|t|}{T} & |t| < T \\ 0 & |t| \geq T \end{vmatrix}$	$\sqrt{3}$	3	$W_1(f) = T\left(\dfrac{\sin 2\pi fT}{2\pi fT}\right)^2$
Parzen	$w_2(t) = \begin{vmatrix} 1 - 6\left(\dfrac{|t|}{T}\right)^2 + 6\left(\dfrac{|t|}{T}\right)^3 & |t| \leq \dfrac{T}{2} \\ 2\left(1 - \dfrac{|t|}{T}\right)^3 & \dfrac{T}{2} < |t| < T \\ 0 & |t| \geq T \end{vmatrix}$	1.93	3.71	$W_2(f) = \dfrac{3}{4}T\left(\dfrac{\sin 2\pi fT}{2\pi fT}\right)^4$
Hann	$w_3(t) = \begin{vmatrix} 0.5\left(1 + \cos\dfrac{\pi t}{T}\right) & |t| < T \\ 0 & |t| \geq T \end{vmatrix}$	1.63	2.67	$W_3(f) = 0.5W_0(f) + 0.25W_0\left(f + \frac{1}{2T}\right) + 0.25W_0\left(f - \frac{1}{2T}\right)$
Hamming	$w_4(t) = \begin{vmatrix} 0.54 + 0.46\cos\dfrac{\pi t}{T} & |t| < T \\ 0 & |t| \geq T \end{vmatrix}$	1.59	2.52	$W_4(f) = 0.54W_0(f) + 0.23W_0\left(f + \frac{1}{2T}\right) + 0.23W_0\left(f - \frac{1}{2T}\right)$
Blackman	$w_5(t) = \begin{vmatrix} 0.42 + 0.5\cos\dfrac{\pi t}{T} + 0.08\cos\dfrac{2\pi t}{T} & |t| < T \\ 0 & |t| \geq T \end{vmatrix}$	1.81	3.28	$W_5(f) = 0.42W_0(f) + 0.25W_0\left(f + \frac{1}{2T}\right) + 0.25W_0\left(f - \frac{1}{2T}\right) +$ $0.04W_0\left(f + \frac{1}{T}\right) + 0.04W_0\left(f - \frac{1}{T}\right)$
Max, Fauque, Berthier	$w_6(t) = \begin{vmatrix} \dfrac{\sin \pi t/T}{\pi t/T} & |t| < T \\ 0 & |t| \geq T \end{vmatrix}$	1.49	2.22	$W_6(f) = \begin{vmatrix} 1 & |f| < \frac{1}{2T} \\ 0 & |t| \geq f \end{vmatrix}$

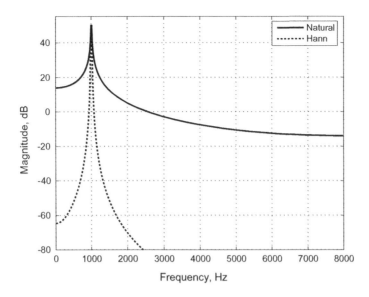

Figure 2.14   Magnitude spectra of signal with natural and Hann weighting windows

frames, a process called "overlap-and-add" is used. The symmetric frame extraction window is moved forward one half of its duration. The frames are extracted and processed normally, then converted back to the time domain. The signal reconstruction includes secondary weighting and combining the first half of the current frame with the second half of the previous frame, as shown in Figure 2.15.

The double weighting and overlap–add technique ensures that there will be no discontinuities in the reconstructed signal. On the down side, decreasing the frame step twice increases the processing time by two times. Each sample in the time domain will be processed twice, as part of two adjacent frames. To keep the same effects from the chosen weighting function $w(t)$, we can use as frame extraction and secondary weighing functions $w_L(t) = w(t)^{1/2}$. For example, if we intend to use the Hann window as weighting function, then we will apply the modified Hann window twice:

$$w_L(t) = \sqrt{\cos\left(2\pi\frac{t}{L}\right)} = \sin\left(\pi\frac{t}{L}\right) \quad \forall t \in \left[-\frac{L}{2}, +\frac{L}{2}\right). \tag{2.36}$$

The equivalence of splitting the weighting process into two and applying the function once is easy to prove.

EXERCISE

Look at the provided MATLAB script *ProcessWAV.m*. It takes as parameters the names of the input and output WAV files and performs the full framing and overlap–add

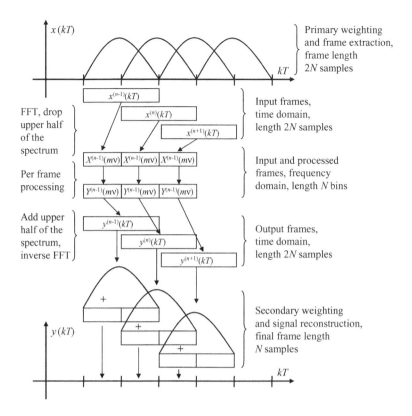

Figure 2.15   Overlap-add-process for processing and reconstruction of signals

process. The script uses the *ConvertFFT.m* script for weighting and conversion to the frequency domain and the *RestoreFFT.m* script for converting back to the time domain and restoring the signal. Note that the second script uses the remembered previous frame. While the script does practically nothing and just returns the same signal, it demonstrates the real-time audio processing chain well.

Modify the provided script to decrease the magnitude of the spectrum two times by multiplying each frequency domain frame by 0.5. Of course, this operation can be done much faster in the time domain, but at least now our script does some processing and demonstrates the linearity of the Fourier transformation.

2.5.6 Spectrogram: Time–Frequency Representation of the Signal

Using shifting frame-extraction weight windows allows us to show the time domain signal in a time–frequency representation. This is the magnitude spectrum as a function of both frequency and time. This presentation is shown as two-dimensional plot in Figure 2.16. The signal shown is a male speaker pronouncing the English word "hello." The pitch and its harmonics are well visible during the vowels. The

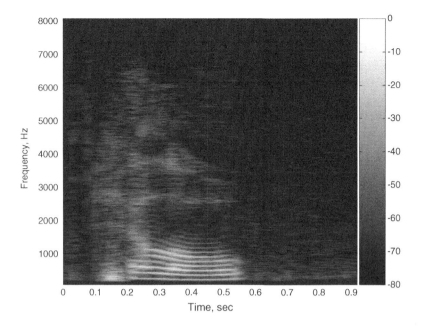

Figure 2.16   Spectrogram of English word "hello"

signal has a duration of one second and a 16 kHz sampling rate. The frame size is 512 samples.

There is a dualism here. The larger frame size gives better resolution in the frequency domain, but decreases the resolution in the time domain owing to the longer frame while increasing the resolution in the time domain decreases the resolution in the frequency domain owing to the shorter frame. Here we can make an interesting analogy with the Heisenberg uncertainty principle in quantum mechanics which applies to the position and momentum of any object and implies that if we continue increasing the accuracy with which one of these is measured, the other will be measured with less and less accuracy.

A factor for consideration here is how fast the spectrum of the observed signals changes. For processing of speech signals for telecommunication purposes the most commonly used frame durations are between 10 and 50 milliseconds. In some cases this is adjusted to be a power-of-two number of samples, in other cases a rounded number of milliseconds. Many audio codecs work with frame durations of 20 ms, which at a 16 kHz sampling rate yields a frame size of 640 samples. As this type of processing targets reconstruction of the signal back into the time domain, the step size is 320 samples. In speech recognition systems the typical frame duration is 25 ms with a frame step of 10 ms. Note that in this case we do not target signal reconstruction.

EXERCISE

Write MATLAB functions that perform Fourier and inverse Fourier transformations according to Equations 2.32 and 2.33. Generate 1024 points of a synthetic sinusoidal signal with a frequency of 1000 Hz and a sampling rate of 16 000 Hz.

- Compute the spectrum of the signal in 1024 points and compare the result with the FFT function in MATLAB.
- Compute the signal spectrum in 8196 points. Convert the magnitudes to decibels and plot. Explain the picture comparing it with Figure 2.14.
- Compute the signal spectrum in 1024 points, convert it back to the time domain in 2048 points. Compare with analytically computed points. Explain.

Plot the short-term spectrum of a speech file (record your own or use the provided) by running *PlotWaveShortTermSpectr.m* from MATLAB. The script parameter is the WAV file name:

```
>> PlotWaveShortTermSpectr('Speech.WAV')
```

In this plot the harmonic structure of the vowels in the speech signal is well visible.

Change the frame size to larger and smaller values and observe the effect on the resolution in frequency and time axes.

The second parameter of the script is the plot type. Default is "S" for the already observed scattered plot. Another way to represent the time–frequency–magnitude processes is by using a direct three-dimensional plot by specifying the second parameter as "D." Run the same script with a second parameter "D," and use MATLAB's abilities to rotate and enlarge the plots to take a closer view of the signal segments.

Plot the time–frequency representation of the white noise signal. Observe the changes in the signal spectrum over time.

2.5.7 Other Methods for Transformation to the Frequency Domain

While Fourier transformation and its CPU-efficient implementation as FFT is used dominantly in audio signal processing, there are alternative transformations that offer certain advantages.

2.5.7.1 Lapped Transformations

Lapped transforms (LT), in general, use weighting functions for frame extraction optimized for various goals [9]. Lapped orthogonal transforms (LOT) are optimized to maximize the transform coding gain in image processing and to reduce the blocking effects. Later the modulated lapped transform (MLT) [10] was developed. The continuous-time equivalent of MLT is called "Malvar wavelets," without the restriction that the window has to be symmetric. MLT is used in the speech coding standard G.722.1; see [11].

MLT was then generalized as modulated complex lapped transforms (MCLT) [12]. The MCLT spectrum has similar properties to the spectrum from a windowed FFT representation, and thus the MCLT can be used for the same applications as windowed FFT: spectral analysis, subband-domain processing, and so on. The main advantage of the MCLT is that there are several simple signal reconstruction formulas, whereas for the FFT there is only one simple reconstruction formula. In particular, with the MCLT you can reconstruct the signal from complex-valued coefficients, but you can also easily reconstruct the signal from just the real parts of the coefficients, making it easier to interface MCLT signal representation to compression systems (since there is half the data in the real parts of the coefficients, when compared to the full complex-valued coefficients).

In its simplest form it can be reduced to using a modified Hann weighing window for frame extraction, a frequency shift to one half the frequency bin, and conversion to the frequency domain using FFT. The first bin of the spectrum is not at 0 Hz, but at a frequency of $F_S/2K = 15.75$ Hz, using the example from Section 2.5.4. The frequency bins are with the same step of $\nu = F_S/K = 31.5$ Hz, and we have no frequency bin exactly at $F_S/2$. Otherwise processing of the MCLT spectrum in this case is the same as in Fourier transformation-based systems. The signal reconstruction consists of an inverse FFT to the time domain and shifting back one half frequency bin. The overlap–add process described above is applied to 50% overlapping frames.

2.5.7.2 Cepstral Analysis

Cepstrum was first defined in [13] as the Fourier transformation of the logarithm of the Fourier image of a real signal with unwrapped phase. Later the definition of the complex spectrum was generalized as the inverse Fourier transformation complex natural logarithm of the Fourier image of the signal:

$$x_{cc}(\theta) = \frac{1}{2\pi} \int\limits_{-\pi}^{\pi} \ln(X(e^{j\omega}))e^{j\omega\theta}d\omega, \tag{2.37}$$

where $\omega = 2\pi f$ and $X(e^{j\omega})$ is the spectrum of $x(t)$. The name cepstrum comes from reversing the order of the first four letters in the word spectrum. Using the same analogy, the independent variable θ is called "quefrency" (derived from frequency). Note that θ is measured with time units. Changes in the log-spectrum are called "liftering" (derived from filtering). The multiplicities of the main quefrency are called "rahmonics" (derived from harmonics).

The real cepstrum, frequently called just cepstrum, is defined as the real part of the inverse Fourier transformation of the natural logarithm of the magnitude of the Fourier image of the signal:

$$x_{cc}(\theta) = \frac{1}{2\pi} \int_{-\pi}^{\pi} \ln(|X(e^{j\omega})|)e^{j\omega\theta}d\omega. \tag{2.38}$$

The difference is that we compute the magnitudes of the signal spectrum and then compute the logarithm and the inverse transformation. Note the similarity between the real cepstrum and the spectrum of the autocorrelation function:

$$x_{cc}(\theta) = \frac{1}{2}\text{FFT}\{\log[X(f)\cdot X(f)^*]\}^{-1} \quad \text{vs.} \quad C_{xx}(\tau) = \text{FFT}\{[X(f)\cdot X(f)^*]\}^{-1}.$$
$$\tag{2.39}$$

Here we applied the Wiener–Khinchin theorem for computing the autocorrelation function from the signal spectrum.

In both cases, computing the cepstrum faces the problem of handling the case when we have frequency bins with values of zero. This singularity can be resolved by adding a small number to the spectrum (equivalent to adding white noise) or just by replacing the logarithm of zero with a certain big number.

Without physical meaning and without modeling real processes, the cepstrum gives information about how fast the spectrum changes. This means that if we have a peak (maximum) at a given quefrency, the spectrum changes with this period. The main advantage of the cepstral analysis compared with spectral analysis is that it converts the multiplicative errors to additive. If we have a observed signal $y(t) = x(t) * h(t)$, which is the actual signal $x(t)$ convolved through a filter $h(t)$, representing the measurement system and the environment, then the spectrum is $Y(f) = X(f) \cdot H(f)$ and the cepstrum is $y_{cc}(\theta) = x_{cc}(\theta) + h_{cc}(\theta)$.

The initial application of the cepstrum was for analysis of shock waves after earthquakes or bomb explosions – that is, processes we observe with substantial changes after passing through an unknown filter. It can be used for radar signals analysis as well. Another common application of the cpestrum is analysis of vibrations. When an internal combustion engine works, every part of it "rings" with its own resonance frequency, modulated by the actual vibration, usually with a lower frequency. Cepstral analysis of the vibration signal gives information about the magnitude and frequency of the modulation frequency.

Figure 2.17 gives an example of the advantages of ceprstral analysis. We have an input signal with a main frequency of 1000 Hz and both frequency and amplitude modulated with period 10 ms. The spectrum of this signal gives us information about the main frequency and the frequency band. The cepstrum clearly shows the modulation period of 10 ms as a peak and the information about main frequency is lost.

In audio processing, cepstral analysis has applications in some algorithms for pitch measurement, reverberation estimation, and frequency correction estimation. Practically all speech recognition systems today have a version of cepstral mean normalization (CMN) in their front ends. The assumption is that if we average the input cepstum we can estimate $h_{cc}(\theta)$ because the speech signal cepstrum in the long term approaches zero, which is a good approximation, but in general not true. Then

$$\hat{h}_{cc}^{(n)}(\theta) = \frac{1}{N} \sum_{n-N}^{n} y_{cc}^{i}(\theta) \qquad (2.40)$$

and

$$\hat{x}_{cc}^{(n)}(\theta) = y_{cc}^{(n)}(\theta) - \hat{h}_{cc}^{(n)}(\theta). \qquad (2.41)$$

Further, this algorithm is improved by better estimation of $h_{cc}(\theta)$ using a-priori knowledge of the speech signal or by estimating the long-term cepstrum of a clean speech signal [14].

2.6 Bandwidth Limiting

The noise signal, as discussed earlier in this chapter, has its energy spread in the entire audible spectrum. This means that following the rule to limit the bandwidth of acquiring and processing to the bandwidth of the desired signal is essential. Using wider frequency band adds nothing but noise.

The upper frequency of the speech signal can vary depending on the speaker gender and language. If we sample a signal coming from a telephone line, it has already been band limited to 3400 Hz. The wideband speech signal is considered to be up to 7000 Hz. While there are some user studies showing that using bandwidth of up to 14000 Hz for speech signals improves the understandability and user satisfaction, doubling the processing time and the necessary bandwidth to transport the speech signal to the recipient seems not to be worth the effort. Another limiting factor is the sampling rate and the low-pass filter in front of the ADC. This filter suppresses all frequencies above one half of the sampling rate to enforce Shannon's theorem. As it cannot be ideal the suppression usually starts at a frequency $0.45 F_S$, which can be considered as the upper limit. This means that for a 16 kHz sampling rate, most commonly used for speech signals, we have the upper limit at 7200 Hz.

The lower limit depends on the acquiring hardware and signal carrier. The signal from a telephone line already has 300 Hz as the lower limit. Earlier in this chapter we mentioned that the pitch main frequency can go down to 50 Hz for deep male voices, but humans have the ability to restore it and to understand speech when the main pitch frequency is missing. The problem is that the lower part of the frequency band is where most of the noise energy is – look at the Hoth noise model. Air-conditioning adds substantially more energy in that part of the frequency band as well. In general, the

(a)

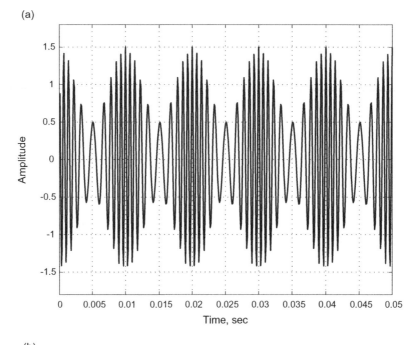

(b)

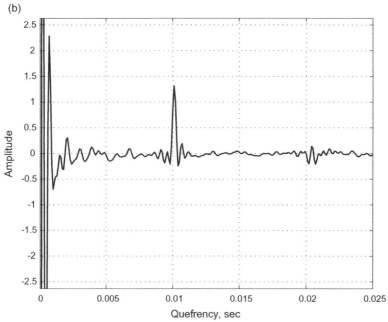

Figure 2.17   (a) FM and AM modulated signal, (b) its cepstrum, and (c) its spectrum

(c)

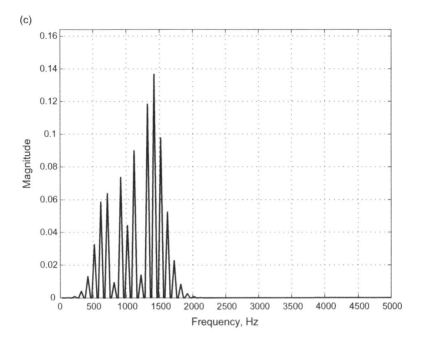

Figure 2.17   (*Continued*)

noise coming from transformers and power lines is 50 Hz and harmonics in Europe and 60 Hz and harmonics in North America. This is why the processing of speech signals usually starts at 200 Hz, which is considered a good lower limit. We would strongly recommend using an analog high-pass filter as part of the microphone preamplifier. The cut-off frequency of this filter at −3 dB should be around 150–180 Hz, and if it is a fourth-order filter we can have a slope of ≈20 dB/octave. This means that at 60 Hz there will be a suppression of 50 dB, which is enough to reduce unpleasant noises.

In summary, we should limit the bandwidth of the processed speech signals to 300–3400 Hz for telephone quality speech, and to 200–7200 Hz for wideband speech. For very-high-quality speech processing systems we can use bandwidths of 100–14 400 Hz, which will require more CPU time and more bandwidth for telecommunication. In addition, such a frequency band requires high-quality loudspeakers and young listeners – the ability of humans to hear high-frequency sounds decreases with age.

EXERCISE

Record you own voice with 32 kHz sampling rate or use the provided *HQSpeech.WAV* file. Modify the provided *ProcessWAV.m* script to perform band-pass filtering for the three bandwidths suggested above. Listen to the script output with a good-quality headset and compare the perceptual quality.

2.7 Signal-to-Noise Ratio: Definition and Measurement

Signal-to-noise ratio (SNR) is one of the most commonly used parameters for signal quality assessment. It is defined as the ratio of the signal and noise energies and is usually measured in decibels (dB). It can be defined in both the time and frequency domains:

$$\Xi = 10\log_{10}\frac{\frac{1}{N_{sig}}\sum_{\substack{signal\\frames}}\sum_{k=0}^{K-1}|x^{(n)}(kT)|^2}{\frac{1}{N_{noise}}\sum_{\substack{signal\\frames}}\sum_{k=0}^{K-1}|x^{(n)}(kT)|^2} = 10\log_{10}\frac{\frac{1}{N_{sig}}\sum_{\substack{signal\\frames}}\sum_{k=0}^{k-1}|X^{(n)}(k\phi)|^2}{\frac{1}{N_{noise}}\sum_{\substack{signal\\frames}}\sum_{k=0}^{k-1}|X^{(n)}(k\phi)|^2}. \qquad (2.42)$$

Here N_{sig} and N_{noise} are the numbers of signal and noise frames, respectively. Further we will define various SNRs in the frequency domain only. The SNR estimation above needs a binary classifier for each frame: signal or noise. These frame classifiers are called "voice activity detectors" (VAD) and will be discussed in Chapter 4. In many cases we cannot be certain of the current frame signal or noise. That is why most VADs are soft decision-based and provide speech presence probability. In this case we can compute the SNR using the speech presence probability for each frame:

$$\Xi = 10\log_{10}\frac{\sum_{n=0}^{N}\left[p^{(n)}\sum_{m=0}^{K-1}|X_k^{(n)}|^2\right]}{\sum_{n=0}^{N}\left[(1-p^{(n)})\sum_{k=0}^{K-1}|X_k^{(n)}|^2\right]} \qquad (2.43)$$

where $p^{(n)}$ is the speech presence probability for the n-th frame. Note the abbreviated denotation $X_k^{(n)}$ for the signal in the k-th frequency bin from the n-th frame. This overall SNR is good for estimating the average signal corruption with noise.

To estimate the momentary SNR, or the SNR for the current frame, we need to know the noise variation σ^2:

$$\Xi^{(n)} = \frac{\sum_{k=0}^{K-1}|X_k^{(n)}|^2}{\sigma^2} \qquad (2.44)$$

which is just the ratio of the current frame power and the average noise power.

We can go further and define the SNR per frequency bin:

$$\xi_k = \frac{|X_k|^2}{\sigma_k^2} \qquad (2.45)$$

where $|X_k|^2$ is the instantaneous power and σ_k^2 is the noise variance, both for the k-th frequency bin. The denotation for the current frame is omitted for simplicity.

Measuring the correct SNR plays an important role not only for signal quality assessment, but it is also an important part of noise suppression and reduction systems. We will look more deeply at these systems in Chapter 4.

2.8 Subjective Quality Measurement

When the processing chain produces an audio signal targeting human ears, using just the SNR as a quality assessment parameter is not enough. In this case it is more about human perception than about improving some technical parameters. Some signals with worse SNRs are perceived as higher quality by humans. Subjective quality measurements play an important role in the design and tuning of practically all audio processing algorithms. They can vary from simple listening to the output (informal evaluation) to time and resource consuming effort with hundreds of listeners involved.

The mean opinion score (MOS) is the most commonly used method. The general approach is that multiple listeners evaluate the quality of audio files on a scale from 1 to 5 by listening to them. See Table 2.5 for the rating description. The set of files each listener evaluates is randomly selected from a large pool and usually contains clean signals, contaminated signals, and sounds processed with various algorithms. After conducting the tests the score for each type of processing is averaged – this is the MOS number. The standard deviation is computed to confirm that we have statistically reliable results. One algorithm can be considered better if at least:

$$MOS_2 > MOS_1 + \sigma_1 + \sigma_2. \tag{2.46}$$

Table 2.5 Mean opinion score perceptual quality rating

Rating	Quality	Description
5	Excellent	Imperceptible, perfect speech signal recorded in a quiet booth
4	Good	Perceptible but not annoying, intelligent and natural-like telephone quality
3	Fair	Slightly annoying, communication quality, but requires some hearing effort
2	Poor	Annoying, low quality, and hard to understand the speech
1	Bad	Very annoying, unclear speech, breakdown

Using MOS methodology is pretty straightforward. It is standardized by ITU-T recommendation P.800 [15] and widely used for evaluation of codecs quality. One of the problems, frequently faced during MOS tests, is that the listener's quality evaluation changes during the tests. After listening to some poor-quality sound files, they tend to give higher scores to previously evaluated records.

Differential MOS tries to fix this problem. Each test consists of listening to two randomly selected records from the pool. The participants do not give absolute scores, but compare the sound quality on a scale from -2 to $+2$. See Table 2.6 for the ratings description. After direct comparison, more reliable data are received when comparing

Table 2.6 Differential perceptual quality rating

Rating	Quality	Description
2	A much better than B	Signal A has much better quality than B
1	A better than B	Signal A has better quality than B
0	A and B are the same	Signals A and B have the same perceptual quality
−1	A is worse than B	Signal A has worse quality than B
−2	A is much worse than B	Signal A has much worse quality than B

two processing algorithms. The differential MOS approach allows the results to be scaled using signals with known MOS quality for the derivation of the final MOS. The potential problem here is that, when the differences in the signal quality are low and the number of listeners not very high, circular results can be obtained; that is, users like A more than B, B more than C, and C more than A. Usually such results are not statistically reliable.

2.9 Other Methods for Quality and Enhancement Measurement

Conduction of MOS tests, while providing precise evaluation of the perceived sound quality, is a long and expensive process. It can and should be done at the end of the algorithm design to prove the achieved improvement; however, doing it frequently during the design process is usually unacceptable. For quick evaluation of the sound quality several methods are designed, which can be implemented as automated tests. They vary from simple computation of a set of parameters to approaches based on machine learning that provide close to actual MOS scores results.

Intrusive methods for objective speech quality assessment require knowledge of both output and clean signals. In some literature sources they are called "double-ended methods." Single-ended or non-intrusive methods use the degraded speech signal only. They are easier to conduct, but the quality assessment reliability is lower. The major criterion for evaluation of the reliability of these speech and audio quality methods is the correlation with MOS test results. Current state-of-the-art algorithms for objective quality measurements are more reliable than MOS obtained from a small listening panel, but still MOS results from large listening panel are considered better.

The algorithms for intrusive measurement of speech quality have preparation and measurement phases. During the preparation phase the contaminated signal is re-sampled if necessary, time aligned, its duration is truncated, and its magnitude adjusted to correspond to the clean input signal. A naive approach is then to compute the average square error; that is, the average difference between the two signals. While this approach gives some information about how close the contaminated signal is to the clean signal, its correlation to MOS scores is low. A better approach is to compute the log-spectral distance between the two signals for the n-th frame:

$$LSD^{(n)}[dB] = \frac{10}{K} \sum_k |\log_{10}(X_{inp}^{(n)}(k)) - \log_{10}(X_{out}^{(n)}(k))|^2 \qquad (2.47)$$

The log-spectral distance can be averaged for all frames of the signal. Using a voice activity detector, the pauses can be removed if we are only interested in the signal frames. This quality (actually similarity) criterion is widely used for quick estimation of the distortions through the channel or the signal enhancement algorithms. It has higher correlation with MOS test results, but still does not account for the psycho-physiological properties of human hearing.

Some sources state that using automatic speech recognition as a quality assessment gives quite a high correlation with MOS results. This and other objective speech quality assessment methods are evaluated and compared in [16].

The block diagram of the most frequently mentioned method for perceptual evaluation of the sound quality (PESQ) [17,18] is shown in Figure 2.18. The reference and degraded signals are level-adjusted and bandpass-filtered. Then they are time-aligned to compensate for the delays in the evaluated system. Each signal is processed with auditory transform blocks and then both are compared in the disturbance processing block. The cognitive modeling block computes the final result, equivalent to the MOS score.

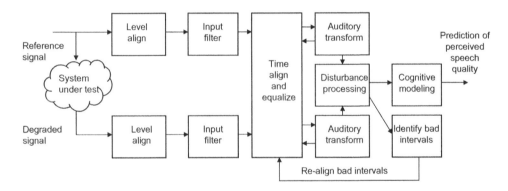

Figure 2.18 Block diagram of perceptual audio sound quality assessment algorithm

Automated measurement of speech and audio quality is standardized in several ITU-T recommendations. They are shown in Table 2.7 with brief descriptions. The standards can be downloaded from the ITU site. They contain reference implementation code in ANSI-C. Using some of these standards for evaluation of the speech enhancement algorithms gives comparable results with other algorithms. Regardless of time savings and the comparability of the results, however, informal listening tests and MOS tests should be performed to confirm the speech quality improvement.

Table 2.7 ITU-T standards for audio and speech quality measurement

Abbreviation	Standard number	Year
Narrowband speech, intrusive		
E-model	ITU-T G.107	2000
PSQM	ITU-T P.861	1996
MNB	ITU-T P.861	1996
PESQ	ITU-T P.862	2001
PESQ-LQ	ITU-T P.862.1	2003
PESQ App. Guide	ITU-T P.862.3	2005
Narrowband speech, non-intrusive		
INMD	ITU-T P.561	1996
CCI	ITU-T P.562	2000
3SQM	ITU-T P.563	2004
Wideband speech		
WB-PESQ	ITU-T P.862.2	2005
Wideband audio		
PEAQ	ITU-T BS.1387	1999

2.10 Summary

This chapter has covered important properties of noise and speech signals: statistical, spectral, temporal, and spatial. This knowledge will be an aid to understanding the speech enhancement algorithms described in subsequent chapters.

Speech enhancement algorithms use different properties of the speech and noise signals. We can distinguish noise-suppression and noise-cancelling algorithms. While the first type reduces the noise based on the corrupted signal only, algorithms of the second group require knowledge of the noise signal. De-reverberation algorithms try to estimate and compensate for the room impulse response, reducing the portion of the captured signal energy coming from reflections. Acoustic echo reduction aims to subtract the signal from loudspeakers captured from the microphones. Similar to the noise-reduction algorithms, there are two major groups: acoustic echo cancellation and acoustic echo suppression.

The analog signal captured from a microphone goes through discretization and quantization on input to a digital computer. These processes have to be performed according to a sampling theorem to prevent information loss. Most of the sound and audio processing algorithms work in the so-called frequency domain. Usually they use Fourier transformation of windowed audio frames and the overlap-and-add process for signal reconstruction.

The signal quality can be evaluated using objective and subjective tests. Subjective quality measurements are based on mean opinion score tests in direct and differential variants. Among the objective quality measurement parameters are measurement of the signal-to-noise-ratio, log-spectral distance, and perceptual estimation of sound quality.

Bibliography

[1] Hoth, D.F. (1941) Room noise spectra at subscriber's telephone location. *Journal of the Acoustical Society of America*, **12**, 499–504.

[2] IEEE (2001) *Draft Standard Methods for Measuring Transmission Performance of Analog and Digital Telephone Sets, Handsets and Headsets*, 269-2001 (revision of 269-1992), IEEE.

[3] Vary, P. and Martin, R. (2006) *Digital Speech Transmission: Enhancement, Coding and Error Concealment*, John Wiley and Sons, Ltd, West Sussex, England.

[4] Martin, R. (2002) Speech enhancement using MMSE short-time spectral estimation with gamma distributed speech priors. Proceedings IEEE ICASSP'02, Orlando, FL.

[5] Halkosaari, T. and Vaalgamaa, M. (2004) Workshop on Wideband Speech Quality in Terminals and Networks: Assessment and Prediction. Mainz, Germany.

[6] ITU-T (1989) *Recommendation G.711: Pulse Code Modulation of Voice Frequencies*, ITU-T, Geneva, Switzerland.

[7] Cooley, J. and Tukey, J. (1965) An algorithm for the machine calculation of complex fourier series. *Mathematics of Computation*, **19**, 297.

[8] Max, J. (1981) *Methodes et techniques de traitement du signal et applications aux mesures physiques*, 3rd edn, vols **1 and 2**, Masson, Paris [in French].

[9] Malvar, H.S. (1986) Optimal Pre- and Post-filters in Noisy Sampled-data Systems, PhD Thesis, MIT, Cambridge, MA.

[10] Malvar, H.S. (1992) *Signal Processing with Lapped Transforms*, Artech House, Boston, MA.

[11] ITU-T (1999) *Recommendation G.722.1: Coding at 24 and 32 kbit/s for Hands-free Operation in Systems with Low Frame Loss*, ITU-T, Geneva, Switzerland.

[12] Malvar, H.S. (1999) A modulated complex lapped transform and its applications to audio processing. Proceedings of IEEE ICASSP'99, Phoenix, AZ, pp. 1421–1424.

[13] Bogert, B.P., Healy, M.J.R. and Tukey, J.W. (1963) The quefrency analysis of time series for echoes: cepstrum, pseudo-autocovariance, cross-cepstrum, and saphe cracking, in *Proceedings of the Symposium on Time Series Analysis* (ed. M. Rosenblatt), Chapter 15, John Wiley & Sons, New York, pp. 209–243.

[14] Acero, A. and Huang, H. (1995) Augmented cepstral normalization for robust speech recognition. Proceedings of the IEEE Workshop on Automatic Speech Recognition. Snowbird, UT.

[15] ITU-T (1996) *Recomendation P.800: Methods for Subjective Determination of Transmission Quality*, ITU-T, Geneva, Switzerland.

[16] Liu, W.M., Jellyman, K., Mason, J.S.D. and Evans, N.W.D. (2006) Assessment of objective quality measures for speech intelligibility estimation. Proceedings of IEEE ICASSP06, Toulouse, France.

[17] Rix, A.W., Beerends, J.G., Hollier, M.P. and Hekstra, A.P. (2001) Perceptual evaluation of speech quality (PESQ): a new method for speech quality assessment of telephone networks and codecs. Proceedings of IEEE ICASSP01, Salt Lake City, UT.

[18] ITU-T (2001) Recommendation P.862: Perceptual Evaluation of Speech Quality (PESQ): An Objective Method for End-to-End Speech Quality Assessment of Narrow-band Telephone Networks and Speech Codecs, ITU-T, Geneva, Switzerland.

[19] Martin, R. and Breithaupt, C. (2003) Speech enhancement in the DFT domain using laplacian speech priors. Proceedings of International Workshop on Acoustic Echo and Noise Control (IWAENC), Kyoto, Japan.

[20] Huang, Y. and Benesty, J. (2004) *Audio Signal Processing for Next Generation Multimedia Communication Systems*, Kluwer Academic, Boston, MA.

[21] Oppenheim, A.V. and Schafer, R.W.D (1999) *Discrete-Time Signal Processing*, Prentice-Hall, Upper Saddle River, NJ, pp. 788–789.

[22] Quckenbush, S.R., Barnwell, T.P. and Clements, M.A. (1988) *Objective Measures of Speech Quality*, Prentice-Hall, Englewood Cliffs, NJ.

3

Sound and Sound Capturing Devices

This chapter defines the sound wave and how it propagates in different media, especially in air. Microphones are the sensors for capturing the sound wave, and several aspects of the microphone will be discussed in detail: principles of operation, parameter definitions, and measurements. Typical microphone models will be derived towards the end of the chapter.

3.1 Sound and Sound Propagation

3.1.1 Sound as a Longitudinal Mechanical Wave

Sound is a waveform consisting of density variations in an elastic medium, propagating away from the source. The propagation medium can be air, water, or a solid material. Sound generation, propagation, and detection are related to the performance and conversion of mechanical work. The sound generator does mechanical work to agitate the medium. For example, a loudspeaker diaphragm initiates movements of air molecules by vibrating in the air. Energy is converted from one form to another – electrical to mechanical. The sound wave, propagating through the air, is nothing but a transition of this mechanical energy using the elastic and inertial properties of the medium. Thus the sound in the air is periodic changes in pressure. On reaching an eardrum or microphone diaphragm, the sound causes microscopic movements or vibrations; that is, the sound does mechanical work on the eardrum or diaphragm.

On the surface of water, a wave propagates by particles moving in a circular manner: up, forward, down, and back. That is why it is called a "circular" wave. When a wave propagates on a rope, the displacement is perpendicular to the direction of the wave motion, so that wave is called "transverse." In contrast, sound consists of "longitudinal" waves, as the medium's particles are displaced *parallel* to the direction of wave

Sound Capture and Processing   Ivan J. Tashev
© 2009 John Wiley & Sons, Ltd

propagation. It is important to note that, in all three cases, the particles of the medium are merely displaced *locally* – they oscillate around a certain position. *It is the wave that travels from source to detector.* It can be estimated that the air particles move less than $1\,\mu m$ (1×10^{-6} m) when sound propagates, and there is no net flow of air. Therefore, sound comprises moving areas of compression and rarefaction, leading to microscopic changes in air pressure, as shown in Figure 3.1.

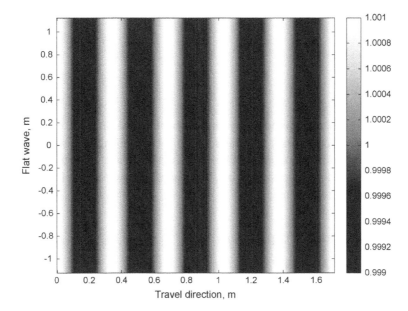

Figure 3.1   Sound wave as changes in the atmospheric pressure

3.1.2 Frequency of the Sound Wave

Whether it is the bang of an explosion, the roar of an earthquake, the steady tone of a whistle, or the sonar pulse of a flying bat, the most fundamental property of sound is its frequency or frequency band. This is measured in cycles per second or hertz (Hz). Sound is a very low-energy phenomenon – humans can hear a 1 kHz tone with an intensity of just 10^{-12} W/m^2. For comparison, daylight intensity is around 1000 W/m^2, a difference of 10^{15} times [1]!

   Devices used to generate and detect sound are called "transducers." To achieve good sensitivity at such low power the transducers should use resonance or resonances. This is why humans and animals use sound within a limited bandwidth.

   We will see later in this book that the sensitivity of human hearing varies with the sound frequency. We are most sensitive to sounds in the range 1000–2500 Hz, and sensitivity to lower and higher frequencies degrades quickly. It is generally assumed that humans can hear sounds with frequencies within the range 20–20 000 Hz. The upper limit is valid for very young people, but the limit reduces with age such that, at the age of 40, it is around 12 000 Hz. This is due to the reduced elasticity of the aging eardrum.

The human hearing range allows grouping of sounds based on their frequencies. Audible sound, or just "sound," has a frequency between 20 Hz and 20 kHz. Sound with a frequency above the upper limit is called "ultrasound," and below the lower limit it is called "infrasound." While ultrasounds are generated and can be heard by many animals (dogs, horses, spiders, bats, dolphins), infrasound is almost never used in nature. With their internal organs, humans can detect, as vibrations, sounds with frequencies as low as 1 Hz. Highly intense infrasound affects the inner ear and can cause dizziness. Figure 3.2 shows the frequency range various animals can hear.

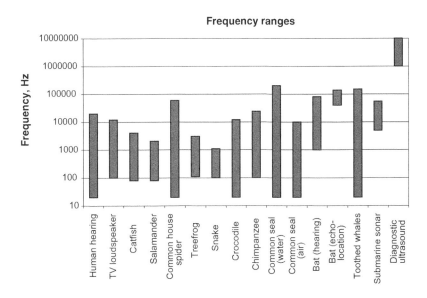

Figure 3.2   Frequency ranges heard by various animals. TV loudspeaker and diagnostic ultrasound are for reference only

An interesting question is how far we can go with the ultrasound frequency. At room temperature and at normal pressure, air molecules have a mean free path of 7×10^{-8} m. This is the average distance they travel between collisions with each other – the way air transmits the changes in temperature and pressure, including sound waves. This mean free path corresponds to half the wavelength of a sound with frequency 2.5 GHz $(2.5 \times 10^{12}$ Hz). We can say that air generates sound at this frequency. It is considered as the boundary where the particulate nature of air becomes noticeable and it ceases to behave as a continuum. This boundary varies with temperature, pressure, and the material.

3.1.3 Speed of Sound

In general, the speed of a longitudinal mechanical wave, like sound, is given by

$$c = \sqrt{\frac{C}{\rho}} \tag{3.1}$$

where C [Pa] is the stiffness of the medium and ρ [kg/m^3] is its density. This means that the sound's speed increases with the stiffness of the material and decreases with the density. Note that this equation assumes that the speed of sound does not depend on frequency. Such media are called "non-dispersive." If the speed of the sound wave does vary with frequency, the material is called "dispersive." With frequencies up to 28 kHz, air is very close to being a non-dispersive medium. We will assume that the speed of audible sound does not depend on frequency. Note that for some ultrasound applications this effect should be accounted for.

For solid materials, the stiffness is given by Young's modulus E. For sheet steel, $E = 200$ GPa, $\rho = 7850$ kg/m^3, and the speed of sound is $c = 5047$ m/s. Note that Young's modulus is defined when the length of the material is much larger than its lateral dimensions. The speed of sound in steel is different if this is not the case.

For the propagation of sound waves in liquids, things are slightly different. In most cases we can assume that liquids are incompressible; that is, their density does not depend on pressure. Obviously the existence of sound in a medium requires density fluctuation, and so at first sight propagation of sound in a medium that is assumed incompressible seems self-contradictory. Sound does propagate in liquids, but it is not a simple wave, described by a linearized wave equation. There are relatively very small changes in the fluid density and we can use the same assumptions for the longitudinal mechanical wave. In this case the stiffness is replaced by the adiabatic bulk module K. For distilled water $K = 2.2 \times 10^9$ Pa and $\rho = 998.2$ kg/m^3 at a temperature of 20 °C. This gives the speed of the sound wave as $c = 1484.6$ m/s. In seawater that is free of bubbles and sediments, sound travels at approximately 1500 m/s. This depends on the temperature (a change of 4 m/s for each degree Celsius), the salinity (a change of 1 m/s for each part per thousand), and the pressure (hence the depth). At 25 °C, a salinity of 35 parts per thousand, and a depth of 1000 m, the speed of sound in seawater is 1550.74 m/s. There are numerous empirical approximations for computing the speed of sound in seawater since the precision of sonars depends on it and it affects the study of the behavior of marine animals that use sounds for orientation and finding their prey. For more information see [2].

For a gas, the bulk module B (equivalent to the stiffness in solids C) is given by $B = \gamma p$, and the speed of sound becomes

$$c = \sqrt{\gamma \frac{p}{\rho}} \tag{3.2}$$

where p is the pressure, ρ is the density, and γ is the ratio of the heat capacity of the gas at a constant pressure to that at a constant volume – frequently called the "adiabatic index." For completely adiabatic compression and expansion (i.e., without heat exchange), $\gamma = 1.4$ for diatomic gas molecules. If the compression and expansion of the gas occurs at a constant temperature (i.e., complete temperature exchange with the gas surroundings assumed to be an infinite reservoir), then $\gamma = 1.0$. Experiments show that the speed of sound in air agrees with $\gamma = 1.4$, which just means that the sound wave propagation is adiabatic. However, with increasing frequency the distance between compressions and rarefactions decreases and the heat conduction between them increases. Under standard conditions of pressure and temperature at frequencies of 7×10^8 Hz, γ becomes 1.0 and the speed of sound approaches $c = (p/\rho)^{1/2}$. This frequency is far above anything used in nature and technology. For audible sound, we can assume that the sound wave propagates adiabatically. As both the pressure and density of the air depend on its temperature (they are exactly proportional for ideal gas and the air behaves very closely to this assumption), we can go one step further and replace their ratio:

$$c = \sqrt{\frac{\gamma k T}{m}} \tag{3.3}$$

where $k = 1.3806504 \times 10^{-23}$ is Boltzmann's constant, T is the temperature in kelvins, and m is the mass of a single molecule (or average mass for gas mixtures) in kilograms. For the mixture of nitrogen, oxygen, carbon dioxide, and other gases we call air, $m = 1.5936 \times 10^{-23}$ kg. Recalling that T [K] $= 273.15 + T$ [°C], we can compute the speed of sound in air at the freezing point (0 °C) to be $c = 331.3$ m/s. More commonly used is the speed of sound in air at room temperature (20 °C), which is 343.2 m/s.

The speed of sound is one characteristic of the medium. Another important property of the medium is the "acoustic impedance." The concept of impedance is common in physics and refers to the ratio of a driving force to the velocity response. In Ohm's law, for example, impedance is the ratio of a driving force (voltage) to the velocity response (current), and it is called "resistance" in simple cases. In acoustics, impedance is given by the ratio of the acoustic pressure amplitude (driving force) and velocity of the particles in the medium (velocity response). After several substitutions[1], we find the expression for acoustic impedance:

$$Z = \rho c. \tag{3.4}$$

The specific acoustic impedance of a material is the product of the density and the speed of sound in that material, measured in [kg m^{-2} s^{-1}]. Table 3.1 shows the acoustic parameters of various materials.

Table 3.1 Acoustic parameters of some materials

Material	Density (kg/m³)	Speed of sound (m/s)	Specific acoustic impedance (kg/m²·s)
Air	1.2	330	400
Water	1000	1480	1.48E + 06
Steel	7850	5050	3.95E + 07
Aluminum	2700	6400	1.70E + 07
Brass	8500	449	3.80E + 07

3.1.4 Wavelength

The length of the sound wave is defined as the smallest distance measured along the direction of the wave propagation between two points having identical pressure. As this is the distance traveled by the sound wave for one period, the wavelength can be computed as

$$\lambda = \frac{c}{f} \tag{3.5}$$

where λ [m] is the wavelength, c [m/s] is the speed of sound, and f [Hz] is the frequency of the sound. As mentioned earlier, humans can hear sounds with frequencies in the range of 20–20 000 Hz. At room temperature the wavelength will be between 17.16 m for 20 Hz and 1.7 cm for 20 kHz. For comparison, the wavelength of visible light (from dark red to violet) is between 740 and 360 nm (1 nanometer $= 1 \times 10^{-9}$ meters). There are two major differences.

The first is that the wavelength of visible light is many times smaller than any object in the surrounding world. This is why light propagates in a straight line and we see sharp shadows behind objects. It is difficult to demonstrate the wave properties and effects of visible light (interference or diffraction). On the other hand, the wavelength of audible sound is comparable with the size of surrounding objects. For example, in a large room with size $5 \times 4 \times 3$ m, the sound will reflect, diffract, and interfere with its own delayed and attenuated copies. The sound will be audible behind objects and its frequency content will be distorted due to interference and diffraction. The second major difference is the wavelength range. The ratio between the largest and smallest wavelength of visible light (740 nm for red, 360 nm for violet light) is approximately 2.1 times. The same ratio for audible sounds is 1000 times. This means that we need simultaneously to deal with different behaviors of a sound wave for different frequencies. A sound wave with frequency 100 Hz will completely diffract around an object with a size of 3 m, but we will detect acoustical shadows for a sound with a frequency of 10 000 Hz.

The wideband character of sound impacts human hearing as follows. The head's diameter (and hence the distance between the ears) is in the range 10–20 cm. For frequencies below 100 Hz, both ears hear practically the same sound and we cannot detect the direction the sound comes from. For frequencies around 1000 Hz, humans use the phase difference between the sound, detected by the two ears, to determine the direction of arrival; and for frequencies above 3500 Hz we rely mostly on the sound energy envelope. Note that the sound wave tends to diffract around objects with a size smaller than the wavelength and reflects from objects with a size larger than the wavelength. Echo location is used by many mammals. Bats, for example, use a chirp signal with a frequency between 30 and 80 kHz. The wavelength of this signal is between 11 and 4 mm – the typical size of insects in the diet of this animal. Toothed whales use frequencies up to 200 kHz. The reason for this is that sound propagates in the water almost five times faster and sound with a frequency of 200 kHz has a wavelength of 7 mm, which is the resolution of the sonar of this animal. To achieve a resolution better than one half of a millimeter, ultrasound medical equipment uses 2.5 MHz as the sound propagates through human tissue practically with the same speed as in water.

3.1.5 Sound Wave Parameters

3.1.5.1 Intensity

The "acoustic intensity" is defined as the rate at which energy in the wave crosses a unit area perpendicular to the direction of propagation. The measurement unit is watts per square meter: [W/m^2]. Two similar waves can be compared by the ratio of their powers, and this is the foundation of the *decibel scale*. To express the intensity of a signal in the decibel scale it should be compared to a reference level. In air the reference standard is taken at a level of $I_{ref} = 10^{-12}$ Wm^{-2}, which is approximately the threshold intensity for normal human hearing at 1000 Hz. Then the intensity level (IL) will be

$$\text{IL} = 10 \log_{10}\left(\frac{I}{I_{ref}}\right). \tag{3.6}$$

3.1.5.2 Sound Pressure Level

The reference intensity in equation (3.6) corresponds to an acoustic pressure amplitude of 28.9 μPa for plane and spherically traveling waves. The sound pressure level (SPL) is usually taken as the root mean square (RMS) of the sinusoidal wave, which is the amplitude divided by $\sqrt{2}$. The result of 20.4 μPa is rounded to 20 μPa, which is the reference level for measuring the sound pressure level:

$$A_{SPL} = 20 \log_{10}\left(\frac{P}{P_{ref}}\right). \tag{3.7}$$

Because of this rounding, SPL is almost, but not exactly, equal to the IL for plane and spherical waves. For other media, different reference levels can be used. For underwater acoustics, for example, levels of $20 \mu Pa$, 1 μbar, and 1 μPa, equivalent to intensities of 2.70×10^{-16}, 6.76×10^{-9}, and $7.76 \times 10^{-19} Wm^{-2}$, respectively, are used. For more complex waveforms the measurements of IL and SPL can disagree.

The advantage of this logarithmic scale is that it can represent the vast range of intensities to which the human hearing can respond. An additional advantage is that the human sensory perception of loudness is logarithmic and we judge one sound to be so many times louder than another. Figure 3.3 shows the existing sound pressure level in various conditions [3].

Sound levels in various conditions

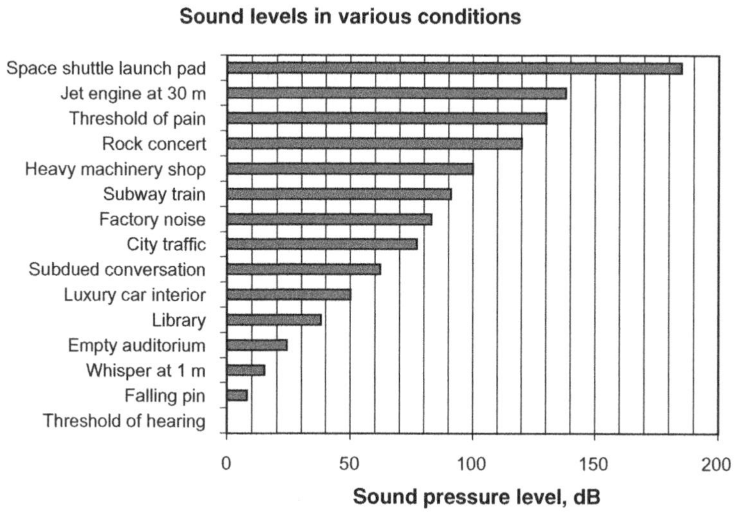

Figure 3.3   Sound pressure levels in various conditions

3.1.5.3 Power

The power of a sound source can be estimated from the intensity through a surface S completely surrounding it:

$$W = \int_S I \cdot dS. \tag{3.8}$$

If we assume S to be a sphere with radius r and a sound source in the center, then

$$I(r) = \frac{W}{S} = \frac{W}{4\pi r^2}. \tag{3.9}$$

This is known as the inverse square law for sound propagation, which accounts for the fact that sound becomes weaker as it travels in an open space away from the source, even if viscous effects of the medium are disregarded. Note that while the intensity decreases with the inverse square of the distance, the sound pressure level decreases with the inverse distance to the sound source.

3.1.5.4 Sound Attenuation

Propagation of sound in an isotropic medium leads to energy dissipation. The wave intensity decays not only due to the inverse square law, but also due to certain energy loses, caused by internal friction. A good model for this process is the exponential decay of the amplitude and intensity of the wave, propagating towards the x axis:

$$
\begin{aligned}
A(x) &= A_0 e^{-bx} \\
I(x) &= I_0 e^{-2bx}
\end{aligned}
\tag{3.10}
$$

where A_0 and I_0 are the amplitude and intensity at point $x = 0$ and b is the sound attenuation ratio, measured in $[\text{m}^{-1}]$. For transverse waves in gases and liquids:

$$
b = \frac{(2\pi f)^2}{2\rho c^3} \left[\frac{4}{3}\eta + \xi + K\frac{c_p - c_V}{c_p c_V} \right],
\tag{3.11}
$$

where f and c are the frequency and the speed of the wave; ρ, η, ξ, and K are the density, shear viscosity, second viscosity, and heat conduction coefficient, respectively; and c_p and c_V are the specific heat capacities at constant pressure and at constant volume respectively. In air, all of these parameters depend on atmospheric pressure, temperature, and humidity. As the exponential attenuation is linear in the logarithmic scale, more frequently it is measured in decibels per meter [dB/m]:

$$
B = \frac{20\log_{10}A(x_1) - 20\log_{10}(A(x_2))}{x_2 - x_1} = \frac{20b}{\log_{10}e}.
\tag{3.12}
$$

Figure 3.4 shows the attenuation of a sound wave at room temperature (20 °C), normal atmospheric pressure (1.013 MPa), and 50% humidity as a function of the frequency. The sound attenuation is negligible for distances comparable to a normal room size. From 1 to 3 meters, for example, the sound pressure level of sound with a frequency 1000 Hz will decrease by 9 dB owing to the inverse distance law and by 0.009 dB due to absorption. In this book we will not consider sound attenuation due to friction and energy loses.

3.1.6 Huygens' Principle, Diffraction, and Reflection

Christiaan Huygens (Dutch mathematician, physicist, and astronomer, 1629–1695) stated a principle by which wave motion can be approximated: each point of an

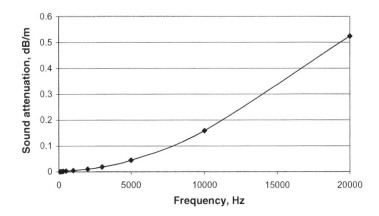

Figure 3.4 Attenuation of sound due to energy losses at normal atmospheric pressure, 20 °C temperature, and 50% humidity

advancing wave can be treated as a point source of secondary wavelets, and so the position of the wavefront a short time later can be found as the envelope of these wavelets.

Using this principle is easy to explain the *diffraction* of a sound wave: the effect when the sound wave diffracts around obstacles and is audible behind the object, or when inside a room with an open door all sounds in the hallway seem to sound as if their source is the door. Simply, the large hole can be approximated by a collection of many small holes, so each is practically a point source.

Refraction occurs when a plane sound wave reaches the border between two media with different speeds of wave propagation. The plane wave changes its direction as described by Snell's law:

$$\frac{\sin \theta_I}{c_1} = \frac{\sin \theta_T}{c_2} \tag{3.13}$$

where θ_I and θ_T are the incident and transmitted angles when the wave crosses the boundary between media with wave speeds of c_1 and c_2, respectively.

Reflection is part of the same process. A portion of the sound wave, reaching the boundary between two media, is transmitted and refracted, a portion is reflected, and a portion is dissipated and lost. The angle of the reflected wave is equal to the incident angle, according to the reflection law:

$$\sin \theta_I = \sin \theta_R. \tag{3.14}$$

Reflection causes a reversal in normal wave velocity, and in pressure (i.e., a compression is reflected as rarefaction). This simply means that the reflection process reverses the phase of the sound wave. The ratio of transmitted to reflected energy depends on the ratio of the acoustic impedances of the media. If they are equal the wave will be transmitted completely, and the two media are said to be "impedance matched." In the case of a large difference in the impedances (air and water, for example), most of the energy is reflected for incidences from any of the boundary sides. Because the acoustic impedance of air and water differ by a factor of nearly 4000, any mechanism designed to couple to one medium is unlikely to couple to the other. This might perhaps be why the common seal appears to employ two different hearing mechanisms, one for aerial hearing and one for aquatic, with maximum audible frequencies of 12 kHz and 160 kHz, respectively [4,5].

3.1.7 Doppler Effect

The Doppler effect is the change of frequency and wavelength when the transmitting and receiving points are moving. It is named after Christian Doppler, Austrian mathematician and physicist, who in 1842 published his famous study on changes of the stars' light colors. For waves that propagate in a medium, such as sound waves, the velocities of the observer and the source are reckoned relative to the medium in which the waves are transmitted. When we stand in a train station and a whistling and moving train passes, we hear the whistle tone change – it is higher when the train approaches, then changes to lower when the train leaves the station. Meanwhile, for a passenger in the train, the whistle's tone remains constant. The total Doppler effect may therefore result from either motion of the source, or motion of the observer, or both. The frequency change is given by

$$f = \frac{f_0}{1 - \frac{|v_s - v_o|}{c} \cos \theta} \tag{3.15}$$

where f_0 is the original frequency, f is the observed frequency, v_s and v_o are the speed vectors of the source and observer respectively, c is the speed of sound, and θ is the angle between the resulting speed direction and the line connecting the source and the observer. The formula above is valid when each speed is lower than the speed of sound. Discussing the effects of supersonic boom is outside the scope of this book.

There are many practical applications of the Doppler effect. In astronomy it allows the measurement of the speed and temperature of stars. Note that, according to Huygens' principle, a reflection can be modeled as a new sound source. From this perspective, sound or wave reflection from a moving object is the same as a moving source. A Doppler radar sends a high-frequency radio wave, which is reflected from the moving object (car, plane) and sent back with a frequency changed by the Doppler effect. The difference in the frequencies $|f - f_0|$ is relatively easy to measure and it is trivial from the difference to compute the speed of the moving object. The same principle is used to measure the flow in blood vessels and pipes using ultrasound.

3.1.8 Weighting Curves and Measuring Sound Pressure Levels

In the definition of the sound pressure level, P is the root mean square of the pressure changes; that is, the RMS of the sound wave magnitude. This way of measuring the SPL includes all frequencies – from infrasound with very low frequency to ultrasound with high frequency. A passing cloud causes infrasound with an SPL of 190 dB, but this is of interest more to the meteorologists than to audio-related measurements. In most cases we are interested in measuring audible sounds, which means that we have to at least remove the frequencies below 20 Hz and above 20 kHz. This is the reason for standardizing the frequency response of the measurement devices – so-called "weighting curves" for measuring the sound pressure level. Later the practice introduced several weighting curves.

One of the most commonly used is A-weighting. The sensitivity is very low in the lower part of the frequency band, has its peak around 2500 Hz, and then goes down again. This weighting is used to simulate the human sensitivity – we have lower sensitivity for low and high frequencies. A human will sense a two times increase in the noise level when the sound pressure level meter shows that this increase is $+6$ dB. To explicitly state that A-weighting is being applied during SPL measurement, the units are denoted as "dBA." We will look at how humans hear different sounds later in this book; here we will just note that A-weighting matches the human perception at ambient noise levels around 40 dB SPL. Widely used interpolation of this curve in s-domain is given by:

$$G_A = \frac{k_A s^4}{(s+129.4)^4(s+676.7)(s+4636)(s+76655)^2} \tag{3.16}$$

where $k_A = 7.39705 \times 10^9$ and $s = j2\pi f$.

In order to find a better correspondence to human evaluation, B-weighting and C-weighting curves are standardized, suitable for louder sounds (70 dB SPL and 90 dB SPL, respectively). They have a wider flat part in the middle of the audible frequency band. C-weighting is used practically to cover the entire audible range of frequencies as it has slopes of decreased sensitivity below 30 Hz and above 15 000 Hz. To underline that C-weighting is applied, the measurement units are denoted as "dBC." The notation "dBB" for B-weighted SPL measurements is used as well. Interpolations in the s-domain for these two curves are given by

$$G_B = \frac{k_B s^3}{(s+129.4)^2(s+995.9)(s+76655)^2} \tag{3.17}$$

and

$$G_C = \frac{k_C s^2}{(s+129.4)^2(s+76655)^2}, \tag{3.18}$$

where $k_B = 5.99186 \times 10^9$ and $k_C = 5.91797 \times 10^9$. All three types of weighting are standardized in ANSI document S1.42.

During the 1960s, with the work on Dolby-B noise suppression algorithms for tape and cassette recorders, it was found that humans are highly sensitive to (or annoyed by) noise in the area around 6 kHz. The research, driven by the British Broadcasting Corporation, lead to Recommendation 468 (ITU-R 468-4). It includes the weighting and the exact methodology for measurement. Later it was simplified by Dolby Laboratories and become known as ITU-R ARM (average response meter). From the frequency response perspective, the curve is just shifted down 6 dB, as the 0 dB reference is moved from 1 kHz to 2 kHz). The shapes of these weightings are shown in Figure 3.5.

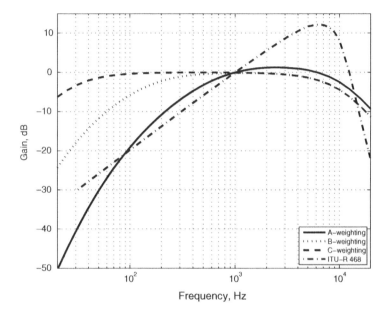

Figure 3.5   Weighting curves for noise measurements: A-, B-, and C-weightings and ITU-R-488

There are known, but less frequently used, D-weighting curves (similar to B, but with a "hump" around 3 kHz to underline the noise), and G-weighting (for infrasound measurements). In the literature, Z-weighting can be found as well, which is an abbreviation of zero weighting and is nothing but a completely flat frequency response and equal emphasis of all frequencies.

In telecommunications, standardized filters for measuring the noise in telephone lines are used. The most common is standardized by ITU-T O.41; in North America C-message weighting is frequently used, standardized by Bell System Technical Reference 41 009. These weightings model telephone line bandwidth and are shown

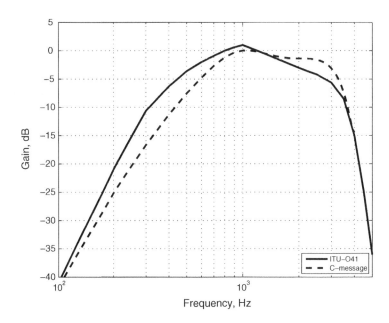

Figure 3.6   Weighting curves for noise measurements in telephony: ITU-O41 and Bell Labs C-message

in Figure 3.6. For some measurements, just a limited frequency band can be specified, which corresponds to a rectangular weighting window; for example: "the proposed algorithm achieved a 4.5 dB improvement of the signal-to-noise ratio (SNR) in the band 300–3400 Hz".

Considering the fact that most of the ambient noise energy is concentrated in the lower part of the frequency band, using some of these weightings we can see much "better" results from some algorithms (signal-to-noise ratio improvement, for example). This is not cheating if the proper weighting is used and explicitly stated when presenting the results. When measuring noise coming from a point noise source, besides specifying the SPL and the weighting, the distance between the microphone and the noise source should be specified. The standard distance is one meter and if it differs it should be explicitly stated. For measuring ambient noise levels (a street, inside an airplane) the distance is meaningless.

3.2 Microphones

3.2.1 Definition

The microphone is a sensor for capturing the small changes in air pressure we call sound and converting them into an electrical signal. The need to convert the sound to an electrical signal arose with the first telephone devices.

3.2.2 Microphone Classification by Conversion Type

Almost every microphone first converts the changes in air pressure to mechanical movements with a diaphragm – a flexible membrane that bends under forces caused by the difference in air pressure on the two sides. These mechanical movements are then converted to an electrical signal. The way different microphones do this conversion allows us to classify them into various groups.

Carbon microphones were the first widely used microphones. The diaphragm movements apply pressure to carbon dust, changing its electrical conductivity. Air pressure changes are thus converted to changes in the microphone's electrical resistance. A simple circuit consisting of a 60-volt battery, a carbon microphone, and a telephone capsule (to convert voltage to sound by applying variable magnetic field to a metal membrane) can be used to transmit sound over large distances. Carbon microphones were used in most telephones for more than half a century. They have a frequency response barely covering the telephone line quality and a low dynamic range.

In *piezoelectric microphones* the diaphragm bends a piezocrystal, which converts the mechanical deformations into an electric voltage. The effect was first used in electrical gramophones. These microphones provide higher sound capture quality, but the output voltage is in the range 50–100 mV and they have a high output impedance. This requires an electronic circuit for amplification and impedance conversion, which is why these microphones became usable only after the invention of electronic vacuum tubes.

Electrodynamic microphones convert the movement of the diaphragm to an electric voltage by moving a coil, attached to the diaphragm, in a magnetic field. As there is no mechanical contact to carbon dust or piezoelectric crystal, the diaphragm can be made small and light. This is why for many years these microphones were used for professional radio broadcasting and in the recording industry. The output voltage is of the order of several millivolts and the output impedance is low. This requires good shielding of the cables and electronic amplification of the signal. An electrodynamic microphone is constructed in a very similar way to a loudspeaker. This means that if a small voltage signal is applied to the coil it will make the diaphragm vibrate and produce sound, and any dynamic loudspeaker can be used as a low-quality microphone.

Condenser microphones convert the movements of the diaphragm into a variable capacitance by means of a non-movable membrane placed parallel to the moving diaphragm. The two plates thus form a capacitor. Changes in capacitance can be sensed by applying a constant voltage (48 V) and measuring the charging current. Very precise circuits can include a generator of high frequency and direct measuring of the condenser microphone's impedance. For more information about condenser microphones see [6].

For a long time, condenser microphones were used for calibration purposes and in acoustic measurement devices owing to their high price and complex electronics required. Things changed with the invention of the electret. Condenser microphones could be built smaller, which required a polarization voltage in the range 1–5 V. A small FET (field-effect transistor), integrated into the microphone, converts the charge changes into an electric voltage. Today nearly every microphone used in telephones, consumer electronics, the recording industry, and radio broadcasting are electret condenser type microphones.

Recently, *miniature electro-mechanical system microphones* (MEMS microphones) have become popular. They are manufactured from a silicon crystal with the same technology used for manufacturing integrated circuits. This technology allows the formation of a diaphragm and holding it at a certain distance from the base. Conversion of the movements of the diaphragm to an electrical signal is based on varying the capacitance between the diaphragm and the base; that is, MEMS microphones are condenser microphones. The advantages of MEMS microphones are their small size ($3 \times 3 \times 1$ mm), the same packaging as surface-mounted components (so the same printed board manufacturing robots can be used for soldering them), and lower manufacturing tolerances (important for their use in microphone arrays). In addition, the accompanying electronic circuitry (preamplifiers, frequency correction, even part of the ADC – so-called "digital MEMS") can be integrated into the microphone chip using the same manufacturing process. The main disadvantages of MEMS microphones are still a higher price and higher self noise. The last is due to the smaller size of the diaphragm.

3.3 Omnidirectional and Pressure Gradient Microphones

3.3.1 Pressure Microphone

Most microphones use a flexible diaphragm as a sensing device, which measures not the air pressure directly, but the difference in pressure on the two sides of the diaphragm. These differences cause the diaphragm to bend, which is converted later to an electrical signal. To make the diaphragm measure changes in air pressure, on one side we have to have constant air pressure.

A closed cylindrical capsule with a diaphragm on one of the sides, as shown in Figure 3.7a, forms a classic pressure microphone. Usually there are very small holes on the back of the capsule to equalize changes in atmospheric pressure. Since the size of the diaphragm is much smaller than the wavelength of the highest frequency we want to capture, we can say that the pressure microphone samples the air pressure at a single point. This microphone is *not* sensitive to the direction of arrival of the sound wave, so it is called "omnidirectional."

The diameter of the diaphragm for the popular electret omnidirectional microphones is 5–9 mm, while the wavelength of an acoustic wave with frequency of 20 kHz is 17 mm. This allows such microphones to have a practically flat frequency response for the entire range of audible frequencies from 20 Hz to 20 kHz.

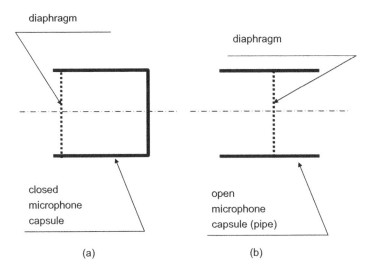

Figure 3.7   Microphone types: (a) pressure, and (b) pressure gradient

3.3.2 Pressure-gradient Microphone

If the diaphragm is placed into a small pipe open at both ends, as shown in Figure 3.7b, the new microphone measures the difference in air pressure at the two entrance points of the pipe. This is called a "pressure-gradient microphone" because it measures the air pressure gradient towards its axis. This microphone is sensitive to the direction of travel of the sound wave. For a sound wave that propagates parallel to the axis of the microphone, it will detect the first derivative of the air pressure. For a sound wave coming perpendicular to the main axis, marked with a dashed-dotted line, the air pressure on both sidess of the diaphragm will be the same at any time and there will be no output signal. This microphone has a directivity pattern in the shape of a figure 8 – sensitive to sounds coming from the front or back, but not to sounds coming from the sides. The same sound, coming from the back, will have an opposite phase from the one coming from the front. We can say that, while the pressure microphone is an acoustic *mono*pole, the pressure-gradient type is an acoustic *di*pole.

Another distinct characteristic of the pressure-gradient microphone, besides its specific directivity pattern, is its frequency response. As it captures the difference of the air pressure at two points, the output signal for a far-field sound wave should be the difference between the signal at one of the diaphragm sides and the same signal, delayed due to the microphone length l and incident angle θ as it arrives at the other side of the diaphragm:

$$U(f,\theta) = 1 - \exp\left(-j2\pi f \frac{l\cos\theta}{c}\right) \qquad (3.19)$$

where c is the speed of sound.

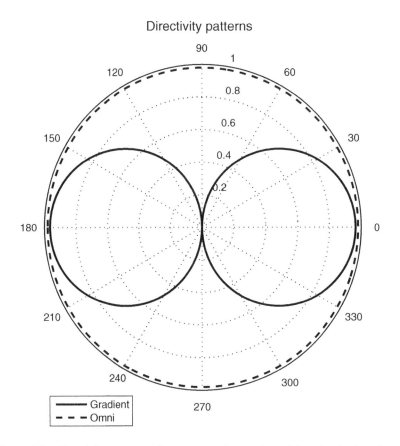

Figure 3.8 Directivity patterns of pressure-gradient and omnidirectional microphones

Figure 3.8 shows a plot of the gain (i.e., the ratio of the output and input signals) of a microphone with length $l = 5$ mm as a function of the incident angle θ for 1000 Hz, which confirms the figure-8 directivity pattern. Owing to the different wavelength, the derivative's magnitude changes and this microphone has -6 dB/octave slope towards the lower part of the frequency band. To compensate for this, usually a first-order low-pass filter is used:

$$U(f,\theta) = \frac{Z_C}{Z_C + R}\left(1 - \exp\left(-j2\pi f\frac{l\cos\theta}{c}\right)\right), \qquad (3.20)$$

where $Z_C = -j/2\pi f C$ is the capacitor impedance and $\tau = RC$ is the filter's time constant. Figure 3.9 shows the frequency response towards the main response axis, corrected with a first-order low-pass filter with a time constant of 1 ms. With this time constant the lowest part of the frequency band is undercompensated – this slight slope is presented in the majority of the microphone capsules, like these used in laptops, speakerphones, and so on. This frequency correction does not affect the directivity pattern.

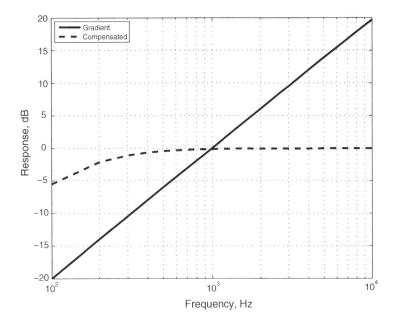

Figure 3.9 Frequency response of pressure-gradient microphone towards MRA: non-compensated and compensated

EXERCISE

Write a MATLAB® script that models the pressure gradient microphone directivity and frequency response, according to Equation 3.19. As parameters, use the frequency and the cosine of the incident angle:

```
GradientMicrophoneDirectivity(Freq, CosTheta).
```

Plot and compare the directivity pattern for various frequencies for a full circle – the result should look like a figure 8.

Add to the script above the first-order frequency correction according to Equation 3.20. Plot the frequency response towards the main response axis to verify the proper compensation and compare with Figure 3.9.

3.4 Parameter Definitions

3.4.1 Microphone Sensitivity

Microphone sensitivity is the ratio for converting changes in air pressure to electrical voltage. It is measured in [V/Pa] and allows estimation of the output voltage, given the sound field intensity. Since using decibels and the logarithmic scale is more convenient, 1 V/Pa is widely used as a reference level, or 0 dB.

For example, a normal conversation has a sound pressure level of 65 dB. With a reference level of 20 μPa, this means that the RMS amplitude of the sound wave is 35.5 mPa. If the sensitivity of the microphone we use is given to be −40 dB, then the RMS voltage on the microphone output will be 0.355 mV.

3.4.2 Microphone Noise and Output SNR

Regardless of the principle of converting the air pressure to an electrical signal, this process is accompanied by the introduction of some electrical noise. Its spectrum is typically flat (i.e., white noise), except with frequency-equalized unidirectional microphones when it will have increased magnitude towards the low frequencies (i.e., pink noise).

Manufacturers usually define the self noise of a microphone as its signal-to-noise ratio at a certain level of input sound. A typical value for a good electret microphone is an output SNR of 60 dB with sound at 96 dB SPL. A normal human voice at one meter generates sound with a 65 dB SPL. This means that under complete absence of ambient noise we will have an output SNR of 29 dB. While it seems to be much lower than the specified value, the ambient noise will still be dominant. In normal office conditions its level is around 50 dB SPL. This means that the SNR of the input acoustical signal is 15 dB and the microphone self noise is masked by the dominant ambient noise.

Another way to measure the microphone's self noise is to compare it with acoustical noise that will generate the same output noise level. In the numerical example above, the self noise of this microphone is equivalent to an external noise of 36 dB SPL. This is a very quiet condition, close to the noise level in specialized rooms for acoustic measurements.

The microphone self noise in good microphones is quite low and can be disregarded when capturing sound in normal conditions with a single microphone. It should be accounted for when capturing sound in very quiet conditions or trying to cover a very large dynamic range. This is the case when there is a loudspeaker nearby – in the speakerphone, for example. Another application when we should account for the microphones noise is in microphone arrays, when the self noise is a limiting factor to how directive we can make the array. This problem will be discussed in detail later in this book.

3.4.3 Directivity Pattern

The directivity pattern of a microphone is its sensitivity as a function of the incident angle. It is frequency sensitive, so usually the frequency is stated as an additional parameter. The incident angle is defined as the angle between the microphone's main response axis (usually the direction of highest sensitivity) and the direction of arrival of the sound. As the microphone can have a different phase shift for sounds coming from various directions and at various frequencies, the directivity pattern in general is a complex function. Usually the directivity pattern is normalized; that is, its magnitude is assumed to be 1 and it has no phase shift at 1000 Hz for sounds coming from the direction of the main response axis. In this book we will denote it as $U(f, d)$, where d is the direction. In the general case, $d = \{\varphi, \theta, \rho\}$ contains the direction, elevation, and distance to the sound source – that is, its coordinates in a radial coordinate system. This makes the directivity pattern a function of four parameters, which is difficult to specify and verify, or even represent.

In most cases we are interested in far-field sound capture, i.e. when the distance approaches infinity and the sound propagates as a flat wave. This model is good for distances above one meter and can be used in most cases. Then for a given frequency we can represent the directivity pattern as a 3-dimensional plot like the one in Figure 3.10a. The coordinate system is dimensionless and each point from the surface represents the microphone gain towards that direction by its distance to the coordinate system origin. This particular directivity pattern is the figure-8 directivity of the pressure-gradient microphone. It is more difficult to represent the phase response using 3D plots.

In addition, we can assume that the directivity pattern is symmetric around the main response axis. Then we can omit one of the angular dimensions of the direction. This allows us to create 3D plots representing the directivity pattern of the microphone as a function of the frequency and the incident angle. This type of plot is shown in Figure 3.10b. The same approach can be used to show the phase response. The particular directivity pattern, shown on this figure, is for a cardioid microphone, described later in this chapter.

Either of the two plots above can be created for a specific distance or plane in 3D space (i.e., given a proportion of φ and θ). Frequently used are polar plots, similar to that in Figure 3.8. Usually such plots contain the directivity pattern of the microphone for several frequencies, plotted with different line patterns.

3.4.4 Frequency Response

The frequency response is a section of the surface shown in Figure 3.10b. It is a 2-dimensional plot representing the magnitude or phase response as a function of the frequency for a given incident angle. The most commonly used is 0 degrees; that is, the direction towards the main response axis. Some microphone manufacturers add a $-180°$ line to illustrate the microphone suppression for sounds coming from the back.

3.4.5 Directivity Index

The directivity plots above present a large amount of numerical data. The directivity index (DI) summarizes it as one number. The DI for a given direction is defined as the ratio of the received power from this direction to the average received power

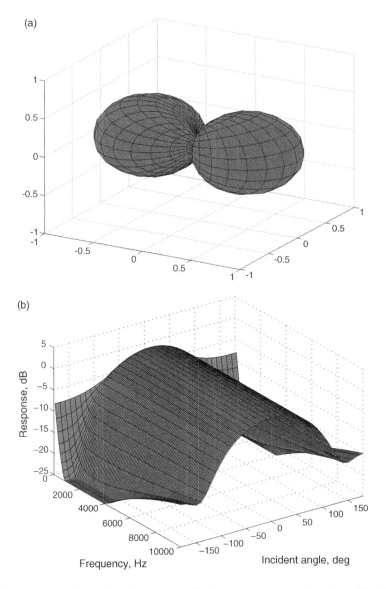

Figure 3.10   Representation of directivity pattern: (a) for one frequency as function of the 3D incident angle; (b) as a function of the frequency and the incident angle

from all directions:

$$\mathrm{DI}(f) = 10.\log_{10}\left(\frac{P(f,\varphi_T,\theta_T)}{\frac{1}{4\pi}\int\limits_0^\pi \mathrm{d}\theta \int\limits_0^{2\pi} \mathrm{d}\varphi \cdot P(f,\varphi,\theta)}\right) \tag{3.21}$$

where $P(f, \varphi, \theta) = |U(f, d)|^2$ and $\rho = \rho_0 = $ constant. Usually it is represented logarithmically and measured in decibels. It can be easily integrated towards the frequency axis to receive one single number, representing the total directivity of the microphone:

$$DI_{tot} = 10.\log_{10}\left(\int_0^{f_{max}} \frac{P(f, \varphi_T, \theta_T)}{\frac{1}{4\pi}\int_0^\pi d\theta \int_0^{2\pi} d\varphi \cdot P(f, \varphi, \theta)} df\right). \tag{3.22}$$

3.4.6 Ambient Noise Suppression

The total directivity index is a frequently used parameter to measure and compare the spatial selectivity of microphones. However the same DI can represent quite different noise suppression capabilities. The reason for this is that the ambient noise energy is not equally spread – it is concentrated mostly in the lower part of the frequency band. This means that a microphone with a higher DI in the lower part will suppress more of the ambient noise than a microphone with higher DI in the upper part of the frequency band, even if these two microphones have the same total DI. To measure the ambient noise suppression capabilities of the microphones the "noise gain" is defined, which is nothing but $DI(f)$ weighted with the noise spectrum before to be integrated towards the frequency-domain axis. It is normalized with the total noise energy; that is, the energy an ideal omnidirectional microphone will capture:

$$G_N = 10.\log_{10}\left(\int_0^{f_{max}} \frac{(|N_A(f)|^2)DI(f)^2}{(|N_A(f)|^2)} df\right) \tag{3.23}$$

where $N_A(f)$ is the ambient noise spectrum. Note that the noise gain represents the amount of noise a directional microphone will capture, compared to an omnidirectional microphone under the same conditions. It does not say anything about how humans will perceive the output noise. As humans have a higher sensitivity around 2000 Hz, the ambient noise spectrum can be additionally weighted with the human sensitivity curve. In this case we will have an estimation of the psychoacoustic noise gain. Based on the target device, the ambient noise can be weighted with another function as well. If we design the sound capture block of a speakerphone, it is a good idea to use C-message or ITU-T O.41 weighting to adjust the noise gain to what we will eventually measure at the end of the telephone channel.

3.4.7 Additional Electrical Parameters

Each microphone has additional electrical parameters defining its limits, or specifying how it has to be connected to the electric circuitry. These parameters vary with different microphone types, manufacturers, and working conditions. Here we will enumerate a few, without any claims that the list is complete.

Bias voltage is a critical parameter for condenser and electret microphones. This is the voltage used to provide power to the integrated FET. For condenser microphones a typical value is 48 V, for the electret it is from 1 to 5 V.

Output impedance of the microphone affects the design of input circuitry. If the output impedance is high, the microphone preamplifier should have a high input impedance too. Dynamic microphones usually have output impedance in tens or hundreds of ohms, while a crystal microphone has hundreds of kiloohms up to several megaohms. The optimal load for condenser and electret microphones is specified by the manufacturer and is usually in the range of tens of kiloohms.

Maximum sound pressure level defines the loudness of the sound the microphone can capture before reaching a certain level of nonlinear distortion. For most microphones it is of the order of 96 dB SPL.

Total harmonic distortion specifies the nonlinearity of the microphone as an electrical sensor for changes in air pressure. Modern microphones are quite good in that, the nonlinear distortion they introduce is below 0.1% in the normal range of speech levels.

3.4.8 Manufacturing Tolerances

All the parameters mentioned above vary from microphone to microphone of the same type, even when purchased from the same manufacturer. The reason for this is manufacturing tolerances which is the total effect of the differences, all in acceptable boundaries, of all components of the microphone. The model of manufacturing tolerances can be extremely complex, accounting for tolerances in the directivity pattern for every frequency and every direction for both magnitude and phase. This is usually unnecessary and simplified models are used.

The manufacturers usually specify the tolerances of the sensitivity towards the main response axis. This is denoted in the microphone specifications as a sensitivity of -47 ± 4 dB, for example. For single microphone systems this is enough to design the necessary limits for adjusting the preamplifier gain.

For designing an array of microphones, a Gaussian distribution of the parameters and the manufacturing tolerances is used, which reduces the model to the standard deviation of the sensitivity towards the main response axis as a function of the frequency, $\sigma_{Mm}(f)$, and the standard deviation of the phase response towards the main response axis as a function of the frequency, $\sigma_{Mf}(f)$. These deviations are assumed to be the same for every incident angle. In reality, if a microphone is 3 dB more sensitive at the main response axis, most probably it will have 3 dB better sensitivity at $90°$ and the shape of the directivity pattern will be the same, causing lower relative variations in the phase response. This is because the sensitivity depends mostly on the electret and the diaphragm, subject to higher tolerances, and the directivity pattern depends mostly of the mechanical shape and back holes size, subject to more precise manufacturing.

Figures 3.11a and b show the sensitivity and phase maximum difference and standard deviations measured from a batch of 30 electret cardioid microphones of the same type and from the same manufacturer. They are higher in the lower part of the frequency band and decline towards the higher frequencies. These deviations heavily depend on the manufacturer. Microphones of the same type and with more or less the

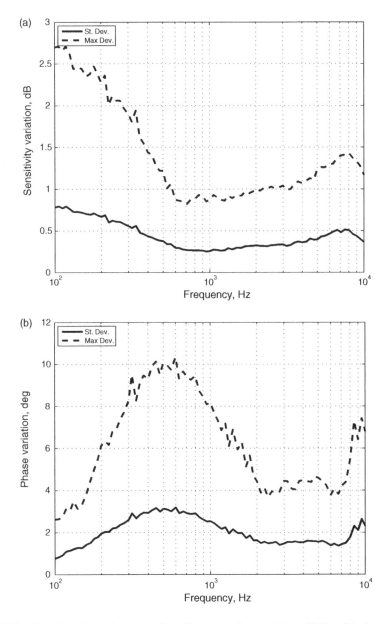

Figure 3.11   Manufacturing tolerances of cardiod microphones: (a) sensitivity; (b) phase response

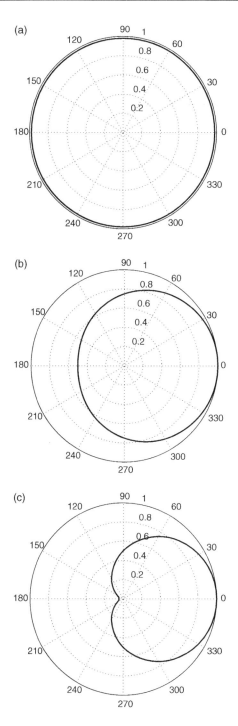

Figure 3.12 Directivity patterns of first-order microphones: (a) omnidirectional; (b) subcardiod; (c) cardiod; (d) supercardioid; (e) hypercardiod; (f) figure-8

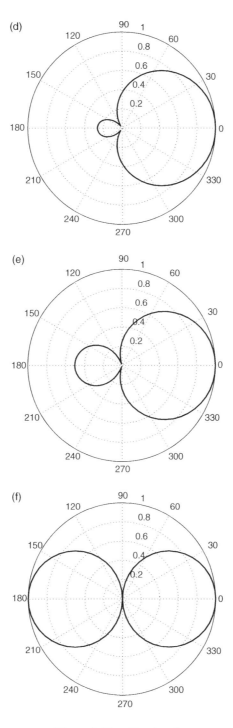

Figure 3.12 *(Continued)*

same parameters in the specification can have quite different manufacturing toler-ances. Microphone characteristics with specified tolerances are in general a good sign – at least the manufacturer looked at the parameter tolerances and guarantees that they fit within a certain window.

3.5 First-order Directional Microphones

A pressure omnidirectional microphone and a pressure-gradient microphone can be combined in a way to form various directivity patterns. This can be generalized as subtraction of the delayed signal of one microphone from the signal of a second microphone, placed at a certain distance l. If the delay time is τ, then

$$U(f, \theta) = 1 - \exp\left(-j2\pi f\left(\tau + \frac{l\cos\theta}{\nu}\right)\right)$$

$$U(f, \theta)_{\text{Norm}} \approx \alpha + (1 - \alpha)\cos\theta$$

(3.24)

where α is a parameter. The family of directivity patterns forms the first- order directional microphones. There are several values of α, that form directivity patterns optimal in one way or another. They are summarized in Table 3.2 and the correspond-ing directivity patterns are plotted in Figure 3.12.

Table 3.2 First-order directivity microphones

Type	α	DI	Note
Omnidirectional	1.00	0.00	No directivity, acoustic monopole
Subcardioid	0.75	2.40	
Cardioid	0.50	4.80	Zero at $180°$
Supercardioid	~ 0.35	5.70	Highest front-to-back ratio, zeros at $\pm 125°$
Hypercardioid	0.25	6.00	Highest DI, zeros at $\pm 109°$
Figure-8	0.00	4.80	Zeros at $\pm 90°$, acoustic dipole

When $\alpha = 1$, the gradient microphone is disabled and the directivity pattern is omnidirectional. It has a directivity index of $0\,\text{dB}$ and is the reference level for computing the noise gain.

The name of the *cardioid* pattern comes from the Greek word for heart and reflects the specific shape of the pattern. It does not rotate the phase for signals coming from any direction and has a null at $180°$. This microphone has DI of $4.8\,\text{dB}$.

Further decreasing the value of α leads to a *supercardiod* directivity pattern. This pattern has the highest front-to-back ratio. This makes it suitable for placing at a theatrical scene, capturing the sounds on the stage and suppressing sounds coming from the audience (clapping, coughing). The microphone directivity pattern has two zeros at $\pm125°$. Note that in the small lobe at the back the phase of the acoustic signal is inverted. This microphone has a DI of 5.7 dB.

The *hypercardiod* directivity pattern has the highest directivity index of 6 dB and the highest noise suppression capabilities. Such microphones are used for sound capturing in noisy places – live radio or TV broadcasting, for example. The directivity pattern has two zeros at $\pm109°$, and the signal phase in the back lobe is inverted.

At $\alpha=0$, the pressure microphone is disabled and we have the figure-8 directivity pattern of the pressure-gradient microphone. There are two zeros in the directivity pattern at $\pm90°$, and the signal phase in the back lobe is inverted. This directivity pattern is usable in radio and TV broadcasting, and when properly held can capture the voices of the interviewer and the interviewee simultaneously. In addition it suppresses noises coming from the sides.

The first-order directivity patterns can be achieved in several ways. There are expensive professional microphones that actually contain two separate microphone capsules – pressure and pressure gradient. This allows the operator to adjust the directivity pattern during recordings; one such microphone can have any of the first-order directivity patterns by proper mixing of the signals from the two microphones. More common is placing the diaphragm in a small capsule, as shown in Figure 3.7a, and adding holes in the back side. This is acoustical combination of the pressure and pressure-gradient microphone. By varying the size of the holes, the ratio of the pressure and pressure gradient signals is adjusted, achieving different directivity patterns. The directivity pattern is fixed after manufacture. This approach requires frequency correction to flatten the frequency response, which is usually done inside the microphone capsule. Figure 3.13 shows omnidirectional and unidirectional electret microphone capsules. The holes at the back are visible. In the omnidirectional case they are smaller and used only to keep the pressure inside the capsule equal to atmospheric pressure. For unidirectional microphones the holes are bigger and play a role in forming the directivity pattern.

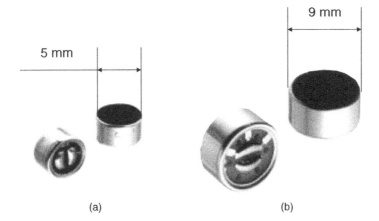

5 mm   9 mm

(a)   (b)

Figure 3.13   Omnidirectional and unidirectional electret microphone capsules

By using more microphones better directivity and noise reduction can be achieved; see [7]. These microphones have so-called second-, third- and higher order directivity patterns. The design of these is generalized in the chapter on microphone arrays later in this book.

EXERCISE

Create a MATLAB script *MicrophoneDirectivity.m* to compute the complex microphone gain for a given frequency, cosine of the incident angle, and type of the microphone (0 – omni, 1 – subcardioid, 2 – cardioid, 3 – supercardioid, 4 – hypercardioid, 5 – figure of 8):

```
MicGain = MicrophoneDirectivity(Freq, CosTheta, Type)
```

Use the already created script *GradientMicrophoneDirectivity.m* and the values for α from Table 3.2.

3.6 Noise-canceling Microphones and the Proximity Effect

The microphone models considered so far have assumed a far-field sound and a flat wavefront. While this model is valid for most sound capture applications when the microphone is at a distance of one meter or more, the model is not correct for headset applications when the sound source is close to the microphone (2–15 cm). Then the microphones have different characteristics. In addition, the ambient noise is still a far field and this allows us to benefit from this difference.

Consider again the pressure-gradient microphone as shown in Figure 3.14. The microphone is a pipe with diameter δ and length l, both much smaller than the shortest

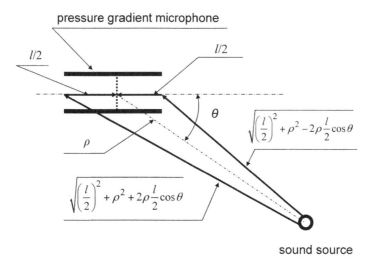

Figure 3.14 Another look at the pressure gradient microphone for close sound sources

wavelength in the frequency band. The center of the coordinate system is the center of the diaphragm. For sound at distance ρ, the distances the sound travels to reach the front and back sides of the diaphragm are

$$
\rho_f(\theta, \rho) = \frac{l}{2} + \sqrt{\left(\frac{l}{2}\right)^2 + \rho^2 - 2\rho\frac{l}{2}\cos\theta}
$$

$$
\rho_b(\theta, \rho) = \frac{l}{2} + \sqrt{\left(\frac{l}{2}\right)^2 + \rho^2 + 2\rho\frac{l}{2}\cos\theta}
$$

(3.25)

respectively, and the directivity pattern of the differential microphone is

$$
U(f, \theta, \rho) = \frac{1}{\rho_f(\theta, \rho)}\exp\left(-j2\pi f\frac{\rho_{f(\theta, \rho)}}{c}\right) - \frac{1}{\rho_b(\theta, \rho)}\exp\left(-j2\pi f\frac{\rho_b(\theta, \rho)}{c}\right). \quad (3.26)
$$

Figure 3.15a shows the frequency response of a pressure-gradient microphone towards the main response axis for distances of 2 cm and 2 m. The size of the microphone is assumed to be $\delta = l = 5$ mm. We have a different frequency response for far-field and near-field sounds because of the different attenuation due to the inverse distance law. This causes differences not only in the phase, which we have for far-field sound, but in the magnitudes as well.

If this is a close-talk microphone, the frequency response should be compensated to flatten the curve for a distance of 2 cm. For far-field sounds – that is, for unwanted ambient noise – the resulting frequency response will be with substantial suppression in the lower part of the frequency band where most of the noise energy is concentrated, as shown in Figure 3.15c. This is why the pressure-gradient microphone with frequency compensation for close distances is called a noise-canceling microphone.

For far-field sound capture, the frequency response should be compensated so as to flatten the curve for a distance of 2 m. If a person approaches such a microphone and starts to talk very close, the lower part of the frequency band will be overcompensated, as shown in Figure 3.15b. Underlying the basses causes the voice to sound deeper. This is called the "proximity effect."

EXERCISE

Modify the MATLAB script *GradientMicrophoneDirectivity.m* to account for the distance: add the additional parameter $\texttt{Dist}$ for distance and replace the implementation of equation with the implementation of Equation 3.26:

```
GradientMicrophoneDirectivity(Dist, Freq, CosTheta)
```

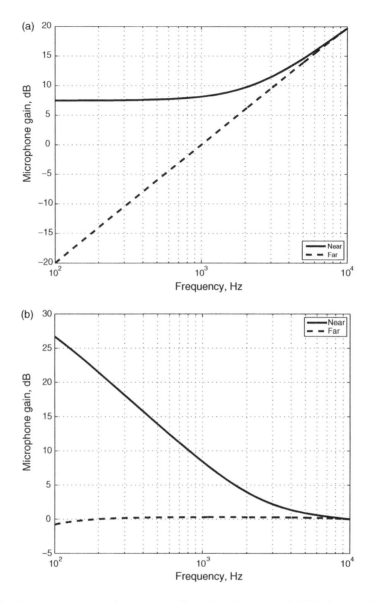

Figure 3.15 Frequency response of pressure-gradient microphone towards MRA for near (2 cm) and far (2 m) sound sources: (a) non-compensated; (b) compensated for far-field sources; (c) compensated for near-field sources

Plot and compare the frequency responses for 1 cm and 1 m. Modify *Microphone-Directivity.m* to require and pass the distance to the new *GradientMicrophone-Directivity.m* script.

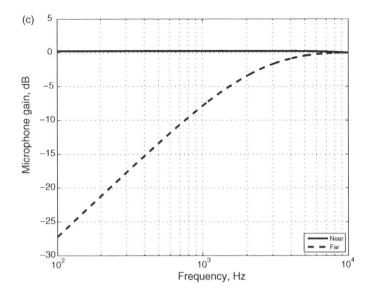

Figure 3.15   *(Continued)*

3.7 Measurement of Microphone Parameters

3.7.1 Sensitivity

The sensitivity of a microphone can be measured by simultaneous capturing of sound with a reference microphone, or by capturing sound with known amplitude.

In the first case, the magnitudes of the synchronously captured signals are compared and the unknown sensitivity derived from the sensitivity of the reference microphone. Reference microphones are expensive and usually of the capacitive type. The majority of them are omnidirectional. If the measured microphone is directional, the measurement should be done with a point noise source (loudspeaker) in a specialized chamber with low noise and reverberation. The noise source should be placed on the main response axis of the measured microphone. In all cases, the two microphones should be positioned as close to each other as possible.

Another approach is to use a sound source with a known intensity. Such a sound source is shown in Figure 3.16. It generates sound with 96 dB SPL and the measured microphone should be placed tightly to the source. Then the magnitude of the signal from the microphone is measured and the sensitivity computed directly. The measurement should be done in a quiet environment and requires well-coupled sound source and microphone.

3.7.2 Directivity Pattern

To measure the directivity pattern of a microphone, it has to be placed in a low-noise and low-reverberation chamber. A good-quality speaker can be used as a sound source. It has to be placed far enough away to avoid close proximity variations in the sound

Figure 3.16   Sound source for measurement of a microphone's sensitivity

field. As a test signal, a sweep or any chirp signal which covers the measured band can be used. A reference microphone, positioned as close as possible to the measured, should be used to exclude the loudspeaker's frequency response from the measurement. It has to be on the rotational axis to avoid phase shifts. The measured microphone should be placed on a turntable, controlled either manually or by a computer. The rotation axis should be perpendicular to the main response axis. The measured microphone and the loudspeaker should be placed at the same height, which guarantees that at some rotation angle the main response axis of the microphone will point right to the speaker. Such a setup is shown in Figure 3.17.

The recording process consists of positioning the microphone at a certain angle and simultaneously recording the signals from the measured and the reference microphones while the loudspeaker plays the chirp signal. For phase-response measurements, the sampling rate of the recordings should be the same and the ADCs should be synchronized to avoid sampling rate drift. Then the measured microphone is rotated and the entire process repeated. A typical rotation angle is $10°$, which means that 36 records should be made. Even under the assumption that the directivity pattern is symmetric, making measurements around a full circle makes sense as they can be used for averaging. This is not obligatory; any number of angles can be used. The sampling rate should be the highest available for better phase/frequency-response estimation. Use a 48 kHz sampling rate if possible, even when the work band is up to 8 kHz.

The first step of processing the records is trimming and aligning. Using the cross-correlation function and quadratic interpolation for the maximum and the two neighboring points, the delay between the reference and measured microphones is computed for each pair. The delays from all angles are averaged and rounded to the nearest integer number. This is the delay in samples used for aligning each pair of

Figure 3.17   Setup in an anechoic chamber for measurement of a microphone's directivity

recordings. After the alignment, the test signal position is recognized by processing the reference signals in each pair and both signals are trimmed to contain the test signal plus 10% of its duration silence before and after.

The second step is to compute the frequency response of the measured microphone for each incident angle. The easiest way to do this is by computing the spectra of the reference and measured microphone signals and then comparing them. As we specify the test signal, there should be no problems with zeros in the working band. After finishing the processing for each measurement angle, under perfect measurement conditions we should have the microphone directivity pattern as a function of the incident angle and frequency. Real measurements show that it can be noisy and non-symmetric. In addition it has a resolution towards incident angles based on the measurements and towards frequencies based on the length of the test signal and the sampling rate.

The third step is re-sampling and smoothing. This should be done for the entire directivity pattern together. The number of measurements in the incident angle axis is doubled by adding the opposite measurement, owing to the assumption that the directivity pattern is symmetric. Then the spectrum of this 2-dimensional complex signal is computed with irregular sampling intervals by using the classic formula for Fourier transformation (see Equation 2.30). This allows us to have a spectrum with desired resolution. The 2-dimensional spectrum is smoothed by applying a low-pass filter:

$$H(\Phi, T) = \frac{1}{4} \left(\cos^{N_\Phi} \left(\frac{2\pi\Phi}{f_S} \right) + 1.0 \right) \left(\cos^{N_T} \left(\frac{1}{T} \right) + 1.0 \right) \qquad (3.27)$$

where Φ and T are the frequency axes corresponding to f and θ, respectively. The smoothing depends on N_Φ and N_T and can be adjusted based on how noisy the measurements are. When they approach infinity the directivity pattern obtained will always be omnidirectional. Typical values vary between 0.5 (low level of smoothing) to 64 (high smoothing). After applying the smoothing filter, the actual directivity pattern is computed by taking the inverse Fourier transformation and pruning it for the frequency range we need and incident angle values between $0°$ and $180°$. If directivity pattern values between the known points are needed, they can be obtained by doing linear interpolation for the magnitude and phase separately.

Figure 3.18 shows the directivity pattern of a cardioid microphone capsule, measured using the methodology described above. The first chart shows the magnitudes; the second chart shows phase as function of the frequency and incident angle.

3.7.3 Self Noise

The noise from the microphone can be measured in the low-noise and low-reverberation chamber – see Figure 3.17. It is already in the recording made for measuring the directivity pattern. We need one additional recording with a reference sound source – see Figure 3.16. Computing the ratio of the noise and reference signal RMS gives the SNR at the intensity of the sound source. It is easy to convert it if a different reference sound source intensity is desired.

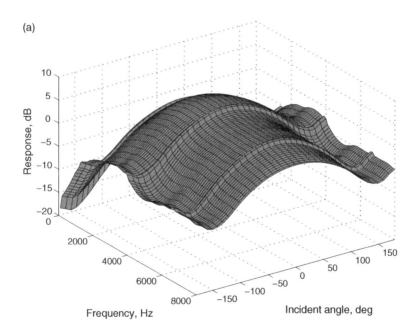

Figure 3.18   Measured directivity pattern of a cardiod microphone: (a) magnitude; (b) phase

(b)

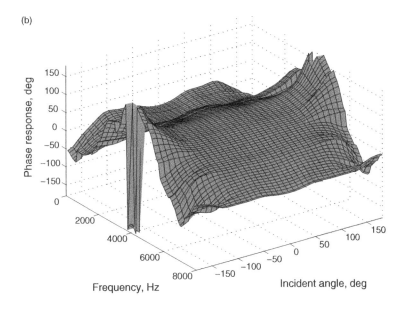

Figure 3.18 *(Continued)*

It is a good idea to convert and store the spectrum of the microphone noise for subsequent use. It will be further used during the design of microphone arrays. Figure 3.19 shows the self-noise spectrum of the cardioid microphone from Figure 3.18. The second plot represents the noise magnitudes at normal room conditions and is for comparison purposes only.

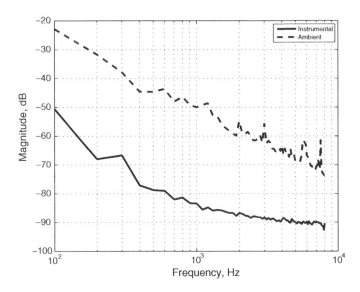

Figure 3.19 Magnitude spectrum of microphone noise compared with the magnitude spectrum of room noise under normal conditions

EXERCISE

Look at the MATLAB script *InstrumentalNoise.m* which returns the magnitude of the microphone noise for a given frequency:

```
instNoise = InstrumentalNoise(Freq)
```

Plot the microphone noise spectrum for frequencies from 100 Hz to 10 000 Hz.

There is another MATLAB script modeling the ambient noise in a room under normal conditions: *AmbientNoise.m*. The script returns the noise magnitude for given frequency:

```
ambNoise = AmbientNoise(Freq)
```

Plot both instrumental and ambient noise spectra and compare them.

Write a MATLAB script that writes the data file for the scripts above from a recorded .WAV file. Record, with the microphone you use, the noise in your room and generate the data file. In very silent conditions (open air during the night, noiseless chamber, etc.) record the instrumental noise of your microphone. Generate the data file for the script above. Plot and compare the results.

3.8 Microphone Models

Microphone models vary, and their complexity should be limited to the necessary precision. The model should provide the directivity pattern function as a function of the frequency and sound source position $U(f, d)$, where $d = \{\rho, \varphi, \theta\}$. For building such models, Equations 3.25 and 3.26 should be used. A simpler and computationally less expensive model can be obtained by assuming axis symmetry and far-field sound; then Equation 3.24 should be used.

The microphone self noise can be measured directly in a sound chamber. The average spectrum can be computed from this recording and the magnitude for each frequency obtained by using linear interpolation.

These directivity pattern and noise models have enough precision for the design of multi-microphone systems. They are partially idealized and do not include mechanical resonances, electrical circuitry details and models, and so on. For more detailed models, see [8]. This precision of the models is necessary for the design of the microphones themselves, but not for designing and understanding the processes in sound capture systems.

3.9 Summary

In this chapter we noted that sound is nothing more than moving compressions and rarefactions in the air, which form a longitudinal mechanical wave. Sound has a frequency and a wavelength. The former is how many cycles per second of compression and rarefaction a point experiences, while the latter is the smallest distance between areas with the same pressure in the direction of propagation of the sound wave. Another very important characteristic of sound is its speed of propagation. For air, this depends on the temperature and pressure, and for normal conditions it is 343.2 m/s. The speed of sound is practically the same for all audible frequencies; that is, the air is a non-dispersive medium. The sound wave magnitude is characterized by its intensity and sound pressure level. The intensity decreases with the square of the distance (the inverse square law), and the sound pressure level decreases with the inverse of the distance. For sound capture purposes, the sound attenuation due to energy loses in the air is negligible and can be ignored. As the wavelength of the audible sound waves is comparable with the objects in our surroundings, typical wave effects take place: interference, diffraction, reflection, refraction, and the Doppler effect.

The sensors that convert sound to an electrical signal are called microphones. Most of them use diaphragms to convert pressure changes to mechanical vibrations. They can be grouped by the way these mechanical vibrations are converted to electrical signals: carbon, piezoelectric, electrodynamic, condenser, and electret. There are two major ways the diaphragm is exposed to the air pressure and two basic types of microphone: pressure and pressure gradient. They have omnidirectional and figure-8 directivity patterns, respectively. Combining the signals from two of these microphones in a certain proportion allows formation of the directivity patterns of first-order directivity microphones: omnidirectional, subcardioid, cardioid, supercardioid, hypercardioid, and figure-8. The directivity pattern of a real microphone and its frequency response depend on the distance as well. Microphones with frequency compensation for far-field sounds underline basses when talked to from a close distance – the proximity effect. Microphones with frequency compensation for near-field sounds suppress the far-field noises – a noise-canceling microphone. The most important parameters of a microphone are the sensitivity, the directivity index, and the signal-to-noise ratio (SNR) (indirect measure for self noise).

In this chapter, we have built the directivity and noise models of the microphones we shall use in this book to design microphone arrays.

Bibliography

[1] Leighton, T.G. (1999) *The Acoustic Bubble*, Academic Press, San Diego, CA.

[2] Dushaw, B.D., Worcester, P.F., Cornuelle, B.D. and Howe, B.M. (1993) On equations for the speed of sound in seawater. *Journal of the Acoustical Society of America*, **93**, 255–275.

[3] Raichel, D. (2006) *The Science and Applications of Acoustics*, 2nd edn, Springer, New York.

[4] Wolski, L.F., Anderson, R.C., Bowles, A.E. and Yochem, P.K. (2003) Measuring hearing in the harbor seal (Phoca vitulina): comparison of behavioral and auditory brainstem response techniques. *Journal of the Acoustical Society of America*, **113**, 629–637.

[5] Babushina, Ye.S., Zaslavskii, G.L. and Yurkevich, L.I. (1991) Air and underwater hearing characteristics of the northern fur seal: audiograms, frequency and differential thresholds. *Biophysics*, **36**, 900–913.

[6] Wong, G.S. and Embleton, T.F. (1995) *AIP Handbook of Condenser Microphones*, AIP Press, New York.

[7] Huang, Y. and Benesty, J. (2004) *Audio Signal Processing for Next Generation Multimedia Communication Systems*, Kluwer Academic, Boston, MA.

[8] Eargle, J. (2001) *The Microphone Book*, Focal Press, Boston, MA.

4

Single-channel Noise Reduction

This chapter deals with noise reduction of a single channel. We assume that we have a mixture of a useful signal, usually human speech, and an unwanted signal – which we call noise. The goal of this type of processing is to provide an estimate of the useful signal – an enhanced signal with better properties and characteristics.

The problem with a noisy speech signal is that a human listener can understand a lower percentage of the spoken words. In addition, this understanding requires more mental effort on the part of the listener. This means that the listener can quickly lose attention – an unwanted outcome during meetings over a noisy telephone line, for example. If the noisy signal is sent to a speech recognition engine, the noise reduces the recognition rate as it masks speech features important for the recognizer.

With noise-reduction algorithms, as with most other signal processing algorithms, there are multiple trade-offs. One is between better reduction of the unwanted noise signal and introduction of undesired effects – additional signals and distortions in the wanted speech signal. From this perspective, while improvement in the signal-to-noise ratio (SNR) remains the main evaluation criterion of the efficiency of these algorithms, subjective listening tests or objective sound quality evaluations are also important. The perfect noise-reduction algorithm will make the main speaker's voice more understandable so that it seems to stand out, while preserving relevant background noise (train station, party sounds, and so on). Such an algorithm should not introduce noticeable distortions in either foreground (wanted speech) or background (unwanted noise) signals.

Most single-channel algorithms are based on building statistical models of the speech and noise signals. In this chapter we will look at the commonly used approaches for suppression of noise, the algorithms to distinguish between noise and voice (called "voice activity detectors"), and some adaptive noise-canceling algorithms. Exercises with implementation of some of these algorithms will be provided for better understanding of the processes inside the noise suppressors.

4.1 Noise Suppression as a Signal Estimation Problem

Let the speech signal be $x(t)$. This signal is captured after it has been mixed with noise $d(t)$. We can assume that these two signals are statistically independent. The capturing process is linear, so the captured signal is $y(t) = x(t) + d(t)$. The goal of the noise reduction is to estimate the speech signal $x(t)$ using the observed signal $y(t)$ and some known properties of both speech and noise signals. This process is shown in Figure 4.1. We have the unobservable part when the speech and noise signals are mixed. In the observable part we have only the observed signal and some a-priori knowledge about the character of the signals. The estimation process is optimal in one way or another; that is, it satisfies a certain criterion.

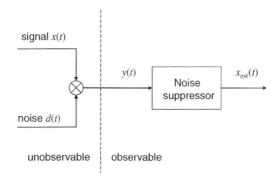

Figure 4.1   Noise reduction as a signal estimation problem

As we saw in Chapter 2, the majority of the audio processing algorithms work with audio frames, obtained by the overlap–add procedure, described in the same chapter. This is why we will look at most of the noise suppression algorithms from the perspective of processing in the frequency domain, after short-time spectral conversion. Since the transformation is linear, we have $Y_k^{(n)} = X_k^{(n)} + D_k^{(n)}$ where k is the frequency bin index and n is the frame index. Under these conditions it is a common assumption that the values of the frequency bins are statistically independent for both noise and speech signals. This allows derivation of the speech signal estimation algorithms for each frequency bin independently, which greatly simplifies the algorithms and the corresponding equations.

4.2 Suppression Rules

4.2.1 Noise Suppression as Gain-based Processing

The early work of Norbert Wiener [1] has an impact on signal processing algorithms by providing an original look from a statistical point of view. He derived an optimal filter for estimation of a signal corrupted by noise in the time domain. Later derivations for representation in the frequency domain found that this optimal estimator applies a real-valued gain to the complex spectral vector. In general, human perception of speech is insensitive to the phase of the signal [2], and the same is true for automated speech recognizers. In later work, Ephraim and Malah [3] proved that the optimal phase estimator for a signal corrupted with noise just takes the phase of the noisy signal. In this manner, noise reduction can be viewed as an application of a time-varying, non-negative, real-valued gain $H_k^{(n)}$ to each frequency bin k of the observed signal $Y_k^{(n)}$ to obtain the estimate $\hat{X}_k^{(n)}$ of the original signal spectrum:

$$\hat{X}_k^{(n)} = H_k^{(n)} \cdot Y_k^{(n)}. \tag{4.1}$$

The time-varying real-valued gain H_k is called the *suppression rule* and it is estimated for each frame. Note that the complex value of the signal estimation for this frequency bin $\hat{X}_k^{(n)}$ keeps the phase of the observed complex signal $Y_k^{(n)}$ because $H_k^{(n)}$ is real-valued.

4.2.2 Definition of A-Priori and A-Posteriori SNRs

In most algorithms for estimation of the suppression rule, a-priori and a-posteriori SNRs are involved. They were defined for the first time in [4]. The authors model the elements of the noise spectrum as independent, identically distributed Gaussian variables with a zero mean and variances $\lambda_d(k)$:

$$D_k \sim \mathbb{N}(0, \lambda_d(k)). \tag{4.2}$$

In the same paper they model the signal as a stationary sum of sinusoidal signals with magnitude A_k which is an estimation of the signal magnitude for this frequency bin. In general we can say that $\lambda_d(k) \triangleq E\{|D_k|^2\}$ and $\lambda_x(k) \triangleq E\{|X_k|^2\}$. Then we can define the a-priori SNR ξ_k and a-posteriori SNR γ_k as

$$\begin{aligned} \xi_k &\triangleq \frac{\lambda_x(k)}{\lambda_d(k)} \\[2mm] \gamma_k &\triangleq \frac{|Y_k|^2}{\lambda_d(k)}. \end{aligned} \tag{4.3}$$

Note that while ξ_k is an average (statistical) SNR, γ_k can be viewed as a momentary SNR.

4.2.3 Wiener Suppression Rule

The Wiener estimator for a discrete signal corrupted by noise was initially derived in the time domain as an N-tap FIR (finite impulse response) filter; that is:

$$\hat{x}(n) = \sum_{i=0}^{N-1} h_i^* \cdot y(n-i).$$

(4.4)

Then the estimation error will be

$$e(n) = x(n) - \hat{x}(n)$$

(4.5)

and the goal is to find filter $\mathbf{h}_{\text{opt}}$ with coefficients h_i that minimizes the estimation error:

$$E\{|e(n)|^2\} = E\{e(n)e^*(n)\}.$$

(4.6)

After taking first derivatives, the filter that zeroes them is

$$\mathbf{h}_{\text{opt}} = \mathbf{R}_{yy}^{-1} \cdot \mathbf{r}_{yy}(0)$$

(4.7)

where $\mathbf{r}_{yy}(n)$ is the autocorrelation vector

$$\mathbf{r}_{yy}(n) = [r_{yy}(n), \ r_{yy}(n-1), \cdots, \ r_{yy}(n-N+1)]^{\text{T}}$$

(4.8)

and $\mathbf{R}_{yy}$ is the autocorrelation matrix

$$\mathbf{R}_{yy} = \begin{bmatrix} r_{yy}(0) & r_{yy}(1) & \cdots & r_{yy}(N-1) \\ r_{yy}^*(1) & r_{yy}(0) & \cdots & r_{yy}(N-2) \\ \vdots & \vdots & \ddots & \vdots \\ r_{yy}^*(N-1) & r_{yy}^*(N-2) & \cdots & r_{yy}(0) \end{bmatrix}.$$

(4.9)

Considering the stationarity of the signals, and the fact that they are statistically independent, in the frequency domain the optimal filter derivation is much simpler. The filter that minimizes the derivatives is

$$\mathbf{S}_{ss}^* - \mathbf{H}_{\text{opt}} \cdot \mathbf{S}_{yy} = 0$$

(4.10)

where $\mathbf{S}_{ss}$ is the power spectral density of the signal and $\mathbf{S}_{yy}$ is the power spectral density of the observation. Due to statistical independence, $\mathbf{S}_{yy} = \mathbf{S}_{ss} + \mathbf{S}_{dd}$ and the optimal filter is

$$H_{\text{opt}} = \frac{S_{ss}}{S_{dd} + S_{ss}} = \frac{\lambda_s}{\lambda_d + \lambda_s}. \tag{4.11}$$

Equation 4.11 can easily be converted in terms of a-priori SNR for each frequency bin as

$$H_k = \frac{\xi_k}{1 + \xi_k} \tag{4.12}$$

which is the *Wiener suppression rule*. This rule minimizes the mean square error of the estimated signal.

4.2.4 Artifacts and Distortions

The derived elegant solution for the optimal suppression rule stumbles on some difficulties when it is applied to real noise-reduction systems – estimation of the a-priori SNR. While obtaining the noise variation $\lambda_d(k)$ is relatively easy, estimation of the signal variation is not trivial. The logical step of using the a-posteriori SNR γ_k to estimate ξ_k leads to the approximate solution

$$H_k^{(n)} = \frac{\xi_k}{1 + \xi_k} \approx \frac{\gamma_k - 1}{\gamma_k} = \frac{|Y_k^{(n)}|^2 - \lambda_d(k)}{|Y_k^{(n)}|^2} \tag{4.13}$$

which is easier to implement in practice. By definition, $H_k^{(n)}$ is non-negative and real; but, for values of $|Y_k^{(n)}|^2$ smaller than $\lambda_d(k)$, Equation 4.13 can have negative values. The common approach is to limit the values of the suppression rule to be non-negative. Another practical problem is potential division by zero for frequency bins where the input signal has a zero value. This is solved by adding a small number ε, and we finally have the Wiener suppression rule widely used in practice:

$$H_k^{(n)} = \frac{\max[0, |Y_k^{(n)}|^2 - \lambda_d(k)]}{|Y_k^{(n)}|^2 + \varepsilon}. \tag{4.14}$$

This approximate rule provides good noise suppression and improves the SNR of the output signal. On the down side, in the output signal some artifacts and distortions may be audible.

The artifacts appear during the silence segments when a very specific type of noise, called "musical noise," can be heard. Investigations have shown that this is due to the fact that some frequency bins are zeroed during these segments owing to the way suppression gain is estimated via Equation 4.14. During silence segments, where we only have a noise signal with variation $\lambda_d(k)$, a substantial number of cases $|Y_k^{(n)}|^2$ will be smaller than $\lambda_d(k)$ and the corresponding frequency bin will be zeroed. This "patchy" spectrogram, converted to the time domain, has the specific musical noise sound.

The Wiener filter is a minimum mean-square estimator, which provides an approximate value of the output signal. Introducing some distortions when compared with the original signal is inevitable. The problem with distortions, audible as metallic and unnatural sound of the estimated speech signal, increases in signals with low SNR owing to the approximation in Equation 4.13.

4.2.5 Spectral Subtraction Rule

The musical noises and distortions in the output signal stimulated the search for better estimators. Considering that less suppression means less musical noise and less distortion, the spectral subtraction rule [5] was introduced. It is defined as

$$H_k^{(n)} = \sqrt{\frac{\gamma_k - 1}{\gamma_k}} \tag{4.15}$$

and frequently used with the approximation

$$H_k^{(n)} \approx \sqrt{\frac{\max[0, |Y_k^{(n)}|^2 - \lambda_d(k)]}{|Y_k^{(n)}|^2 + \varepsilon}}. \tag{4.16}$$

The rule is optimal in the sense of estimation of the speech magnitude spectrum. The overall noise suppression is less, but the spectral subtraction suppression rule has a lower distortion of the estimated speech signal; that is, the output sounds better to the human ear. The problem with the musical noise remains, as many frequency bins are still zeroed during silence periods.

4.2.6 Maximum-likelihood Suppression Rule

McAulay and Malpass [4] proposed a new suppression rule, optimal in the maximum-likelihood sense. They modeled the speech signal as a deterministic waveform of unknown amplitude and phase, and the noise as a random Gaussian signal. Under these conditions the maximum-likelihood suppression rule is

$$H_k^{(n)} = \frac{1}{2} + \frac{1}{2}\sqrt{\frac{\left|Y_k^{(n)}\right|^2 - \lambda_d(k)}{\left|Y_k^{(n)}\right|^2}}.$$ (4.17)

The first thing to notice is that this suppression rule never becomes zero. This completely eliminates the musical noise. A second fact to notice is that its minimal gain value is actually quite high and the rule never goes below 0.5. This substantially reduces the noise-suppression capabilities of the maximum-likelihood suppression rule; it has the lowest noise suppression among the suppression rules we discuss in this book. The same practical measures to limit the value under the square root to be non-negative and to prevent division by zero as in Equation 4.16 should be taken here as well.

Figure 4.2 shows the three suppression rules, discussed so far, as a function of the a-posteriori SNR. It is obvious that Wiener filtering suppresses the most, and maximum-likelihood suppresses the least, of the signal energy. For performance comparison of the different suppression rules, see later in this chapter. We will return to the work of McAulay and Malpass in the section about suppression with the uncertain presence of a speech signal.

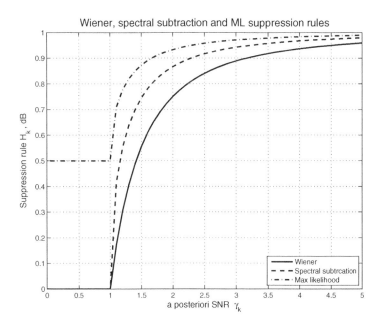

Figure 4.2 Wiener, spectral subtraction, and maximum-likelihood suppression rules as functions of the a-posteriori SNR

4.2.7 Ephraim and Malah Short-term MMSE Suppression Rule

Ephraim and Malah [3] model both speech and noise signals as zero-mean random Gaussian signals. Under the conditions of short term-spectral estimation, they derive a suppression rule, known in the form

$$H_k = \frac{\sqrt{\pi \nu_k}}{2\gamma_k} \left[(1 + \nu_k) I_0 \left(\frac{\nu_k}{2} \right) + \nu_k I_1 \left(\frac{\nu_k}{2} \right) \right] \exp \left(\frac{\nu_k}{2} \right). \qquad (4.18)$$

Here $I_0(\cdot)$ and $I_1(\cdot)$ denote the modified Bessel functions of zero- and first-order, respectively, and

$$\nu_k \triangleq \frac{\xi_k}{1 + \xi_k} \gamma_k. \qquad (4.19)$$

The spectral magnitude estimator, given by Equation 4.18, is optimal in the MMSE sense. It provides good noise suppression comparable to that of the Wiener filter, while maintaining lower distortions and artifacts. For the first time, the suppression rule is defined as a function of both a-priori SNR ξ_k and a-posteriori SNR γ_k: $H_k(\xi_k, \gamma_k)$. Figure 4.3 shows the shape of the Ephraim and Malah suppression rule as a function of these two parameters.

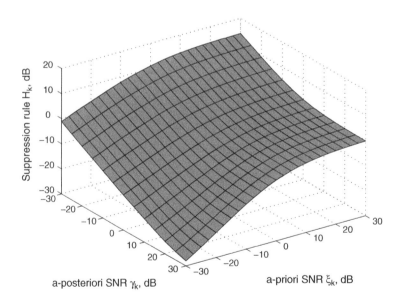

Figure 4.3   Short-term minimum mean-square estimator (MMSE) suppression rule as a function of a-priori and a-posteriori SNRs

4.2.8 Ephraim and Malah Short-term Log-MMSE Suppression Rule

Considering the fact that humans hear in a logarithmic scale of sound pressure level (i.e., magnitudes), Ephraim and Malah [6] derived another suppression rule that is optimal in the MMSE log-spectral amplitude sense. The suppression rule is simpler than the MMSE spectral amplitude estimator:

$$H_k = \frac{\xi_k}{1 + \xi_k} \left\{ \frac{1}{2} \int\limits_{v_k}^{\infty} \frac{\exp(-t)}{t} \, dt \right\} \tag{4.20}$$

but unfortunately it contains an integral that has to be computed in real time. As this is one of the best performing suppression rules, numerous interpolations and approximations for fast computation of this integral have been designed. The integral is a function of one variable and can be easily tabulated and interpolated in real time.

Regardless of the quite different optimization criterion and analytic form, the short-term log-MMSE suppression rule is surprisingly close to the short-term MMSE suppression rule. Figure 4.4 shows the shape of a log-MMSE suppression rule and the difference between this rule and the MMSE suppression rule. The mean of the difference is 1.12 dB, and the maximum difference is only 1.46 dB for $\xi_k \in [-30, +30]$ dB and $\gamma_k \in [-30, +30]$ dB.

4.2.9 More Efficient Solutions

The Wiener filter approach relies on second-order statistics only. Therefore it makes fewer assumptions about the shapes of the probability densities involved. The suppression rules from Ephraim and Malah take explicitly into account the probability density functions of the speech and the noise signals. The MMSE and log-MMSE optimal solutions lead to integrals, exponents, and Bessel functions that are difficult to compute.

Wolfe and Godsill [7] looked for computationally more efficient alternatives. They derived three additional suppression rules, using different criteria for optimality: the 'joint maximum a-posteriori spectral amplitude and phase' (JMAP SAP) estimator, the 'maximum a-posteriori spectral amplitude' (MAP SA) estimator, and the 'minimum mean-square-error spectral power' (MMSE SP) estimator. They assume both speech and noise signals to be Gaussian random processes. The three criteria they use and the corresponding suppression rules are shown in Table 4.1. The table shows the mean and maximum difference between these suppression rules, and Ephraim and Malah's short-term MMSE estimator.

All three rules are much faster to compute in real time as they do not contain Bessel functions and exponents. In the same paper the authors compare these three rules with the short-time MMSE suppression rule, derived by Ephraim and Malah. The average

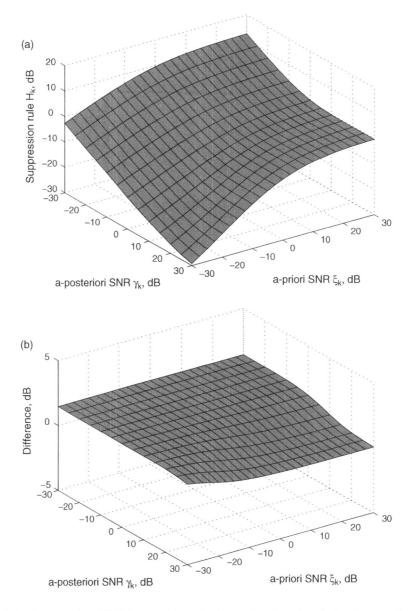

Figure 4.4 Short-term log-MMSE suppression rule: (a) as a function of a-priori and a-posteriori SNRs; (b) difference between MMSE and log-MMSE suppression rules

difference for $(\gamma_k - 1, \xi_k) \in [-30, 30]$ dB is of the order of 1 dB, as Table 4.1 shows. This means that all four rules should have approximately the same noise suppression. Figure 4.5 shows the difference between Ephraim and Malah's suppression rule and the MAP SA estimator – the rule with highest difference.

Table 4.1 More efficient suppression rules

Optimality	Suppression rule	Mean diff (dB)	Max diff (dB)
Joint maximum a-posteriori spectral amplitude and phase (JMAP SAE) estimator	$H_k = \dfrac{\xi_k + \sqrt{\xi_k^2 + 2(1+\xi_k)\frac{\xi_k}{\gamma_k}}}{2(1+\xi_k)}$	0.522	+1.77
Maximum a-posteriori spectral amplitude (MAP SA) estimator	$H_k = \dfrac{\xi_k + \sqrt{\xi_k^2 + (1+\xi_k)\frac{\xi_k}{\gamma_k}}}{2(1+\xi_k)}$	1.261	+4.70
MMSE spectral power estimator	$H_k = \sqrt{\dfrac{\xi_k}{1+\xi_k}\left(\dfrac{1+\nu_k}{\lambda_k}\right)}$	0.685	−1.05

4.2.10 Exploring Other Probability Distributions of the Speech Signal

As was noted in Chapter 2, a speech signal does not have a Gaussian probability distribution. With suppression rules taking into account the actual PDF of the speech signal, the next logical step is to derive suppression rules with better probabilistic models of the speech signal. Martin [8] derives three new suppression rules, under the assumption of Gaussian noise and Gaussian, gamma, and Laplace PDFs of the speech signal – see Equations 2.1, 2.3, and 2.5, respectively. All three rules are optimal in amplitude MMSE sense. For the first time, here the real and imaginary parts in each frequency bin are estimated separately, which may eventually lead to better estimation of the phase; that is

$$E\{S|Y\} = E\{S_R|Y_R\} + jE\{S_I|Y_I\} \tag{4.21}$$

where Y_R and Y_I are the real and imaginary parts of the input signal, S_R and S_I are the real and imaginary parts of the estimated speech signal, all in the corresponding frequency bins.

Assumption of Gaussian noise and Gaussian speech PDFs leads directly to the Wiener estimation rule in Equation 4.12.

With Gaussian noise and gamma distribution of the speech signal, the suppression rule is

$$E\{S_R|Y_R\} = \frac{\sqrt[4]{1.5}}{2\pi\sigma_n p(Y_R)} \int_{-\infty}^{+\infty} S_R|S_R|^{-0.5}.\exp\left(-\frac{Y_R^2}{\sigma_n^2} + \frac{2Y_R S_R}{\sigma_n^2} - \frac{S_R^2}{\sigma_n^2} - \frac{\sqrt{3}|S_R|}{\sqrt{2}\sigma_s}\right)dS_R. \tag{4.22}$$

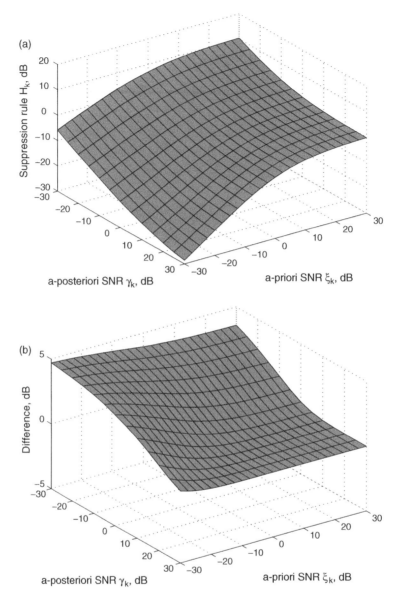

Figure 4.5 Maximum a-posteriori spectral amplitude estimator (MAP SA) suppression rule: (a) as a function of a-priori and a-posteriori SNRs; (b) difference between MMSE and MAP SA suppression rules

Here σ_n and σ_s are the variations of the noise and speech signal. After solving the integral, the suppression rule takes the form

$$E\{S_R|Y_R\} = \frac{\sigma_n}{2\sqrt{2}Z_{GR}} \left\{ \exp\left(\frac{G_{R-}^2}{2}\right) D_{-1.5}(\sqrt{2}G_{R-}) - \exp\left(\frac{G_{R+}^2}{2}\right) D_{-1.5}(\sqrt{2}G_{R+}) \right\}$$

$$(4.23)$$

where G_{R+}, G_{R-} and Z_{GR} are given by

$$G_{R+} = \frac{\sqrt{3}\sigma_n}{2\sqrt{2}\sigma_s} + \frac{Y_R}{\sigma_n} = \frac{\sqrt{3}}{2\sqrt{2}\sqrt{\xi}} + \frac{Y_R}{\sigma_n} \qquad (4.24)$$

$$G_{R-} = \frac{\sqrt{3}\sigma_n}{2\sqrt{2}\sigma_s} - \frac{Y_R}{\sigma_n} = \frac{\sqrt{3}}{2\sqrt{2}\sqrt{\xi}} - \frac{Y_R}{\sigma_n} \qquad (4.25)$$

$$Z_{GR} = \exp\left(\frac{G_{R-}^2}{2}\right)D_{-0.5}(\sqrt{2}G_{R-}) + \exp\left(\frac{G_{R+}^2}{2}\right)D_{-0.5}(\sqrt{2}G_{R+}). \qquad (4.26)$$

In the equations above, $D_p(z)$ denotes a parabolic cylinder function. The computational complexity is obvious, the suppression rule includes numerous exponents and parabolic cylinder functions. Note that in real time the estimation of this rule has to be done twice for each frequency bin – once for the real part and once for the imaginary part. Figure 4.6 shows the shape of this suppression rule. While we already had suppression gain values above 0 dB in previous suppression rules, here in high a-priori

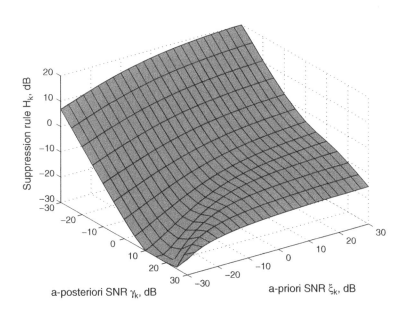

Figure 4.6 Suppression rule under the assumption of Gaussian noise and gamma speech distributions as function of a-priori and a-posteriori SNRs

and low a-posteriori SNRs they reach $+20\,\mathrm{dB}$. As we will see later, this can cause some instability of the entire noise suppressor, which is not desirable.

Under the assumptions of Gaussian noise and Laplace PDF, one more suppression rule is derived, which is even more complex than Equation 4.23. The experimental results with a speech signal corrupted with Gaussian and car noise show a slight advantage with the Gaussian/gamma suppression rule – in the range of 0.1–$0.4\,\mathrm{dB}$ better suppression than the Wiener suppression rule.

4.2.11 Probability-based Suppression Rules

Assume that we have a zero-mean Gaussian noise with magnitude variance λ_d and a speech signal with Gaussian distribution and magnitude variance λ_x. If we have a complex signal with independent and identically distributed real X_R and imaginary X_I parts, both modeled as Gaussian noise $\mathbb{N}(0, \sigma^2)$, then the magnitude of this noise, $|X| = (X_R^2 + X_I^2)^{1/2}$, will have the Rayleigh distribution

$$p(|X||\sigma) = \frac{|X|}{\sigma^2} \exp\left(-\frac{|X|^2}{2\sigma^2}\right) \tag{4.27}$$

where σ is the only parameter. The maximum-likelihood estimator for the parameter σ is $\sigma^2 = \frac{1}{2N} \sum_{i=0}^{N-1} |X_i|^2$ which leads to $\sigma_d = \lambda_d/2$ and $\sigma_x = \lambda_x/2$. Then the noise and speech signals will have the following distributions:

$$
\begin{aligned}
p_d(|Y|) &= \frac{2|Y|}{\lambda_d} \exp\left(-\frac{|Y|^2}{\lambda_d}\right) \\
p_x(|Y|) &= \frac{2|Y|}{\lambda_x} \exp\left(-\frac{|Y|^2}{\lambda_x}\right).
\end{aligned}
\tag{4.28}
$$

Now assume that we have two hypotheses:

- H_d: the noise signal dominates in this frame and frequency bin;
- H_x: the speech signal dominates in this frame and frequency bin.

Then the probability of the second hypothesis is given by

$$P(H_x||Y|) = \frac{p_x(|Y|)P(H_x)}{p_x(|Y|)P(H_x) + p_d(|Y|)P(H_d)} \tag{4.29}$$

where $P(H_d)$ and $P(H_x)$ are the a-priori probabilities for the corresponding hypotheses, and $p_d(|Y|)$ and $p_x(|Y|)$ are the distributions, defined in Equation 4.28. We will return to this equation later, but for now we just factorize $P_x(|Y|)$. In this case the probability of the speech signal to be dominant for a given magnitude $|Y|$ will be

$$P(\mathrm{H}_x||Y|) = P_x(|Y|)\frac{p_x(|Y|)}{p_x(|Y|) + \frac{P_d(|Y|)}{P_x(|Y|)}p_d(|Y|)} \tag{4.30}$$

which after some substitutions can be expressed in terms of a-priori and a-posteriori SNRs:

$$P(\mathrm{H}_x||Y|) = P_x(|Y|)\frac{\exp\left(-\frac{\gamma}{\xi}\right)}{\exp\left(-\frac{\gamma}{\xi}\right) + \exp(-\gamma)} = P_x(|Y|)\frac{1}{1 + \exp(-\gamma)/\exp(-\gamma/\xi)}. \tag{4.31}$$

The probability of the speech signal dominating the frequency bin can be used as a suppression rule:

$$H_k = \frac{1}{1 + \dfrac{\exp(-\gamma_k)}{\exp(-\gamma_k/\xi_k)}}. \tag{4.32}$$

We will return later to discuss why the prior probability of signal presence $P_x(|Y|)$ was removed from the suppression rule. The shape of the derived 'most probable amplitude' (MPA) estimator is shown in Figure 4.7a. It is quite different from the suppression rules plotted above. One of the differences is that this suppression rule never goes above 1.0, as it is a probability. This guarantees the stability of the entire noise-reduction system, as we will see later in this chapter. Figure 4.7b shows the suppression rule as a function of the a-posteriori SNR γ (i.e., the magnitude $|Y|$ for given λ_d) for 5 dB and 15 dB a-priori SNRs ξ. The Wiener and maximum-likelihood suppression rules are plotted for comparison. The second obvious difference is that the rule never goes to zero – again because it is a probability. This eliminates the problem with musical noise. The MPA estimator has interesting behavior in the area of very low a-priori SNR, where it actually suppresses the high amplitudes and lets the low amplitudes go unattenuated. In the area of 0 dB a-priori SNR, the suppression rule is constant and equals 0.5 – we cannot separate a mixture of two Gaussians with the same variation, the best we can do from the probabilistic standpoint is to attenuate the magnitude to one-half.

This probabilistic approach can be easily adopted for other than Gaussian PDFs of the speech signal and the noise. There are many studies regarding the speech signal probability distribution, but it is commonly accepted that for short-time audio frames (10–50 ms) a Laplace distribution models the speech signal best [9]. For periods of speech in the range of one to two seconds, the gamma distribution provides better

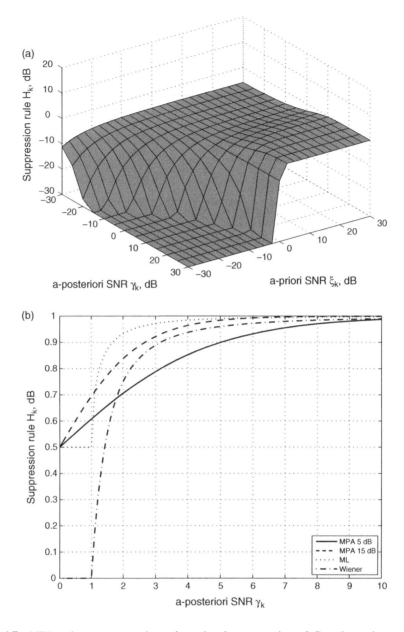

Figure 4.7   MPA estimator suppression rule under the assumption of Gaussian noise and speech distributions: (a) as a function of a-priori and a-posteriori SNRs; (b) comparison between Wiener, ML, and MPA suppression rules as a function of the a-posteriori SNR (MPA is given for 5 and 15 dB a-priori SNRs, respectively)

modeling. Considering the fact that the gamma distribution is more generic (the Laplace distribution is a particular case of the gamma distribution), we will use the gamma distribution. Given Gaussian complex noise with magnitude variance λ_d and Rayleigh distribution of the amplitudes (the same as above), and a speech signal with magnitude variance λ_x and gamma distribution, we have

$$p(x|k,\theta) = \frac{|x|^{k-1}}{\theta^k \Gamma(k)} \exp\left(-\frac{|x|}{\theta}\right) \tag{4.33}$$

where k is the shape and θ is the scale parameter. In our case $k = 1$, and the magnitude of the speech signal is exponentially distributed:

$$p(|x||\theta) = \frac{1}{2\beta} \exp\left(-\frac{|x|}{2\beta}\right). \tag{4.34}$$

The exponential distribution parameter is the magnitude variance $\beta^2 \approx \lambda_x$. Under the same assumption of the two hypotheses above, the suppression rule for gamma speech and Gaussian noise is

$$H_k = \frac{1}{1 + 4\sqrt{\frac{\gamma_k}{\xi_k}} \dfrac{\exp(-\gamma_k)}{\exp\left(-\frac{1}{2}\sqrt{\frac{\gamma_k}{\xi_k}}\right)}}. \tag{4.35}$$

This suppression rule is plotted in Figure 4.8. Note the similarity in the shape of the rule for low a-priori/high a-posteriori SNRs with Figure 4.6 – the MMSE solution for gamma speech and Gaussian noise – and the similarity in the shape in low a-posteriori/ high a-priori SNRs with the shape of the previous probabilistic rule.

4.2.12 Comparison of the Suppression Rules

To compare the effectiveness of the suppression rules alone, an experiment was conducted in a controlled environment. The speech signal, recorded with a close-talk microphone and high SNR, was mixed with noise, recorded in normal office conditions, to generate signals with 0, 10 and 20 dB SNRs. For each experiment, all three signals – clean speech, the noise signal, and the mixture – were available. Audio frames with 512 samples and 50% overlapping frames, weighted with a Han window, were used for conversion to the frequency domain and synthesis back to the time domain. The entire overlap–add process was discussed in detail in Chapter 2, where a MATLAB® script called *ProcessWAV.m* was provided. All three signals were converted to the frequency domain and precise estimations for a-priori SNR ξ_k and a-posteriori SNR γ_k were available for each frame. They were estimated as

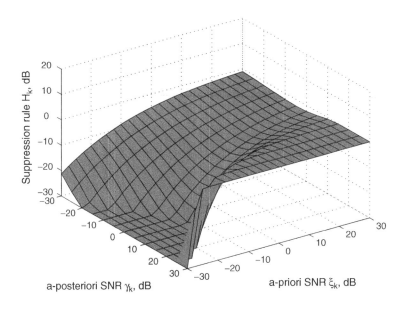

Figure 4.8 MPA estimator suppression rule under the assumption of Gaussian noise and Laplace speech distributions as a function of a-priori and a-posteriori SNRs

$$\lambda_d^{(n)}(k) = (1-\alpha)\lambda_d^{(n)}(k) + \alpha|D_k^{(n)}|^2$$

$$\xi_k^{(n)} \triangleq \frac{|X_k^{(n)}|^2}{\lambda_d(k)} \qquad\qquad\qquad (4.36)$$

$$\gamma_k^{(n)} \triangleq \frac{|Y_k^{(n)}|^2}{\lambda_d^{(n)}(k)}$$

for $\alpha = 0.02$. Here $D_k^{(n)}$, $X_k^{(n)}$, and $Y_k^{(n)}$ are the noise, clean speech, and mixed signals for the k-th frequency bin in the n-th frame. Note that while the noise variance λ_d is averaged, the speech signal variance is taken as $|X|^2$; that is, as momentary variance. Under any circumstances in a real scenario, when we have access only to the mixed signal, we cannot have such a precise estimation of the signal and noise variances and SNRs.

The criteria for comparison were the mean square error (MSE), log-spectral distance (LSD, see Equation 2.47), improvement in the SNR (as difference between the SNR after and before the processing, measured in decibels), and mean opinion score (MOS), measured with the implementation of objective quality measurement algorithm PESQ-W.

The results are shown in Table 4.2. There are sections for each SNR separately and averaged values for each algorithm.

From the MSE perspective, the best performers are Wiener (which is optimal in exactly this sense), closely followed by the entire group of efficient alternatives, and

Table 4.2 Comparison of various suppression rules

Algorithm	Equation	0 dB				10 dB				20 dB				Average			
		MSE	LSD	SNRI	MOS	MSE	LSD	SNRI	MOS	MSE	LSD	SNRI	MOS	MSE	LSD	SNRI	MOS
Do nothing	(4.54)	0.0429	0.885		2.377	0.0136	0.536		3.129	0.0043	0.300		3.906	0.0203	0.574		3.137
Wiener	(4.14)	0.0066	0.458	22.23	3.311	0.0028	0.405	20.27	3.875	0.0012	0.291	13.53	4.217	0.0035	0.384	18.68	3.801
Wiener DDA	(4.12)	0.0052	0.767	39.94	3.462	0.0024	0.475	29.34	4.029	0.0012	0.267	16.97	4.298	0.0029	0.503	28.75	3.930
Maximum-likelihood	(4.17)	0.0229	0.694	5.76	2.728	0.0075	0.421	5.56	3.443	0.0026	0.258	4.67	4.082	0.0110	0.457	5.33	3.418
MMSE	(4.18)	0.0057	0.317	29.39	3.631	0.0026	0.279	23.27	4.032	0.0013	0.211	13.77	4.276	0.0032	0.269	22.14	3.980
Log-MMSE	(4.20)	0.0055	0.332	30.71	3.677	0.0025	0.295	24.44	4.064	0.0013	0.221	14.71	4.288	0.0031	0.283	23.29	4.010
Spectral subtraction	(4.16)	0.0059	0.322	28.50	3.604	0.0027	0.286	22.55	4.001	0.0013	0.223	13.10	4.259	0.0033	0.277	21.38	3.955
JMAP SAE	Table 4.1, 1	0.0056	0.325	30.79	3.658	0.0026	0.280	24.01	4.038	0.0013	0.210	13.92	4.280	0.0031	0.272	22.91	3.992
MAP SAE	Table 4.1, 2	0.0054	0.370	33.17	3.717	0.0025	0.308	25.76	4.075	0.0012	0.220	15.08	4.295	0.0030	0.300	24.67	4.029
MMSE SAE	Table 4.1, 3	0.0058	0.314	28.44	3.592	0.0027	0.274	22.39	4.003	0.0013	0.208	13.05	4.265	0.0033	0.265	21.29	3.953
Prob. Gauss-Gauss	(4.32)	0.0055	1.187	38.80	3.008	0.0025	0.746	29.79	3.692	0.0012	0.333	17.00	4.214	0.0031	0.755	28.53	3.638
Prob. Laplace-Gauss	(4.35)	0.0055	1.233	40.04	2.958	0.0025	0.822	30.10	3.623	0.0012	0.379	17.61	4.180	0.0031	0.811	29.25	3.587

probabilistic rules. The maximum-likelihood suppression rule is definitely worst in this sense.

From the LSD perspective, the front runners are MMSE and log-MMSE (which is optimal in the log-MMSE sense). Good results are shown by the entire group of efficient alternatives. Note that Wiener and probabilistic rules are worse from this perspective, which means that they do not deal well with low levels of noise and speech.

The best average SNR improvement definitely has Wiener and probabilistic rules, followed by the efficient alternatives and spectral subtraction. The maximum-likelihood rule, as expected, has the lowest improvement in SNR. It is outperformed by the approximate Wiener suppression rule.

The highest MOS score and the best sound is achieved by log-MMSE and MAP SAE, followed closely by the group of efficient alternatives. The maximum-likelihood suppression rule sounds worse owing to a substantial amount of noise.

Figure 4.9 shows the relationship between the average improvement of SNR and the average MOS score – the last two columns in Table 4.2. It is clear that, to a certain degree, the noise suppression helps, and the signals with more suppressed noise achieve better perceptual sound quality. Enforcing the noise suppression further actually decreases the sound quality, regardless of the better SNR. This is good evidence that, when evaluating noise-suppressing algorithms, improvement in the SNR should not be used as the only criterion, and even not as a main evaluation criterion. Ultimately the goal of this type of speech enhancement is to make the output signal sound better for the human listener. From this perspective, the MOS is a much better criterion. When targeting speech recognition, the best criterion is, of course, the highest recognition rate.

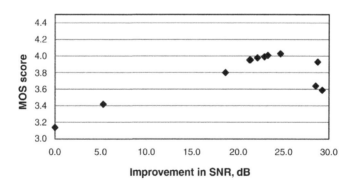

Figure 4.9   MOS results as a function of the SNR improvement for various suppression rules

On comparing the overall performance of the suppression rules, spectral subtraction and the entire group of efficient alternatives of Ephraim and Malah's MMSE (and that very same rule, of course) definitely stand out. To reiterate, this evaluation has been done under the best possible conditions and precise SNR estimates. In real conditions, the parameters for computation of the suppression rules will be estimated and have certain errors. Thus it will be important how robust each one of these suppression rules is to those errors.

EXERCISE

Look at the MATLAB script *SuppressionRule.m* which returns the suppression rule values for the given vectors of a-priori and a-posteriori SNRs:

```
Gain = SuppressionRule(gamma, xi, SuppressionType)
```

The argument `SuppressionType` is a number from 0 to 9 and determines which suppression rule is to be used. The script contains implementation of most of the suppression rules discussed so far. Finish the implementation of the rest of the suppression rules.

Write a MATLAB script that computes the suppression rules as a function of the a-priori and a-posteriori SNRs in the range of ± 30 dB. Limit the gain values in the range from -40 dB to $+20$ dB and plot the rules in three dimensions using the `mesh` function.

4.3 Uncertain Presence of the Speech Signal

All the suppression rules discussed above were derived under the assumption of the presence of both noise and speech signals. The speech signal, however, is not always presented in the short-term spectral representations. Even continuous speech has pauses with durations of 100–200 ms – which, compared with the typical frame sizes of 10–40 ms, means that there will be a substantial number of audio frames without a speech signal at all. Trying to estimate the speech signal in these frames leads to distortions and musical noise.

Classification of audio frames into "noise only" and "contains some speech" is in general a detection and estimation problem [10]. Stable and reliable work of the *voice activity detector* (VAD) is critical for achieving good noise-suppression results. Frame classification is used further to build statistical models of the noise and speech signals, so it leads to modification of the suppression rule as well.

4.3.1 Voice Activity Detectors

Voice activity detectors are algorithms for detecting the presence of speech in a mixed signal consisting of speech plus noise. They can vary from a simple binary decision (yes/no) for the entire audio frame to precise estimators of the speech presence probability for each frequency bin. Most modern noise-suppression systems contain at least one VAD, in many cases two or more. The commonly used algorithms base their decision on the assumption of a quasi-stationary noise;

that is, the noise variance changes more slowly than the variance of the speech signal. This allows one to build a model of the noise and track the changes. Then the decision is based on the assumption that the presence of a speech signal means an increase in energy – for the entire frame and per frequency bin. Such VADs work reliably for SNRs down to 0 dB. There is a separate group of approaches for detecting the presence of the speech signal in very low SNR conditions, or when the noise is highly non-stationary and changes as fast as, or faster than, the energy envelope of the speech signal.

One of the most commonly used and cited VADs is described in [11]. The purpose of this VAD is to detect the silent periods and improve the work of the G.729 codec. It is frequently used as a baseline to compare the performance of improved algorithms for VADs.

4.3.1.1 ROC Curves

Simple VADs produce a binary decision for the presence or absence of a speech signal in the current audio frame. In signal detection theory, such binary classifiers are characterized by the so-called *receiver operating characteristic* (ROC), or simply ROC curve. This is a graphical plot of probabilities of true positives versus false positives. In general, the binary classifier makes a decision that can be interpreted in four ways:

- true positive TP – a correct decision for presence was made = hit;
- true negative TN – a correct decision for absence was made = correct rejection;
- false positive FP – a decision for presence was made when a speech signal is not present = false alarm;
- false negative FN – a decision for absence was made when a speech signal was present = miss.

If we have a total number of positives N_P and a total number of negatives N_N, the true positive rate P_{TP}, or sensitivity, or recall, is defined as the proportion of true positives and all positives:

$$P_{TP} = \frac{N_{TP}}{N_P} = \frac{N_{TP}}{N_{TP} + N_{FN}} \qquad (4.37)$$

and the false positive rate P_{FP}, or false-alarm rate, is defined as the proportion of false positives and all negatives:

$$P_{FP} = \frac{N_{FP}}{N_N} = \frac{N_{FP}}{N_{FP} + N_{TN}}. \qquad (4.38)$$

Another important parameter of this type of detectors is the accuracy P_A, defined as the proportion of the correct decisions (i.e., true positives and true negatives) to all decisions:

$$P_A = \frac{N_{TP} + N_{TN}}{N_P + N_N}. \tag{4.39}$$

Energy-based binary classification (or VAD) for the absence/presence of speech signals can operate using a certain threshold. That is, if the energy of the current frame is greater than the threshold, then we have a speech signal, otherwise the speech signal is not present. The optimal value of the threshold can be estimated using the ROC curves. A typical ROC curve in this case will look like the one in Figure 4.10. For each threshold there is a corresponding point on the chart. For higher thresholds we have fewer false positives, but more false negatives. With a lower threshold the detection rate of the speech signal will be higher, but with the price of more noise frames detected as speech. The diagonal line shows the random-decision approach. The ROC of our classifier should be above this line (otherwise we can just negate the decisions made by the classifier). The threshold, corresponding to the point closest to the upper left corner, minimizes the sum of the squares of false positives and false negatives, $P_{FN}^2 + P_{FP}^2$. It is

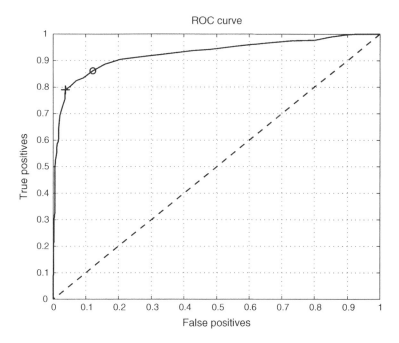

Figure 4.10 Receiver operating characteristics (ROC) curves for various thresholds of energy-based voice activity detectors. A " + " marks the highest accuracy point, an "o" marks the closest to the upper left corner point, where the sum of the squares of false positives and false negatives is minimal

marked with a small circle. The point where the accuracy from Equation 4.39 is highest is marked with a plus sign. Under some special circumstances the optimal point can be different. For estimation of the noise statistical parameters, for example, we may want to specify some true positive rate at the price of reducing the false positive rate, when energy from the speech signal will be averaged as noise statistical parameters.

ROC analysis was first used in World War II for the analysis of radar signals before being employed in signal detection theory. Currently it is widely used in biology and medicine. ROC curves find application in some machine-learning techniques as well.

EXERCISE

Write a MATLAB script to mix a clean speech signal (record or use *Speech.WAV*) with noise (record or use *NoiseHoth.WAV*) to achieve $\sim$5 dB SNR. Write the output into *NoisySpeech.WAV*.

Create a text file containing the beginnings and ends of each speech segment – this is going to be used as a ground truth.

Write a MATLAB script to build the ROC characteristics of a fixed-threshold VAD. The script reads the *NoisySpeech.WAV* and the text file with the ground truth. Then use a set of thresholds for classifying the frames to speech (RMS of the frame above the threshold) or noise (RMS of the frame below the threshold). Compute the false positives and false negatives. Plot the ROC curve – it should look like Figure 4.10.

4.3.1.2 Simple VAD with Dual-time-constant Integrator

Using a fixed threshold in binary VAD limits the system to a certain level of noise in the signal. In real VAD systems, the noise floor is usually adaptively tracked and the decision for absence/presence is made based on the average noise floor level. Assuming that the noise floor changes more slowly than the speech envelope, we can track the noise floor level with two different time constants: one low when the current level is higher than the estimate, and one high when the level is lower than the estimate:

$$L_{min}^{(n)} = \left| \begin{array}{ll} \left(1 - \dfrac{T}{\tau_{up}}\right) L_{min}^{(n-1)} + \dfrac{T}{\tau_{up}} L^{(n)} & L^{(n)} > L_{min}^{(n-1)} \\[3mm] \left(1 - \dfrac{T}{\tau_{down}}\right) L_{min}^{(n-1)} + \dfrac{T}{\tau_{down}} L^{(n)} & L^{(n)} \leq L_{min}^{(n-1)}. \end{array} \right. \tag{4.40}$$

Here $L_{min}^{(n)}$ is the estimate of the noise floor for the n-th frame, $L^{(n)}$ is the estimated signal level for the same frame, T is the frame duration, and τ_{down} and τ_{up} are the time constants for tracking the noise floor level when the level goes lower or higher than the current estimate. As the speech-plus-noise signal has a higher level than just the noise signal, $\tau_{\text{down}} \ll \tau_{\text{up}}$. For estimation of the signal level, weighted RMS is usually used:

$$L^{(n)} = \sqrt{\frac{1}{K} \sum_{k=0}^{K-1} (W_k \cdot |Y_k^{(n)}|)^2}. \tag{4.41}$$

Here $\mathbf{W}$ is a weighting function. The idea is to give more weight to the frequency bins with a higher difference between the signal and the noise (i.e., with higher SNR), which will allow easier differentiation. As the noise energy is usually concentrated in the lower part of the frequency band, this can just be a high-pass filter. Considering the fact that the speech energy decreases in the higher parts of the frequency band, it is a good idea to suppress the higher frequencies as well, in case of unexpected noises there. A standard weighting is often used – C-message or ITU-T Recommendation O.41 for modeling the telephone channel bandwidth. These weightings increase the signal SNR, which makes the detection easier.

The decision for presence/absence of the speech signal can be made based on the estimated noise floor $L_{\text{min}}^{(n)}$, the signal level $L^{(n)}$, and the previous value of the voice activity flag V ($V=0$ for noise, $V=1$ for speech):

$$V^{(n)} = \begin{vmatrix} 0 & \text{if} & \dfrac{L^{(n)}}{L_{\text{min}}^{(n)}} < T_{\text{down}} \\[2ex] 1 & \text{if} & \dfrac{L^{(n)}}{L_{\text{min}}^{(n)}} > T_{\text{up}} \\[2ex] V^{(n-1)} & & \text{otherwise.} \end{vmatrix} \tag{4.42}$$

Note that the flag V switches unconditionally to state "noise" if the proportion of the current and minimal level is below the threshold T_{down}. To switch to "speech" state, this proportion should go above the threshold T_{up} ($T_{\text{down}} < T_{\text{up}}$). To switch back, the proportion should fall below T_{down}, and so on. This hysteresis stabilizes the work of the VAD, decreases the false positives, and the VAD switches back to its "noise" state after the end of the word. The downside is that the VAD switches to "speech" state with a small delay and can cut off the beginning of the word if it starts with a low-level consonant.

This simple, energy-based VAD with binary decision is a software implementation of the well-known hardware VAD used in a countless number of amateur radio stations – see Figure 4.11. The first two RC groups, $R_1 C_1$ and $R_2 C_2$, form the band-pass filter. The low cut-off frequency is $f_{\text{low}} = 1/(2\pi C_1 R_2)$, the high is $f_{\text{high}} = 1/(2\pi R_1 C_2)$. The diode D_1, the resistor R_3, and the integration capacitor C_3 estimate the envelope

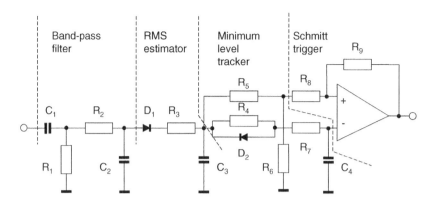

Figure 4.11   Hardware implementation of a binary energy-based voice activity detector

RMS – note that the non-linear VA characteristic of the diode is used as an interpolation of the squaring in Equation 4.41, but the square-root is missing in this analog RMS estimator. The group R_4, D_2, R_7, and C_4 is the minimum level tracker from Equation 4.40. Here $\tau_{up} = (R_4 + R_7)C_4$ and $\tau_{down} = R_7C_4$. Resistors R_5 and R_6 form the absolute threshold, and the operational amplifier, together with R_8 and R_9, forms a Schmitt trigger with a given hysteresis. The Schmitt trigger output is the VAD state.

This VAD has four parameters to adjust, $S = [\tau_{down}, \tau_{up}, T_{down}, T_{up}]$. Typical values used in analog circuitry are $\tau_{down} = 40$ ms, $\tau_{up} = 10$ s, $T_{down} = 1.2$, and $T_{up} = 3$. The performance of this VAD can be improved by maximizing the accuracy (Equation 4.39):

$$S_{opt} = \arg \max_S (P_A(S)) \tag{4.43}$$

which minimizes the sum of false positives and false negatives, regardless of their proportions. The point with maximal accuracy is marked with " $+$ " on the ROC curve in Figure 4.10. If we assume that true positives and true negatives are equally important, we can minimize the minimal distance between the ROC curve and the upper left corner, where the ideal classifier is – zero false positives and all true positives. This optimization criterion is given by

$$S_{opt} = \arg \min_S \left(\min \left(\sqrt{[1-P_{TP}(S)]^2 + P_{FP}(S)^2} \right) \right). \tag{4.44}$$

The optimal from this perspective point is marked with "o" on the ROC curve in Figure 4.10. The accuracy and the corresponding probabilities can be estimated with a large set of manually labeled WAV files with SNR varying in the work range of this

VAD. An easier approach is to use the VAD and clean speech signal to label the frames with the presence and absence of a speech signal, and then to contaminate the clean speech signal with variable amounts of noise. Almost any optimization method can find the solutions above; gradient-based methods will most probably be the most efficient.

Regardless of its simplicity, the energy-based binary decision VAD works surprisingly well. It easily achieves accuracy above 95% for SNRs varying from 5 to 30 dB. A major contributor to this is the weighting in Equation 4.41, which removes most of the noise energy under normal conditions. This weighting is equivalent to band-pass filtering for telephone bandwidth. One of the advantages of this classifier is that it does not need prior knowledge of the noise and speech statistical parameters – the estimated $L_{\min}^{(n)}$ is nothing more than tracking the noise floor. This VAD does not provide soft probability of the speech signal presence in the current frame and per frequency bin.

EXERCISE

Examine the MATLAB script *SimpleVAD.m*. This is an implementation of the voice activity detector above. The function needs the current level, the current time, and a data structure as input parameters, carrying the pre-initialized time constants, thresholds, and so on as described in the initial comment. This data structure has to be initialized at the beginning with the recommended values.

Write a MATLAB script to compute the SNR of the given WAV file, which takes as parameters the file name and returns the signal-to-noise-ratio:

```
SNR = SNRMeasurement(inpFileName)
```

Read the WAV file, initialize the necessary variables (VAD data structure, signal and noise levels, and signal and noise frame counters), organize a loop for processing the file on frames with duration of 20 ms, and compute the RMS level of the current frame. Then use the simple VAD above to do the classification of speech or noise, and add the square of the computed level to the corresponding variable. Increment the corresponding frame counter. When you reach the end of the file, compute the average levels of the signal and noise frames, convert to decibels, and subtract. Display the result.

Modify the MATLAB script above to work in the frequency domain. Use *Process WAV.m* as a template. Pre-compute the weightings for A-, B-, and C-weightings using the approximation formulas from Chapter 3. Organize three sets of noise and speech levels – one for each weighting. At the end, compute and display three SNRs – one for each weighting. For the level provided to the VAD, use either C-message weighting or a rectangular band-pass filter (300–3400 Hz).

Test the final version of the SNR measurement tool and make sure it works well, as this will be a frequently used tool later in the book.

4.3.1.3 Statistical-model-based VAD with Likelihood Ratio Test

Sohn *et al.* [12] use the likelihood ratio test and an effective hangover scheme for classification of the audio frames. Given a speech signal degraded by uncorrelated additive noise, two hypotheses can be considered:

- H_0: speech absent, $Y = D$;
- H_1: speech present, $Y = D + X$;

where D, X, and Y are the K-dimensional complex vectors of the noise, speech and noisy speech. Assuming Gaussian distribution and known variances λ_d and λ_x for the real and imaginary parts of the noise and speech signals, the probability density functions of these two hypotheses are

$$
\begin{aligned}
p(Y|H_0) &= \prod_{k=0}^{K-1} \frac{1}{\pi \lambda_d(k)} \exp\left\{-\frac{|Y_k|^2}{\lambda_d(k)}\right\} \\
p(Y|H_1) &= \prod_{k=0}^{K-1} \frac{1}{\pi [\lambda_d(k) + \lambda_x(k)]} \exp\left\{-\frac{|Y_k|^2}{\lambda_d(k) + \lambda_x(k)}\right\}
\end{aligned}
\tag{4.45}
$$

This yields the likelihood ratio

$$
\Lambda_k \triangleq \frac{p(Y|H_1)}{p(Y|H_0)} = \frac{1}{1+\xi_k} \exp\left\{\frac{\gamma_k \xi_k}{1+\xi_k}\right\}
\tag{4.46}
$$

where ξ_k and γ_k are the a-priori and a-posteriori SNRs. Then a decision rule is established by comparing the geometric mean of the likelihood ratios for all frequency bins:

$$
\log \Lambda = \frac{1}{L} \sum_{k=0}^{K-1} \log \Lambda_k \underset{H_0}{\overset{H_1}{\gtrless}} \eta,
\tag{4.47}
$$

where η is a fixed threshold for deciding which one of the two hypotheses is true.

To account for the timing of the speech signal, Sohn and co-authors use an HMM-based hang-over scheme. It is based on the idea that there is a strong correlation in the consecutive occurrences of speech frames. The sequence is modeled with a first-order Markov process – that is, assuming that the current state depends only on the previous states. The correlative characteristics of speech occurrence can be represented by $P(q_n = H_1 | q_{n-1} = H_1)$ with the following constraint:

$$
P(q_n = H_1 | q_{n-1} = H_1) > P(q_n = H_1).
\tag{4.48}
$$

Here q_n denotes the state of the n-th frame and is either H_0 or H_1. With the assumption that the Markov process is time-invariant, then $a_{ij} \triangleq P(q_n = H_j | q_n = H_i)$; and under the assumption of process staionarity, $P(q_n = H_i) = P(H_i)$, where $P(H_0)$ and $P(H_1)$ are the steady-state probabilities, obtained from $a_{01} P(H_0) = a_{10} P(H_1)$ and $P(H_0) + P(H_1) = 1$. Then the overall process is described with only two parameters, a_{01} and a_{10}. The decision rule is modified as

$$\widehat{\Lambda}_n \triangleq \frac{p(\aleph_n | q_n = H_1)}{p(\aleph_n | q_n = H_0)} = \frac{P(H_0)}{P(H_1)} \cdot \frac{P(q_n = H_1 | \aleph_n)}{P(q_n = H_0 | \aleph_n)} \overset{H_1}{\underset{H_0}{\gtrless}} \eta \qquad (4.49)$$

where $\aleph_n = \{X_n, X_{n-1}, \ldots, X_1\}$ is the set of observations up to the current frame n. The left fraction in Equation 4.49 is the a-priori probability ratio; the right is the a-posteriori probability ratio. Denoting the second one with Γ_n and using the forward procedure described by Sohn and co-authors, we can obtain the following recursive formula:

$$\Gamma_n = \frac{a_{01} + a_{11} \Gamma_{n-1}}{a_{00} + a_{10} \Gamma_{n-1}} \Lambda_n \qquad (4.50)$$

where Λ_n is computed using Equation 4.47. Then the modified decision rule with hangover scheme is compared to the threshold η:

$$\widehat{\Lambda}_n = \frac{P(H_0)}{P(H_1)} . \Gamma_n \overset{H_1}{\underset{H_0}{\gtrless}} \eta. \qquad (4.51)$$

In their practical implementation, the authors used hand-labeled voice data to estimate the probabilities. In their case, $a_{01} = 0.2$ and $a_{10} = 0.1$. With hand-labeled data it is relatively easy to draw the ROC curves for various η and to pick the value with highest accuracy. Taking a more detailed look at Equation 4.50, we find that this is just a more sophisticated time smoothing, replacing the traditional first-order integrator.

4.3.1.4 VAD with Floating Threshold and Hangover Scheme with State Machine

One potential limiting factor of the previous VAD is the fixed threshold. This reduces the range of various SNRs the VAD can operate without significant decrease of the classification accuracy. Theoretically, for each SNR there is an optimal threshold η that can work satisfactorily in a certain range of SNRs around the one it is optimal for. Davis and Nordholm [13] derive a way to estimate the optimal threshold for a given SNR. They compute the log-likelihood ratio using a similar approach as in the previous section. Under the assumption of Gaussian distributions of both noise and speech, the optimal threshold for the acceptable probability for a false alarm P_{FA} is

$$\eta_{opt}(k) = \sqrt{2}\lambda_d(k).\text{erfc}^{-1}(2P_{FA})$$ (4.52)

where $\text{erfc}(u)$ is the complimentary error function [14]. Then the modified decision rule with floating threshold for classification of the current frame is

$$\frac{1}{L}\sum_{k=0}^{K-1} \log \Lambda_k \underset{H_0}{\overset{H_1}{\gtrless}} \text{erfc}^{-1}(2P_{FA}).\sqrt{2}.\frac{1}{L}\sum_{k=0}^{K-1} \lambda_d(k).$$ (4.53)

In the same paper the authors discuss a slightly different hangover scheme. It is a state machine, which requires the decision rule of Equation 4.53 to be in the "speech" state for at least four consequent frames before switching to "speech" state, and the decision rule to be in the "noise" state for ten consequent frames before switching to "noise" state. This state machine effectively delays the decision to gain confidence. The number of frames to delay the decision can be tuned for specific conditions, but in general the first delay reduces the false alarms, and the second delay allows covering the end of the word, which usually has low energy.

4.3.2 Modified Suppression Rule

To make the noise suppressors more efficient, a modified suppression rule was derived by McAulay and Malpass [4]. While comparing the ML suppression rule, derived by them in the same paper, with the Wiener and spectral subtraction rules existing at that time, they concluded that it was apparent that none of the suppression rules adequately suppresses the background noise when the speech is absent, as all of them are derived under the assumption of speech presence. The presence or absence of the speech signal is considered a two-state model:

- H_0: speech absent, $|Y_k^{(n)}| = |D_k^{(n)}|$;
- H_1: speech present, $|Y_k^{(n)}| = |D_k^{(n)} + X_k^{(n)}|$.

Under their assumption of Gaussian noise and the speech signal model as a sum of sinusoidal signals, the modified suppression rule is derived as

$$\tilde{H}_k^{(n)} = P(H_1||Y_k^{(n)}|).H_k^{(n)}$$ (4.54)

where $\tilde{H}_k^{(n)}$ is the modified suppression rule, $H_k^{(n)}$ is the estimated suppression rule (in their case, the ML amplitude estimator), and $P(H_1||Y_k^{(n)}|)$ is the probability to have a speech signal present in this frequency bin and frame. The derivation of Equation 4.54 is generic and does not depend on the specific assumptions for the distributions of the speech and noise signals.

Ephraim and Malah [3] used results from [15] to modify the suppression rule under the uncertain presence of a speech signal as

$$\tilde{H}_k^{(n)} = \frac{\Lambda(Y_k, q_k)}{1 + \Lambda(Y_k, q_k)} \cdot H_k^{(n)} \tag{4.55}$$

where $\Lambda(Y_k, q_k)$ is the generalized likelihood ratio defined by

$$\Lambda(Y_k, q_k) = \mu_k \frac{P(Y_k|H_1)}{P(Y_k|H_0)} \tag{4.56}$$

$\mu_k \triangleq (1-q_k)/q_k$ and q_k is the probability of signal absence in the k-th frequency bin. The two hypotheses H_0 and H_1 for the current frequency bin are the same as above – the absence and presence of a speech signal. For Gaussian speech and Gaussian noise, $\Lambda(Y_k, q_k)$ can be expressed in terms of a-priori and a-posteriori SNRs:

$$\Lambda(Y_k, q_k) = \mu_k \frac{\exp(\nu_k)}{1 + \xi_k}, \tag{4.57}$$

where now ξ_k is defined as the a-priori SNR when speech is present:

$$\xi_k \triangleq \frac{E\{|X_k|^2|H_k\}}{\lambda_d(k)}. \tag{4.58}$$

For convenience and easier estimation, we define

$$\eta_k \triangleq \frac{E\{|X_k|^2\}}{\lambda_d(k)} = (1-q_k)\xi_k, \tag{4.59}$$

which is easier to estimate, and this finalizes the modification of the suppression rule under the uncertain presence of a speech signal. Looking closely at the modifier of the suppression rule in Equation 4.55, we can say that it is actually a generalized probability for the presence of a speech signal, which concurs with the conclusion in Equation 4.54. Note the similarity of Equations 4.46 and 4.57 – they both are actually speech-presence probability estimators.

The modified suppression rule is derived under certain assumptions for the short-term speech and noise probability distributions. As already mentioned, the short-term speech distribution is modeled best with a Laplace distribution, while a Gaussian distribution best models the noise in the majority of cases. On the other hand, the gamma distribution models longer intervals of speech signals (1–2 seconds); see [9].

4.3.3 Presence Probability Estimators

An alternative approach to introduce the uncertain presence of the speech signal is presented in Equation 4.29 with the a-priori probabilities for a dominant presence of speech $P(H_d)$ or noise $P(H_x)$, respectively. The noise and speech probability for the entire frame can be used as a prior for each frequency bin. Using the likelihood, estimated from Equation 4.47 or 4.51, we can estimate the probability for speech presence or absence in the current audio frame:

$$P^{(n)}(H_1) = \frac{\Lambda^{(n)}}{1 + \Lambda^{(n)}}$$
$$\qquad(4.60)$$
$$P^{(n)}(H_0) = 1 - P^{(n)}(H_1).$$

The level $L^{(n)}$, or RMS estimated according to Equation 4.41, has a Rayleigh distribution during the noise frames. The distribution of the levels during speech frames can be assumed to be either Rayleigh (assuming Gaussian distribution of the speech signal) or exponential (assuming Laplace distribution of the speech signal). These assumptions and the hypotheses for a predominant noise or speech signal lead to the probability estimators in Equation 4.31 or 4.35, but this time for the entire audio frame.

Estimation of the speech-presence probability $P_k^{(n)}(H_1||Y_k|)$ for each frequency bin can be done using either of the two methods described above for estimation of the speech-presence probability per audio frame.

4.4 Estimation of the Signal and Noise Parameters

4.4.1 Noise Models: Updating and Statistical Parameters

Estimation of the noise variation λ_d, given the VAD state $V^{(n)}$ for the n-th frame can be done using a simple recursive formula:

$$\lambda_d^{(n)}(k) = \begin{vmatrix} (1-\alpha)\lambda_d^{(n-1)}(k) + \alpha|Y_k^{(n)}|^2 & \text{for} & V^{(n)} = 0 \\ \lambda_d^{(n-1)}(k) & \text{for} & V^{(n)} = 1 \end{vmatrix} \qquad(4.61)$$

where α is the adaptation parameter with typical values between 0.5 and 0.95. This adaptation formula works well in real systems. One thing we may want to fix is that at the end of the word the noise estimation is outdated if the noise variance changed meantime. In addition, speech is quite a sparse signal in the frequency domain. Even with speech present, a lot of frequency bins contain just noise and we can update the noise variance for them even when the speech signal is present in the frame. This leads to the probability-based noise variance estimator

$$\lambda_d^{(n)}(k) = P_k^{(n)}(H_1||Y_k|)\lambda_d^{(n-1)}(k) + P_k^{(n)}(H_0||Y_k|)|Y_k|^2; \qquad (4.62)$$

that is, we use the speech absence and presence probabilities for the frequency bin to modify the adaptation parameter. A small problem here is that, when $P_k^{(n)}(H_1||Y_k|)$ approaches one, which is completely possible, then we will have an estimate that is actually the momentary value of the noise variation. In most cases this is not desirable, as λ_k is a statistical parameter. This leads to combining the two estimators:

$$\lambda_d^{(n)}(k) = (1-\alpha)P_k^{(n)}(H_1||Y_k|)\lambda_d^{(n-1)}(k) + \alpha P_k^{(n)}(H_0||Y_k|)|Y_k|^2 \qquad (4.63)$$

which is nothing more than limiting the adaptation speed to the value of the adaptation parameter α in cases of high probability of noise-only presence.

Another interesting approach for estimation of the noise variance is based on minimum statistics [16]. It is based on the estimation of floating and optimal values of the smoothing parameter α. During the speech-absence frames we want our noise variance estimate λ_d to be as close as possible to the actual σ_d^2. Therefore the goal is to minimize $E\{(\lambda_d^{(n)} - \sigma_d^2)^2|\lambda_d^{(n-1)}\}$. After setting the first derivative to zero, we can find the optimal value for the smoothing parameter:

$$\alpha_{\text{opt}}^{(n)}(k) = \frac{1}{1 + \left(\dfrac{\lambda^{(n-1)}}{\lambda_d} - 1\right)^2}. \qquad (4.64)$$

Since the parameter $\alpha_{\text{opt}}^{(n)}(k)$ is always between 0 and 1, a stable non-negative variance estimation is guaranteed. Looking closely at the denominator, we can say that the term $\tilde{\gamma}_k^{(n)} = \lambda^{(n-1)}/\lambda_d$ can be recognized and estimated as a smoothed version of the a-posteriori SNR $\gamma_k^{(n)}$. The values of the parameter $\alpha_{\text{opt}}^{(n)}$ for $\tilde{\gamma}_k^{(n)}$ varying between 0 and 10 are shown in Figure 4.12. Note that the minimum statistics approach does not require VAD, as the smoothed a-posteriori SNR $\tilde{\gamma}_k^{(n)}$ is estimated for each frame and then the optimal parameter $\alpha_{\text{opt}}^{(n)}(k)$ adapts accordingly.

4.4.2 A-Priori SNR Estimation

Most of the noise suppression rules described so far depend on two parameters: a-priori and a-posteriori SNRs, ξ_k and γ_k, respectively. They have to be estimated for each

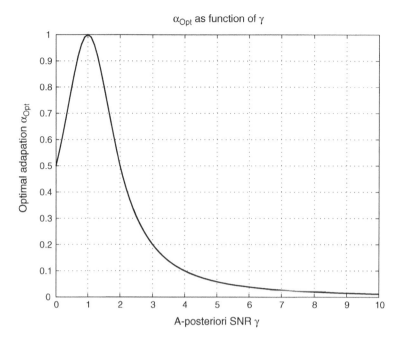

Figure 4.12 The adaptation parameter optimal in a minimum statistics sense as a function of the smoothed a-priori SNR

frame and frequency bin, according to Equation 4.36. With the noise variance estimate $\lambda_k^{(n)}$ and current magnitude $|Y_k^{(n)}|$ in the frequency bin, estimation of the a-posteriori SNR $\gamma_k^{(n)}$ is not a problem. However, estimation of the a-priori SNR requires using the clean speech magnitude $|X_k^{(n)}|$ (which we actually want to estimate). There are several approaches to overcome this problem.

The first is to use the fact that noise and speech are not correlated. Then the maximum-likelihood estimate of the a-priori SNR is

$$\xi_k^{(n)} = \max[0, (\gamma_k^{(n)} - 1)]. \tag{4.65}$$

We already discussed the disadvantages of this approach in the section on suppression rules, and experiments confirmed that the Wiener suppression rule based on this estimate (Equation 4.14) performed worse than the one based on the actual a-priori SNR estimation (Equation 4.12).

A definitely better approach is based on the fact that the speech signal changes more slowly than the normal duration of an audio frame, 10–40 ms. This makes the speech signal in consequent audio frames highly correlated, and we can use the previous clean-speech signal estimate $|\hat{X}_k^{(n-1)}|$ as an approximate value of the clean-speech signal magnitude in the current frame. This is called the *decision-directed approach* (DDA) and was derived by Ephraim and Malah [3]. Analyses showed that this approach contributes as much to the success of their noise suppressor as the suppression rule derived by them. The a-priori SNR is estimated as

$$\xi_k^{(n)} = \beta \frac{|\hat{X}_k^{(n-1)}|^2}{\lambda_d^{(n)}} + (1-\beta).\max[0, (\gamma_k^{(n)}-1)] \qquad (4.66)$$

where β varies between 0.9 and 0.98. The second term is mostly a stabilizer and plays a role at the beginning. Regardless of the fact that this is an approximate solution, it provides much better results with any suppression rule than the estimator from Equation 4.65. On the negative side, the decision-directed approach introduces dependency of the current output on the values of previous outputs; that is, it converts the noise suppressor to a system with feedback. This immediately raises concerns about the stability of the entire system. These concerns increase especially when we have modules with gain higher than 1. This is a case with both suppression rules from Ephraim and Malah and their computationally efficient alternatives. Under certain circumstances and not very well set time constants, these noise suppressors can become unstable and provide output with audible echoes. This is not the case with the Wiener or probabilistic suppression rules.

Another potential problem with a DDA-based estimation of a-priori SNR is that it provides a one-frame-delayed estimation, which in the transition moments (from speech to noise and from noise to speech) is not correct. This problem can be addressed by using forward–backward DDA [17]. Assuming backward off-line processing, we can define the backward DDA-based a-priori SNR estimation as

$$\xi_{Bk}^{(n)} = \beta \frac{|\hat{X}_k^{(n+1)}|^2}{\lambda_d^{(n)}} + (1-\beta).\max[0, (\gamma_k^{(n)}-1)]. \qquad (4.67)$$

Note that, for estimation of the n-th frame, frame $(n+1)$ is used. The final estimation of the a-priori SNR is a linear combination between the forward and backward estimations:

$$\xi_k^{(n)} = \beta_k^{(n)} \xi_{Fk}^{(n)} + \left(1-\beta_k^{(n)}\right)\xi_{Bk}^{(n)} \qquad (4.68)$$

where $\xi_{Fk}^{(n)}$ is the forward estimation according to Equation 4.66, $\xi_{Bk}^{(n)}$ is the backward estimation according to Equation 4.67, and $\beta_k^{(n)}$ is a time- and frequency-dependent adaptation constant. At the beginning of a speech segment, $\xi_{Bk}^{(n)}$ should be preferred, while at the end of the speech segment $x_{Fk}^{(n)}$ is a better estimate. The adaptation constant can be computed using labeled audio files and learning techniques, or interpolated with a smooth function based on the distance after the beginning of the word or before its end. The proposed approach provides a 0.7–1.1 dB better SNR improvement; and, according to tests conducted with human listeners, the output of a forward–backward DDA-based noise suppressor is preferred. Good estimation of $\xi_{Bk}^{(n)}$ requires at least two frames delay, which is not acceptable in most of the cases when the noise suppressor is part of a real-time communication system.

4.5 Architecture of a Noise Suppressor

Having discussed all the components of a noise suppressor it is time to put all of them together. The overall block diagram is shown in Figure 4.13.

One of the most important things in statistical-model-based noise suppression is to have good estimations of the noise statistical parameters. This can be done in two stages:

- Use a simple VAD like the one described in Section 4.3.1.2, and do frame classification and create a rough noise model $\tilde{\lambda}_d^{(n)}(k)$ with adaptation per Equation 4.61.
- With the rough noise model, estimate rough a-priori and a-posteriori SNRs, $\tilde{\xi}_k^{(n)}$ and $\tilde{\gamma}_k^{(n)}$, according to Equations 4.66 and 4.3, and use them in a more sophisticated VAD, like the one described in Section 4.3.1.3.
- With the obtained likelihood ratios for each frequency bin, estimate the speech presence probability $P_k^{(n)}(\mathrm{H}_1||Y_k|)$ according to Equation 4.60. Estimate the speech presence probability $P^{(n)}(\mathrm{H}_1)$ for the entire frame as well.
- Use the speech-presence probability to estimate a more precise noise model $\lambda_d^{(n)}(k)$ according to Equation 4.62 or 4.63.

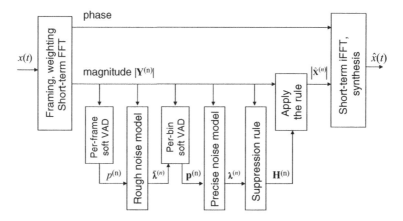

Figure 4.13   Block diagram of noise suppressor

An alternative approach is to use the minimal statistics method, described in Equation 4.64. It can be used for estimation of the final noise model of the first stage only.

The precise noise model $\lambda_d^{(n)}(k)$ can be used for estimation of the final a-priori and a-posteriori SNRs, $\xi_k^{(n)}$ and $\gamma_k^{(n)}$, and the selected suppression rule $H_k^{(n)}$. Spectral subtraction and the log-MMSE estimator are among the most frequently used. The suppression rule can be modified with the speech-presence probability for the entire frame.

Estimate the output signal by applying the suppression rule to the output signal.

The noise suppressor described above is a good practical example of a working system. Properly implemented it sounds good and provides improvement in the overall SNR. The number of potential combinations between various VADs, noise model updaters, and suppression rules is very large and it is not intended here to provide detailed comparison between all variants.

An example of how the noise suppressor works with a real signal is shown in Figure 4.14. Part (a) shows the input signal with SNR = 10 dB in the time domain. Note the noise during the pauses. The same figure shows the speech presence probability, computed later by the voice detector using Equations 4.51 and 4.60. Part (b) shows the spectrogram of the input signal. The noise is easily visible during the pauses and especially in the lower part of the frequency band. There are some constant tones,

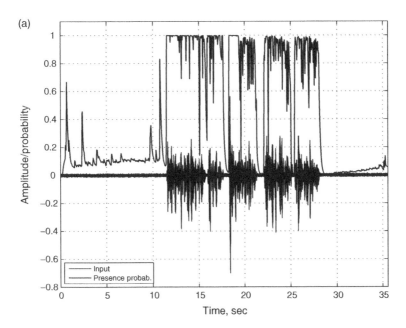

Figure 4.14 Noise suppressor: (a) input signal and speech presence probability per frame; (b) input signal – spectrogram; (c) speech presence probability after the voice activity detector; (d) suppression gain; (e) output signal – spectrogram; (f) output signal in the time domain

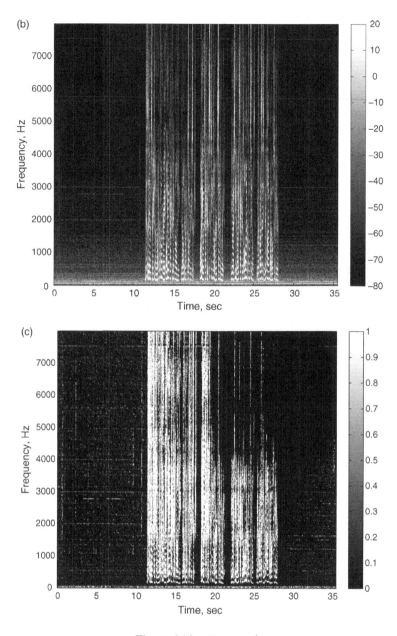

Figure 4.14   *(Continued)*

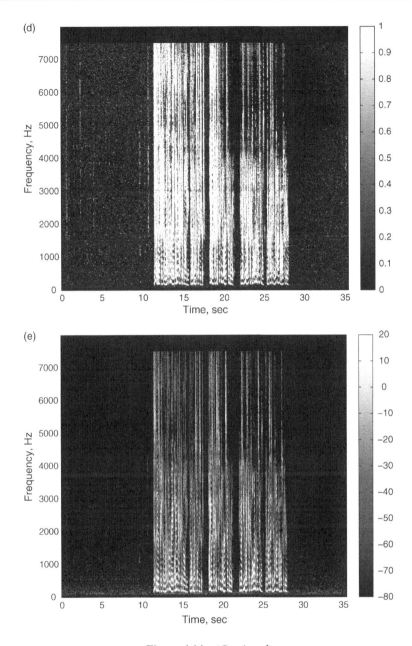

Figure 4.14   *(Continued)*

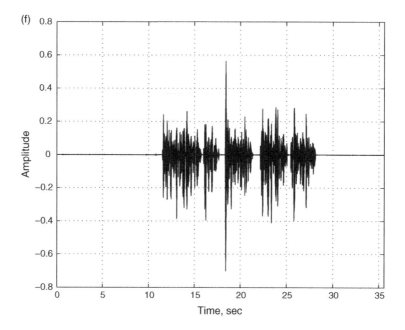

Figure 4.14 *(Continued)*

represented by horizontal lines, caused by computer fans and hard disks. We will discuss a more elegant way to deal with them later in this chapter. Figure 4.14c shows the output of the second voice activity detector. This is the speech presence probability per frequency bin, computed according to Equations 4.46, 4.51, and 4.60. The speech presence probability is used for updating the precise noise model – Equation 4.63. The suppression gain is presented in Figure 4.14c. It is computed according to Equation 4.20 and modified for the uncertain presence of a speech signal in the frame according to Equation 4.55. Note that to the first several frequency bins (below 100 Hz) some minimal gain value is assigned. The noise model there is very volatile and cannot provide a good enough signal estimation. The same is done for the last several frequency bins (above 7500 Hz) to remove aliasing effects. Figure 4.14d shows the spectrogram of the output signal. Note the minimal residual noise in the pause segments and compare this spectrogram with Figure 4.14b. The output signal in the time domain is shown in Figure 4.14e. Compare the noise level with the noise in the input signal – Figure 4.14a.

A comparison of the various suppression rules is presented in Table 4.3. These results are computed using the noise-suppressor implementation above by varying the algorithm for computing the suppression rule. This is similar to the comparison described in Section 4.2.11 and presented in Table 4.2.

From the MMSE perspective, the best results are achieved by the two probabilistic rules. They are apparently more robust to the inevitable errors in estimation of the noise models. At the other end is, again, the ML suppression rule.

Table 4.3 Comparison of various suppression rules for end-to-end noise suppressor

Algorithm	Equation	0 dB				10 dB				20 dB				Average			
		MSE	LSD	SNRI	MOS	MSE	LSD	SNRI	MOS	MSE	LSD	SNRI	MOS	MSE	LSD	SNRI	MOS
Do nothing		0.0429	0.886		2.40	0.0136	0.537		3.16	0.0043	0.301		3.93	0.0203	0.575		3.16
Uncertain presence	(4.54)	0.0184	0.727	7.12	2.62	0.0060	0.484	7.14	3.36	0.0023	0.401	6.85	3.98	0.0089	0.537	7.04	3.32
Wiener, a-posteriori	(4.14)	0.0097	0.516	14.64	2.93	0.0037	0.518	14.48	3.60	0.0018	0.621	13.15	3.96	0.0051	0.552	14.09	3.50
Wiener, a-priori	(4.12)	0.0110	0.497	17.51	2.93	0.0048	0.542	18.09	3.65	0.0022	0.665	16.20	4.05	0.0060	0.568	17.27	3.54
Maximum-likelihood	(4.17)	0.0152	0.655	8.91	2.72	0.0052	0.456	9.04	3.46	0.0021	0.425	8.62	3.99	0.0075	0.512	8.86	3.39
MMSE	(4.18)	0.0116	0.519	14.20	2.85	0.0048	0.468	14.79	3.60	0.0021	0.552	13.90	4.04	0.0062	0.513	14.30	3.50
Log-MMSE	(4.20)	0.0114	0.509	15.15	2.87	0.0049	0.482	15.78	3.62	0.0022	0.578	14.65	4.05	0.0061	0.523	15.19	3.51
Spectral subtraction	(4.16)	0.0115	0.529	13.16	2.88	0.0043	0.469	13.41	3.62	0.0019	0.546	12.62	4.03	0.0059	0.515	13.06	3.51
JMAP SAE	Table 4.1, 1	0.0112	0.513	14.41	2.86	0.0048	0.474	14.96	3.60	0.0021	0.566	14.02	4.04	0.0061	0.518	14.46	3.50
MAP SAE	Table 4.1, 2	0.0119	0.502	15.64	2.89	0.0048	0.496	16.22	3.63	0.0021	0.600	14.95	4.05	0.0060	0.533	15.60	3.52
MMSE SPE	Table 4.1, 3	0.0095	0.532	13.37	2.83	0.0049	0.458	13.91	3.58	0.0021	0.530	13.20	4.04	0.0063	0.506	13.49	3.48
Prob. Gaus-Gauss	(4.32)	0.0085	0.499	19.46	2.90	0.0036	0.559	19.55	3.62	0.0018	0.687	16.50	4.03	0.0049	0.582	18.50	3.52
Prob. Laplace-Gauss	(4.35)	0.0103	0.486	18.67	3.01	0.0034	0.540	18.53	3.68	0.0017	0.670	15.91	4.04	0.0045	0.565	17.70	3.57

From the LSD perspective, best results are achieved by MMSE and log-MMSE rules (optimal in this sense), closely followed by the group of the efficient alternatives. In general all rules achieve quite similar performance.

The two probabilistic suppression rules achieve best noise suppression, followed by the Wiener suppression rule. The ML rule lags behind, as in the clean experiment before.

There is no clear winner as far as the MOS results are concerned, but the two probabilistic suppression rules have a small advantage. Figure 4.15 shows the MOS results as a function of the improvement in SNR. Compare it with Figure 4.9. The first thing to notice is the substantially lower improvement in SNR. Under controlled conditions and precise a-priori and a-posteriori SNR estimations (from the noise-only and speech-only signals, known in this experiment) the improvement in SNR goes above 35 dB in some cases. Here all algorithms are grouped in the range from 9 to 19 dB improvement. Another interesting point is the MOS results. In the controlled experiment we were able to achieve MOS up to 4.1, where under real conditions the average MOS barely approaches 3.6. Note that the sound quality of the output signal does not depend heavily on the suppression rule used. The reliability of the VAD, how precise we can estimate the noise model, and the suppression rule modifier for the uncertain presence of a speech signal are equally important factors for building a good noise suppressor. Another interesting effect to notice is that, while the noise suppressor improves the perceptual quality of the output signal for input SNRs of 0 and 10 dB, the MOS results are much less improved for input SNRs of 20 dB. This is the reason why in some practical implementations the noise suppressor is turned off (or gradually reduced) above a certain input SNR.

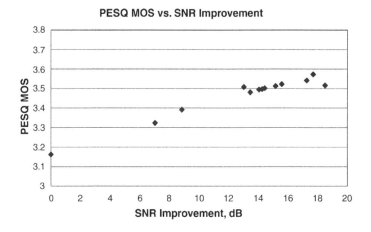

Figure 4.15 MOS results as a function of the SNR improvement for various suppression rules – end-to-end noise suppressor

EXERCISE

Create a MATLAB script for suppression of noise. The script should take the input and output file names as parameters. Use the provided *ProcessWAV.m* script as a template. Follow the steps and formulas above. Use the provided *SimpleVAD.m* for computing the rough noise model, and *SuppressionRule.m* for computing the chosen suppression rule.

Write a MATLAB script to generate a set of WAV files with given SNR (0, 5, 10, 15, 20, and 25 dB, for example) from clean speech and noise (record or use some of the provided). Evaluate the noise suppression script above by processing the generated set of files with noisy speech. Use the provided tool *SNRMeasurement* for evaluation, and, most importantly – listen, listen, listen to the output files.

4.6 Optimizing the Entire System

A good noise suppressor contains many parameters – the adaptation time constants and thresholds seen in many of the equations above. The tuning should start block by block.

The voice activity detectors should be tuned using the approach described at the end of Section 4.3.1.2. This means preparing a set of sound files with SNRs varying over the working range – say, from 0 to 30 dB in steps of 5 dB. This can be done by mixing various noises with a given magnitude to a set of clean speech recordings. Even the simplest VAD works well with a 30 dB SNR. It can be used to label the speech absence and presence frames. Then the mixed files are processed with the algorithm under tuning and the labeled data are used to compute the frame classification accuracy. Maximizing the accuracy can be done completely automatically by using the gradient-descent optimization algorithm.

Once we are sure that each of the blocks works well, it is time to tune the entire system. If the noise suppressor is part of a telecommunication system, the main optimization criterion is how it sounds. This means widely using objective quality measurements (PESQ, for example) to tune the entire system end-to-end. The improvement in SNR should be monitored, and solutions that sound the same but have higher improvement of the SNR should be preferred. The final several variants should go through extensive MOS tests with real human listeners.

If it is expected that speech-enhancement system output will be sent to an automatic speech recognizer, the system should be optimized to maximize the accuracy of the speech recognizer. Note that a speech enhancement system, optimized for speech recognition, may not be optimal for human listeners – and vice versa. If differences are substantial, then a separate processing should be done for the speech recognizer. Note that when doing this the noise-suppression procedure becomes specialized to the particular speech recognizer; that is, it cannot be generic. Different speech recognizers employ various techniques for increasing the noise robustness, and the preceding noise suppressor should not confuse them. If the speech recognizer builds noise models, then applying the modified suppression rule usually does not work well. The reason for this is that the modified suppression rule suppresses more during the pauses – the time the next block will use to build the noise models. This will result in under-estimation of the noise floor and reduction of the efficiency of this block. Comparison of various noise-suppression algorithms from the speech recognition rate standpoint can be found in [18].

The speech recognition system's sensitivity to distortions and artifacts can affect even the suppression rule. In [19], the suppression rule, which is a function of two parameters, ξ_k and γ_k, is represented as a discrete 10×10 matrix. Fine estimation of the suppression rule values is done by bi-linear interpolation. Then the suppression rule can be subject to optimization with the goal of maximizing the speech recognition rate. As a corpus for speech recognition and optimization, the AURO-RA 2 test is used. After 15 iterations, the new suppression rule allows the speech recognizer to achieve a higher recognition rate. The starting point (Ephraim and Malah's MMSE estimator) and the resulting suppression rule after 25 iterations are shown in Figure 4.16. It is clear that, after the optimization, the new rule suppresses less noise. This is just an example of using an external optimization criterion (speech recognition error rate) to make one of the components of a speech enhancement system (the suppression rule) better. This technique is applicable to most of the parameters in the noise-suppression system.

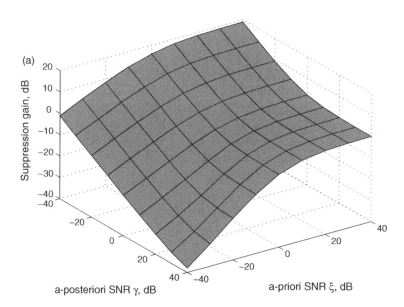

Figure 4.16 Speech-recognition friendly suppression rule: (a) starting point – MMSE suppression rule; (b) after 25 iterations of optimization to maximize the speech recognition rate

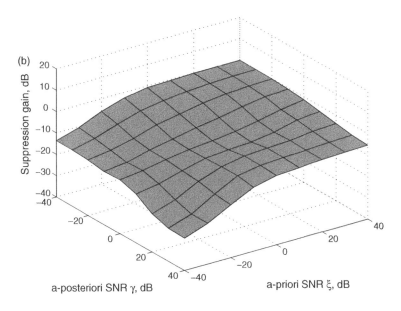

Figure 4.16   (*Continued*)

4.7 Specialized Noise-reduction Systems

4.7.1 Adaptive Noise Cancellation

The noise-suppression techniques described above depend heavily on the accuracy of the noise model. They cannot reconstruct the phase of the original signal, owing to the stochastic nature of the noise. What if we have a distorted copy of the original noise? A good example here is placing a microphone into the engine compartment of a car to pick up the engine noise. The noise, captured from a microphone inside the car cabin, will contain the driver's speech signal and the engine noise, distorted during its propagation to the car cabin. This approach is called "adaptive noise cancellation" and its block diagram is shown in Figure 4.17. The second microphone picks up the noise signal $\mathbf{N}^{(n)}$ – the vector representing the spectrum of this signal in audio frame n. The microphone inside the car cabin acquires the speech signal $\mathbf{X}^{(n)}$ and the engine noise, convolved with the engine-cabin impulse response $\mathbf{H} \cdot \mathbf{N}^{(n)}$:

$$\mathbf{Y}^{(n)} = \mathbf{X}^{(n)} + \mathbf{H} \cdot \mathbf{N}^{(n)}. \tag{4.69}$$

Note that in the frequency domain the convolution becomes multiplication. The active noise canceller tries to estimate the engine–cabin impulse response $\hat{\mathbf{H}}$ and to

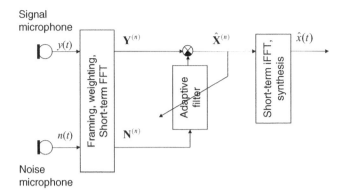

Figure 4.17 Block diagram of adaptive noise cancellation algorithm

calculate the speech signal by subtracting it from the mixture:

$$\hat{\mathbf{X}}^{(n)} = \mathbf{Y}^{(n)} - \hat{\mathbf{H}} \cdot \mathbf{N}^{(n)}. \tag{4.70}$$

Any adaptive filter techniques can be used, but one of the proven effective and most frequently used is the *normalized least-mean-squares* (NLMS). In each silent frame the impulse response is estimated for each frequency bin as

$$H_k^{(n+1)} = H_k^{(n)} + \mu_k \frac{E_k^{(n)} \cdot \bar{Y}_k^{(n)}}{|Y_k^{(n)}|^2 + \varepsilon} \tag{4.71}$$

where ε is a small number to prevent division by zero. $E_k^{(n)} = Y_k^{(n)} - H_k^{(n)} N_k^{(n)}$ is the error that the adaptive filter tries to minimize and that is supposed to be zero during *non-speech* intervals. Here $\bar{Y}_k^{(n)}$ denotes the complex conjugate of the k-th frequency bin in the n-th frame. In general, the NLMS filter is a minimum mean-square-error estimator using the steepest gradient descent. The adaptation time constant μ_k can vary with the frequency bins and the time. It should be $0 < \mu < 1/\lambda_{\max}$, where $\lambda_{\max}$ is the largest eigenvalue of the correlation matrix of the two input signals. In many implementations a fixed value for μ is used.

The adaptation process can be performed on every bin of every frame if we believe that the captured noise $\mathbf{N}$ does not contain a portion of the speech signal $\mathbf{X}$. If there are leaks of $\mathbf{X}$ in $\mathbf{N}$, then it will be better to do the adaptation only during the silence intervals. This will prevent cancellation of portions of the speech signal. If we have estimates of the speech absence probability for the frequency bin $P_k^{(n)}(\mathrm{H}_0||X_k|)$, provided by a VAD, then we can modify the adaptation step $\mu = P_k^{(n)}(\mathrm{H}_0||X_k|)\tilde{\mu}$. Here $\tilde{\mu}$ is the initial (non-variable) adaptation step. In silent frequency bins the adaptation goes with this step, while in frequency bins with speech present the adaptation slows down to prevent cancelling of the speech signal.

The theory of adaptive filters and algorithms is out of the scope of this book. There are numerous papers, books, and book chapters discussing the variable-size adaptation step, other adaptation algorithms, constraints to guarantee convergence, and so on. See [20] for an exhaustive theory of adaptive filters.

Only one dominant noise source is assumed in the adaptive noise cancellation. The delay between the noise-capturing microphone and the speech-capturing microphone should be less than one-quarter of the frame length. For the commonly used frame lengths of 10–50 ms, this means a distance between 3.4 and 17.15 m. In a car environment, for example, the distance between the microphone in the engine compartment and the microphone in the cabin is under 2 m, which means that any frame length above 25 ms will contain practically the same engine noise. If the delay between the noise-capturing and speech-capturing microphones is larger, and the frame length should not be increased for other reasons (latency of a real-time communication system, for example), then a specialized delay-estimation and time-shifting logic should be implemented for the signal from the noise microphone. This logic should delay the noise signal in a way so as to align it with the same noise component in the speech microphone.

The limitations of this type of noise cancelling are due to the presence of other noise sources. In the car example this can be the noise from the tires. In this case we have four additional and independent noise sources. The noise microphone in the engine compartment captures the noise from the engine and from all four tires with a given delay between the engine noise and the noise signal from each tire. The adaptive filter will try to subtract this mixture from the noise, captured by the speech microphone. As the delay from each tire to the speech-capturing microphone is different, there will be a residual noise even after the adaptive filter converges. The adaptive noise-cancellation system cannot cancel noises that are not presented in the noise-only signal – wind noise in the cabin, or other passengers talking. Note that for best results the two ADCs have to have synchronous clocks, otherwise tracking the clock drift will decrease the performance of the adaptive filter.

Overall this technique is linear and does not introduce distortions. It is complimentary to the classic noise-suppression techniques and should precede the non-linear noise suppressor. In many cases, adding a second microphone is justified by the speech enhancement this system can achieve. The general idea of having a noise-only signal, which can be subtracted from the mixture of useful signals and noise using an adaptive filter, will be explored further in the chapter on microphone arrays.

EXERCISE

Create a MATLAB script for adaptive noise cancellation. The script should take as parameters the input and output file names. Perform the processing in the frequency domain. Use the provided *ProcessWAV.m* script as a template. Follow the steps and formulas above. Use the provided *SimpleVAD.m* for controlling the adaptation process.

Evaluate the solution using the provided two-channel WAV file. The first channel is recorded with a microphone inside the car cabin, the second channel is the engine sound only. Measure the improvement in SNR and listen to the output.

4.7.2 Psychoacoustic Noise Suppression

4.7.2.1 Human Hearing Organ

The human hearing organ – the ear – consists of outer, middle, and inner parts. A very schematic diagram of the ear is shown in Figure 4.18. Note that not all parts are present in this figure. Part of the inner ear, for example, is the human vestibular system, which has nothing to do with hearing.

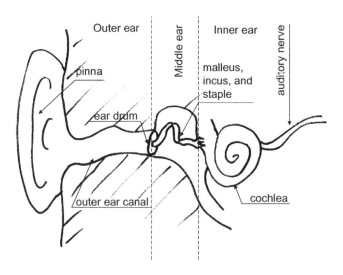

Figure 4.18   Human hearing organ

The outer ear – the pinna and outer ear canal – play a role in forming the directivity of the human hearing, together with the head and the shoulders. In addition, the pinna acts like a funnel, amplifying and directing the sound to the outer ear canal. It increases the sound level 10–15 dB in the frequency range between 1.5 and 7 kHz, where the energy of human speech is concentrated. The outer ear canal is a tube with a length of 30 mm, and a practically uniform diameter of around 7 mm. The ear canal has an acoustical resonance at approximately 3000 Hz, which is the reason for increased sensitivity of humans to these frequencies. The ear canal ends with an ear drum, which converts the changes in the air pressure to mechanical vibrations.

The middle ear is an air-filled cavity and contains the smallest bones in the human body: malleus, incus, and staple. They transfer the mechanical vibrations of the ear drum to the inner ear, which is filled with liquid. These three bones form an acoustical impedance transformer to maximize the energy transmission.

The inner ear is filled with liquid. The staple vibrations are transferred to the cochlea – a tubular organ with a spiral shape (the name comes from the Latin word for snail), and 2.5 turns. The tube is 32 mm long and has a diameter of 0.05 mm at the beginning and 0.5 mm at the end. Inside the cochlea is the Organ of Corti, which consists of thousands of hair cells. They sense the movements of the liquid in the inner ear and cause the neurons connected to them to fire. The auditory nerve transfers the electrical pulses to the brain for further processing. Without going into great detail, the specific shape and construction of the cochlea results in different frequencies agitating different sets of hair cells: for lower frequencies the ones at the end of the Organ of Corti, for higher frequencies the ones at the beginning. This is how humans can distinguish between frequencies.

Detailed description of the physiology and acoustics of the human hearing organ can be found in [21].

4.7.2.2 Loudness

Human hearing is not perfect; the same sound pressure level generates a different audio sensation for different frequencies. The perceived loudness of the sound is a function of both the frequency and sound pressure level. This is why we introduce the loudness level – the sound pressure level of a reference frequency that causes the same subjective loudness. The reference frequency is chosen to be 1000 Hz, and the measuring unit is called a "phon." This means that a sound with magnitude 40 dB SPL and frequency 1000 Hz will have loudness of 40 phons. The human ear is less sensitive towards the lower frequencies, which means that a sound with higher SPL will be necessary to cause the same audio sensation. These equal-loudness curves were studied in the early years by Fletcher and Munson [22] and later were replaced by more precise measurements done by Robinson and Dadson [23]. These measurements became the basis of the standard ISO 226. Later they were revised based on newer and more precise measurements, done by scientists from various countries, and the standard was updated in 2003 as ISO 226:2003.

The equal-loudness curves for several loudness levels are shown in Figure 4.19. The dashed line shows segments with low confidence, or not confirmed by many measurements. The role of the outer ear is clearly visible, as the human hearing has the highest sensitivity in the range 300–7000 Hz. The human sensitivity degrades smoothly for sounds with lower frequencies and rapidly for sounds with frequencies above 15 000 Hz. The upper threshold of human hearing is age-dependent. Young people at the age of 25 years can hear frequencies up to 19–20 kHz, while at the age of 45 years the upper frequency for individuals with normal hearing is down to 15–16 kHz. The curve for 0 phons is actually the threshold of human hearing as a function of the frequency.

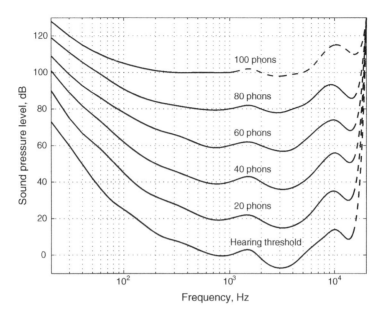

Figure 4.19 Equal-loudness curves for several loudness levels as a function of the frequency and sound pressure level

4.7.2.3 Masking Effects

The human ability to distinguish the frequencies of single tones is quite remarkable – we have a spectral resolution of about 4 Hz for frequencies below 500 Hz, which slowly degrades above this frequency, but remains better than 0.7%. In total, around 640 frequency steps can be distinguished in the audible range for humans.

The situation changes when several sinusoidal signals with different loudness are involved. Experiments and measurements found that the spectral selectivity of human hearing can be modeled as a set of filters with asymmetric shape and a frequency-dependent bandwidth – constant and around 100 Hz in the lower part of the frequency band, decreasing to ∼20% of the center frequency above 500 Hz. This simply means that a tone with frequency 1000 Hz will cause a sensation to a group of hair cells in the cochlea around the ones responsible for detecting this frequency; that is, it will "leak" into the neighboring frequencies. Figure 4.20 shows the excitation of the hair cells in the cochlea for a triple-tone audio signal – 500, 2000, and 8000 Hz – according to [21]. The leakage of the signal in the neighboring frequencies is clearly visible.

To model the non-linear frequency resolution of the human ear, the *Bark scale* is introduced, named after the famous physicist H. G. Barkhausen. It converts the frequencies in hertz into a number in the range from 1 to 25 in a non-linear manner,

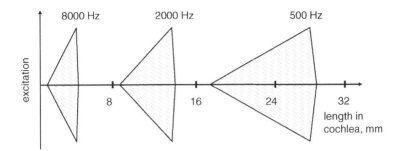

Figure 4.20 Excitation of the hair cells in the cochlea for a triple-tone audio signal – 500, 2000, and 8000 Hz with the same amplitude. Signal leakage towards neighboring hair cells is clearly visible. The horizontal axis is the length inside the cochlea in millimeters, the vertical is the relative magnitude of the excitation

approximated by the formula

$$Z_b(f) = 13 \ \arctan(0.00075f) + 3.5 \ \arctan\left[\left(\frac{f}{7500}\right)^2\right]. \tag{4.72}$$

Figure 4.21 shows the relationship between the linear frequency scale and the Bark scale. The dots mark the frequencies for Bark values of 0.5, 1.5, 2.5, and so on. The leaking of the single-tone frequencies is modeled as a triangle-shaped filter in the Bark scale with slopes $+25$ dB/Bark and -10 dB/Bark, respectively. The more convenient and smoother empirical model for the slope is given in [24]:

$$A(\Delta B) = 15.81 + 7.5(\Delta B + 0.474) - 17.5\sqrt{1 + (\Delta B + 0.474)^2}. \tag{4.73}$$

Here ΔB is the distance from the center frequency in Barks and $A(\Delta B)$ is the attenuation in decibels. This means that the human auditory system can be modeled as a set of overlapping filters with asymmetric triangle-shaped frequency responses. The center frequencies of these filters are chosen to be the dots in Figure 4.21. To simplify the model further, these filters are replaced with a set of rectangular non-overlapping filters with equivalent noise bandwidth; that is, under white-noise excitation they will provide the same average magnitude of the output signal as the corresponding triangle-shaped filter. For the band around 1000 Hz this is shown in Figure 4.22. Instead of using the triangle-shaped filter with a peak at 1000 Hz, the rectangular filter will be used, which passes all frequencies between 920 and 1080 Hz. Table 4.4 shows the center, beginning, and end frequency of the commonly used filter bank of rectangular filters; see [25] for more details. The widely used name for these filters is "critical bands."

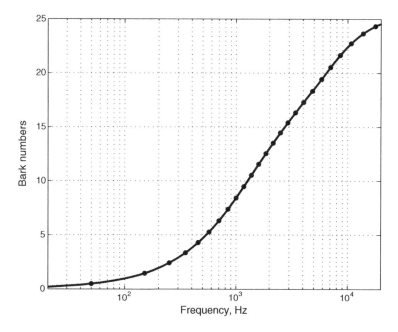

Figure 4.21 Conversion function from frequency to Bark scale. The dots show the positions of integer Bark numbers: 1, 2, 3, and so on

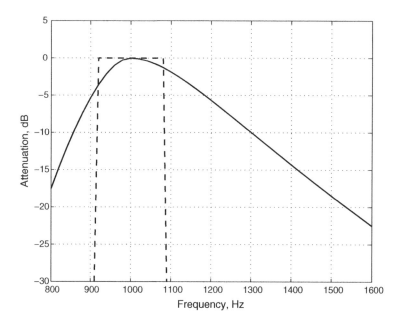

Figure 4.22 Equivalent rectangular bandwidth filter for the critical band with center 1000 Hz. The solid line is the critical band filter, the dashed line the equivalent rectangular bandwidth filter

Table 4.4 Center, beginning, and end of the critical bands (hertz)

Center	Beginning	End
50	0	100
150	100	200
250	200	300
350	300	400
450	400	510
570	510	640
700	630	775
840	765	920
1000	920	1095
1170	1075	1275
1370	1265	1490
1600	1480	1740
1850	1710	2010
2150	1990	2340
2500	2310	2725
2900	2675	3175
3400	3125	3750
4000	3650	4450
4800	4350	5350
5800	5250	6450
7000	6350	7900
8500	7600	9750
10 500	9250	12 250
13 500	11 750	20 000

The masking effects in the frequency domain are caused by the leaking. If we have additional signals with lower amplitude and close frequency, they may be below the hearing threshold. This is shown in Figure 4.23, where we have three signals with close frequencies. The 1000 Hz tone has the highest amplitude. The tone with frequency 1500 Hz and lower amplitude will not be audible, because it is masked by the leakage from the 1000 Hz tone. The third tone with frequency 3000 Hz and the same amplitude as the second one will be audible, since it is far enough to be masked. This simply means that the auditory nerves have a non-linear response to the audio sensation: there is an absolute hearing threshold, and there is a relative threshold, which indicates that the variation in excitation amplitude can be detected only if it is above this relative threshold. The absolute threshold is the 0 phones curve from Figure 4.21. Humans can detect the presence of an additional tone if it is ~8 dB above the masking threshold [26]. Multiple studies of human hearing have shown that a noise can mask a single tone if its level is ~4 dB above the tone, and a tone can mask noise if it has a level of ~24 dB above the noise in that Bark band [27]. These thresholds are used for designing more sophisticated audio compression algorithms.

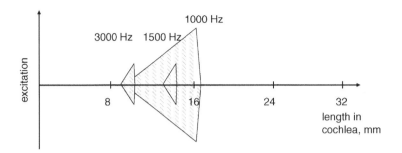

Figure 4.23 Excitation of the hair cells in the cochlea for a triple-tone audio signal – 1000, 1500, and 3000 Hz with different amplitudes. The signal with frequency 1500 Hz will be masked, while the signal with frequency 3000 Hz will be audible. The horizontal axis is the length inside the cochlea in millimeters, the vertical is the relative magnitude of the excitation

In addition to the masking effect in the frequency domain there is masking in the time domain as well, known as "non-simultaneous masking." Pre-masking appears when the masking tone starts some time after the masked tone or noise. This is shown in Figure 4.24. Measurable results registered for 20–30 ms, when the masking threshold is at -50 dB of its simultaneous masking threshold level. The effect is much more sensible in the post-masking case. Signals are still masked 100–200 ms after the masking tone is removed. It is assumed that the masking threshold is at -30, -40, and -45 dB of its simultaneous masking level after 50, 100 and 150 ms, respectively [28].

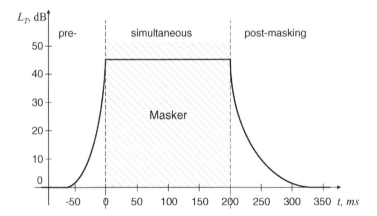

Figure 4.24 Masking in time: pre-, simultaneous, and post-masking effects. The horizontal axis is time, with masking white-noise signal between 0 and 200 ms. The vertical line is the necessary level L_T above the noise level for the audibility of a sinusoidal burst (test tone)

At this point we have all the necessary numbers and models to estimate the masking threshold $M_k^{(n)}$ in the frequency domain, given the input magnitude $|Y_k^{(n)}|$ for given frequency bin and frame. It gives the level below which humans cannot hear tones or noises. If we want to use the post-masking effects, we will have to have the magnitude of several previous frames – that is, $|Y_k^{(n-1)}|$, $|Y_k^{(n-2)}|$, $|Y_k^{(n-3)}|$, and so on. The algorithm is simple and starts with computing the average magnitude in each critical band. Then for each band the masking slopes towards the higher and lower frequencies per critical band are computed. The highest value from all the bands is compared with the hearing threshold. The higher value is the masking threshold for the current frame. Humans cannot hear the tones in each frequency bin if their magnitude is lower than the masking threshold.

The major application of psychoacoustics and masking effects is in audio compression. The principle is "do not store or transmit frequency bins that we cannot hear anyway". All major audio compression techniques today (MPEG-1, MPEG-2, MP3, MPEG-4, Windows Media Audio, Windows Real-Time Voice, etc.) are based on this principle. Discussing audio compression techniques is not in the scope of this book, but we will focus on the psychoacoustic approach for noise suppression.

4.7.2.4 Perceptually Balanced Noise Suppressors

We have seen that more suppression means more distortions in the estimated clean signal. In the psychoacoustic approach for noise suppression, the principle is "do not suppress noise below the level we can hear." In some papers it is defined as the principle of "least processing." Wolfe and Godsil [29] apply a perceptually modified suppression rule as follows:

$$\tilde{H}_k^{(n)} = \left| \begin{array}{ll} \left(1 - \dfrac{M_k^{(n)}}{|Y_k^{(n)}|}\right) H_k^{(n)} + \dfrac{M_k^{(n)}}{|Y_k^{(n)}|} & \text{for} \quad M_k^{(n)} < |Y_k^{(n)}| \\ 1 & \text{otherwise.} \end{array} \right.$$ (4.74)

Here, $\tilde{H}_k^{(n)}$ is the perceptually modified suppression rule and $H_k^{(n)}$ is the suppression rule estimated using any of the algorithms already discussed. In that particular paper the authors use the short-term MMSE rule from Ephraim and Malah. The ratio $M_k^{(n)}/|Y_k^{(n)}|$ is the relative masking level. Note that there is no processing done when the relative masking level is above 1; that is, on the masked frequency bins. The relative masking level is used as a limiting factor to the suppression gain, which does not allow suppression below the audible threshold.

The authors report a marginal decrease in the SNR improvement, which is expected, but increase in the user's preference.

Note that the perceptual-based noise suppressors are not compatible with any psychoacoustic-based audio compressor because they are created on antagonistic principles. It is pointless to have a perceptual-based noise suppressor, followed by a psychoacoustic compressor. The major application of perceptually balanced noise suppressors is for cleaning and restoring high-quality music recordings. By minimizing the intervention only to the hearable part of the noise and limiting the suppression to go no further than the masking threshold, the introduced distortion is minimized and the user perception of the restored records is improved.

4.7.3 Suppression of Predictable Components

Frequently speech recordings are contaminated by stationary tones. They usually come from power wiring, inadequate shielding, or grounding of the microphone cables, or placement of the microphones near power lines or transformers. In those cases the interference frequency is 50/60 Hz or 400 Hz and their harmonics. Other kinds of stationary-tone interference come from microphones positioned near TVs, monitors, or video cameras; the microphones can capture interference at frame or line frequencies acoustically from transformers or electronically from the cables. Yet another source of this kind of interference are noises coming from the acoustical environment, such as fans, computer hard drives, and air conditioning. Because of non-linearities and room reverberation, these signals behave mostly as random zero-mean Gaussian noise, but usually there are still predictable components. The frequencies of the predictable portion of these noises vary depending on the fan or hard drive spindle rotating speed. The common property of these signals is that they are practically stationary. In their time–frequency representations they show up as horizontal lines with constant amplitude.

The most intuitive approach to solve this problem and to clean up the contaminated signal is to apply band-pass filtering or notch filters tuned to the constant tones. These approaches remove speech signal components if the interfering frequencies are within the speech band. If the speech signal is contaminated by single-tone interference, then a notch filter works well and the missing frequency is usually inaudible. If the contaminating signal has multiple harmonics, then a set of notch filters or a comb filter may be needed to achieve significant filtering, and that can substantially distort the speech signal.

Classic noise suppressors, like the ones described above, assume that the noise is a stationary zero-mean Gaussian process and build a statistical model of the noise as a vector of variances per frequency bin. The stationary tones have a probability density function that is usually not Gaussian. Using a Gaussian PDF as a model of these signals and some of the known suppression rules (Wiener, or Ephraim and Malah, etc.) results in complete suppression of the speech signals in these frequency bins; that is, the noise suppressor converts to a notch filter for these frequencies.

The problem of tracking frequencies in a time-frequency representation is well studied. In [30], an ARCAP (autoregressive Capone algorithm) method is used to identify the spectral lines, followed by Kalman filtering to track their movement. The method is illustrated with a processing of avalanche signals. It is sensitive to noise and best results are obtained with a forward–backward Kalman filter, which makes it inapplicable for real-time algorithms where low latency is desired. Improving the algorithm further [31] by adding trajectory smoothing with a Fraser filter still keeps the algorithm good for off-line processing only. The birth/death time estimation of spectral lines is improved in [32] as well. In addition, a particle filter is used to perform optimal estimation in jump Markov systems for detection and tracking of spectral lines. The proposed time-varying autoregressive (TVAR) estimator is evaluated with synthetic signals. It is computationally expensive and sensitive to the times of birth/death of spectral lines. Andia [33] proposes image processing techniques to be used to detect, model and remove spectral lines from the time–frequency representation. All of these approaches solve problems that are more complex than necessary, and are mostly suitable for off-line processing of the contaminated signals.

One of the main properties of the predictable components is that they are stationary, or pseudo-stationary, and can be modeled as a linear combination of sinusoidal signals and a noise component:

$$z(t) = \sum_{i=1}^{L} A_i \sin(2\pi f_i t) + \mathbb{N}(0, \lambda) \tag{4.75}$$

where L is the number of stationary tones, each with frequency f_i. Converting this signal to the frequency domain yields the following model for the n-th audio frame:

$$Z_k^{(n)} = \sum_{i=1}^{L} W_T(k) * A_i e^{-j2\pi n T f_i} + \mathbb{N}(0, \lambda_N) \tag{4.76}$$

where W_T is the Fourier image of the frame weighting function, T is the audio frame step, n is the frame number, and k is the frequency bin.

We note the following:

- Due to the "smearing" of the spectral lines because of the weighting, bins neighboring the central bin (for each contaminating frequency) contain portions of the energy.
- These neighboring bins will rotate in the complex plane (phase shift) from frame to frame with the same speed, which can be different than the speed of each bin's central frequency $\exp(-j2\pi n T f_s / K)$.

These two aspects introduce additional complications in the extrapolation of the signal model for the next frame.

Assuming we have perfect estimation for the n-th frame $\hat{Z}_k^{(n)}$, then the extrapolation for the next frame will be

$$\hat{Z}_k^{(n+1)} = \hat{Z}_k^{(n)} \frac{\sum\limits_{i=1}^{L} W_T(k)*A_i e^{-j2\pi(n+1)Tf_i}}{\sum\limits_{i=1}^{L} W_T(k)*A_i e^{-j2\pi nTf_i}}. \tag{4.77}$$

The second term is a complex number that represents the "speed" of rotation of our complex model from frame to frame. As already noted, this "speed" can be different from the "speed" of the central frequency of the bin. Because $W_T(k)$ decays quickly with increasing k, we can assume that one frequency from the contaminating signal dominates in each frequency bin. In this case

$$\frac{\sum\limits_{i=1}^{L} W_T(k)*A_i e^{-j2\pi(n+1)Tf_i}}{\sum\limits_{i=1}^{L} W_T(k)*A_i e^{-j2\pi nTf_i}} \approx e^{-j2\pi Tf_l} + \mathbb{N}(0, \lambda_E). \tag{4.78}$$

where f_l is the dominant frequency and $\mathbb{N}(0, \lambda_E)$ is an error term, modeled as zero-mean Gaussian noise. As the dominant frequency is unknown, the extrapolation can be presented as

$$\hat{Z}_k^{(n+1)} = \hat{Z}_k^{(n)} \hat{Y}_k^{(n)} \tag{4.79}$$

where $\hat{Z}_k^{(n)}$ is the contaminating signal estimation for the n-th frame, and $\hat{Y}_k^{(n)}$ is the rotating speed of the model towards the next frame. Both components have additive Gaussian noise with variances λ_N and λ_E, respectively.

With the speech signal present, Equation 4.75 takes the form

$$x(t) = s(t) + \sum_{i=1}^{L} A_i \sin(2\pi f_i t) + \mathbb{N}(0, \lambda) \tag{4.80}$$

and the representation in the frequency domain of the n-th frame is

$$X_k^{(n)} = S_k^{(n)} + \sum_{i=1}^{L} W_T(k)*A_i e^{-jnTf_i} + \mathbb{N}(0, \lambda_N). \tag{4.81}$$

In this case our estimation of the speech signal is

$$\hat{S}_k^{(n)} = X_k^{(n)} - Z_{est}^{(n)}(k);$$ (4.82)

that is, we just subtract our estimation of the contaminating signal

$$Z_{est}^{(n)}(k) = \hat{Z}_k^{(n-1)} \cdot \hat{Y}_k^{(n-1)}.$$ (4.83)

The speech signal estimation contains the captured noise $\mathbb{N}(0, \lambda_N)$ and the cancellation adds an additional noise component $\sim \mathbb{N}(0, \lambda_E)$ due to the approximations in the model and estimation errors.

In parallel with the contaminating signal cancellation, we should constantly update the contaminating signal model, which for each frequency bin consists of four elements: $\hat{Z}(k)$, $\hat{Y}(k)$, $\lambda_N(k)$, and $\lambda_E(k)$ (from which only the first two are involved in the constant-tones cancellation process). The contaminating signal model is updated as follows:

$$\hat{Z}_k^{(n)} = (1-\alpha)\hat{Z}_k^{(n-1)} + \alpha\left(p_k^{(n)} X_k^{(n)} + \left(1 - p_k^{(n)}\right)\hat{Z}_k^{(n-1)}\right)$$ (4.84)

where $\alpha = T/\tau_Z$, τ_Z is the adaptation time constant, and $p_k^{(n)}$ is the probability that we have only contaminating signal in $X_k^{(n)}$ – that is, the probability of speech absence. It can be provided by any voice activity detector (VAD), which produces per-bin probability estimation of speech presence. The additive noise variance is updated as follows:

$$\lambda_N^{(n)} = (1-\alpha)\lambda_N^{(n-1)} + \alpha\left(p_k^{(n)}\delta_k^{(n)} + \left(1-p_k^{(n)}\right)\lambda_N^{(n-1)}\right)$$ (4.85)

where $\delta_k^{(n)} = ||X_n^{(n)} - Z_{est}^{(n)}(k)||^2$. The rotating speed estimation is updated in the same way:

$$\hat{Y}_k^{(n)} = (1-\beta)\hat{Y}_k^{(n-1)} + \beta\left(p_k Y_{est}^{(n)}(k) + (1-p_k)\hat{Y}_k^{(n-1)}\right)$$ (4.86)

where

$$Y_{est}^{(n)}(k) = \frac{Y_k}{||Y_k|| + \varepsilon}$$

is the normalized rotation speed estimation

$$Y_k = \frac{X_k^{(n)}}{X_k^{(n-1)} + \varepsilon}$$

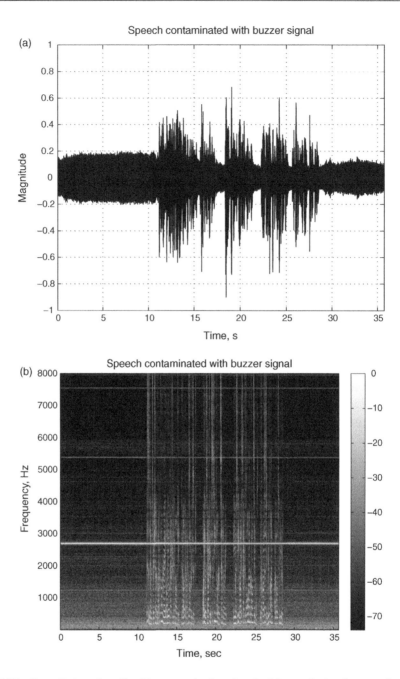

Figure 4.25 Cancellation of predictable contaminating signals: (a) speech signal, contaminated with buzzer signal; (b) spectrogram of the contaminated signal, the buzzer signal being visible as horizontal lines; (c) cleaned speech signal; (d) spectrogram of the cleaned speech signal

(c)

Speech contaminated with buzzer signal – cleaned

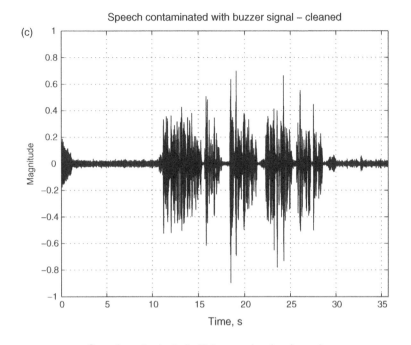

(d)

Speech contaminated with buzzer signal – cleaned

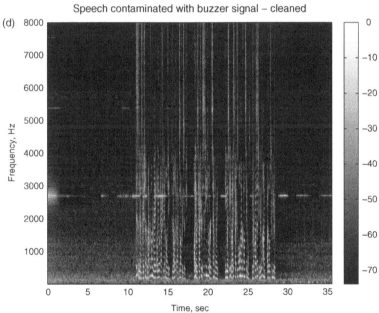

Figure 4.25 *(Continued)*

for the current frame, ε is a small number, $\beta = T/\tau_Y$, and τ_Y is the adaptation time constant.

The proposed method is evaluated in [34] with a speech signal contaminated with several noises: white noise, office noise, and two buzzer sounds with a different number of harmonics. The improvement in SNR is shown in Table 4.5. There is no suppression for white noise and clean speech, as expected. For office noise (three computers with their fans and hard drives, air conditioning) the algorithm improves the SNR by almost 3 dB, removing the predictable components from the noise. The proposed algorithm suppresses the signals from the two buzzers up to 15 dB.

Table 4.5 Improvements in SNR with predictable signals compensation (decibels)

Recording	Input			Output			Improvement
	Signal	Noise	SNR	Signal	Noise	SNR	
White noise		−13.43			−13.43		0.00
Clean speech	−25.37	−60.44	35.07	−25.38	−60.75	35.37	0.30
Office noise	−34.55	−44.62	10.07	−35.02	−47.98	12.96	2.89
Buzzer 1	−21.42	−21.69	0.27	−23.19	−39.35	16.16	15.89
Buzzer 2	−18.56	−20.52	1.96	−24.21	−39.96	15.75	13.79

Figure 4.25 shows the contaminated and cleaned signals and their spectrograms. The contaminating signal is visible as three horizontal lines. This is a real recording in a room where people move, changing the reverberation and interference patterns. After each change the algorithm has to adapt to the new signals. During this time we see the bright traces in the spectrogram. Their magnitude is still much lower than the captured signal, which is visible in the time domain representation of the output signal.

This type of processing is suitable as a pre-processor, before a classic noise suppressor. It removes the predictable part without artifacts and musical noise, leaving the noise suppressor less to suppress, which in general means less musical noise and artifacts. It is computationally inexpensive and even in office conditions reduces almost 3 dB of the noise, which is well audible. It is a safety net when the microphone is accidentally placed near sources of predictable noises. It successfully removes most of the hard drive spindle noise, captured by the microphone in a laptop, but signal and audio processing are not fixes for a not very thoughtful design – the microphones should be kept away from such noise sources.

EXERCISE

Create a MATLAB script for cancellation of predictable components of the noise. The script should take as parameters the input and output file names. Perform the processing in the frequency domain. Use the provided *ProcessWAV.m* script as a template. Follow the steps and formulas above. Use the provided *SimpleVAD.m* for controlling the adaptation process.

Evaluate the solution using the provided *SpeechBuzzer.WAV* file. Measure the improvement in SNR and listen to the output.

4.7.4 Noise Suppression Based on Speech Modeling

The algorithms discussed so far assume one model of the speech signal, usually statistical with Gaussian, gamma, or Laplace distribution. In Chapter 2 we saw that speech is a complex signal and consists of segments with quite different characteristics. Apparently each of the segments can have an optimal suppression filter that is quite different from the optimal filters for the other types of segments.

The idea of different filters for different segments of the speech signal was first proposed by Drucker [35]. In his paper he groups the approximately 40 phonemes in the English language into five broad classes: stops, fricatives, glides, vowels, and nasals. Each phoneme is processed by a separate filter, designed to eliminate the intra-class confusion – that is, the error of assigning a class sound from the same class – and the use of a different filter eliminates interclass confusion. The proposed algorithm works as follows (Figure 4.26). The input speech plus noise is segmented into phonemes; a decision device determines which of the five classes of sounds the phoneme belongs to, and then routes the speech plus noise segment to the appropriate filter. According to the paper, the proposed algorithm increases the intelligibility around 25% in the input SNRs ranging from $-8\,$dB to $0\,$dB; above

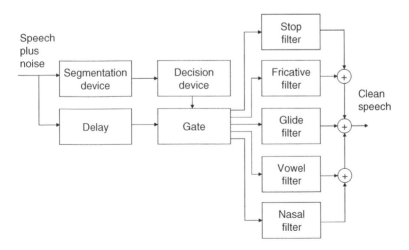

Figure 4.26   Block diagram of a noise suppressor based on speech modeling

0 dB the improvement in the intelligibility decreases to ~7%. The major source of potential problems in this approach is the classifier.

Many researchers later tried either to reduce the classification error, or to synthesize more robust filters to such errors. Algorithms for precise measurement and tracking of the pitch frequency and synthesis of the proper comb filters proved to be very efficient for denoising of the vowels.

In a later work, Ephraim and Cohen [36] proposed parallel processing, based on the probability of classification. In general the processing model with certain probability of speech presence is a multi-class model. From this perspective a useful modeling apparatus can be a *hidden Markov process* (HMP). The classes are not a-priori defined, but they are created in a learning process from training data of clean speech signals. It is in fact a clustering process that can be performed by using vector quantization techniques. Each class may contain spectrally similar vectors of the signal, which can be parameterized as an autoregressive process. Transformations from one spectral prototype to another can be modeled by the HMP. The various noises can be processed in a similar manner. If we have M speech classes and N noise classes, then we have $M \times N$ combinations; that is, estimators for enhancing the speech signals. For a given sequence of noise speech vectors $\mathbf{Y}^t = \{Y^{(n)}, Y^{(n-1)}, \ldots, Y^{(n-t)}\}$, and the probability $p(i, j)|\mathbf{Y}^t)$ of the signal being in state i and the noise being in a state j given $\mathbf{Y}^t$, the MMSE estimator of the clean speech signal $\mathbf{x}$ is

$$\mathrm{E}\{\mathbf{X}^t|\mathbf{Y}^t\} = \sum_{i=1}^{M} \sum_{j=1}^{N} p((i,j)|\mathbf{Y}^t)\mathrm{E}\{\mathbf{X}^t|\mathbf{Y}^t, (i,j)\}. \qquad (4.87)$$

4.8 Practical Tips and Tricks for Noise Suppression

While the mathematical models above are correct and valid to describe the behavior of captured and processed signals, there are some additional tips and tricks that can and should be used for successful implementation of a real noise suppression and reduction system. Audio processing is a science, an art, and a craft. So far we have covered the science part and touched on the art part when talking about human perception. This section is about the craft.

4.8.1 Model Initialization and Tracking

Each time the processing starts there can be different noise levels, or rotational speed. Using some default initial value and adaptation equations like Equation 4.63 leads to slow adaptation in the initial phase. On the other hand, decreasing the time constants for faster adaptation leads to less stable values of the model. To adapt quickly at the beginning and keep the values stable during normal working, we can use a variable time constant in the initial phase:

$$
\lambda_d^{(n)} = \begin{vmatrix} |Y_k^{(n)}|^2 & \text{if} & n = 1 \\[2mm] \left(1 - \dfrac{1}{n}\right)\lambda_d^{(n-1)} + \dfrac{1}{n}|Y_k^{(n)}|^2 & \text{if} & 1 < n < \dfrac{\tau_d}{T} \\[2mm] \left(1 - \dfrac{T}{\tau_d}\right)\lambda_d^{(n-1)} + \dfrac{T}{\tau_d}|Y_k^{(n)}|^2 & \text{for} & n \geq \dfrac{\tau_d}{T}. \end{vmatrix} \qquad (4.88)
$$

In the first frame we initialize with the value of the current frame; the second frame is the average between the first and second frames; and so on until we reach the number of frames big enough to use as an adaptation factor T/τ_d.

4.8.2 Averaging in the Frequency Domain

The noise amplitude in each of the frequency bins varies substantially. To build a more precise model we should increase the adaptation time constant. Unfortunately this leads to slower adaptation to changes in the noise. A good trade-off is to use smoothing towards the neighboring frequency bins as well. Weighting causes leakage, so even a single sinusoidal signal will be represented in several frequency bins. Usually just a moving average of three to seven frequency bins does a sufficient job to stabilize the rough noise model, used in VADs.

4.8.3 Limiting

For robustness to accidental spikes and errors, the values of the measured parameters should be kept within reasonable limits. The standard deviations cannot go below certain minimal values. The mean parameter in a gamma distribution should not go below the standard deviation of the Gaussian noise. Every probability has values between $0 + \varepsilon$ and $1 - \varepsilon$. Likelihood should be limited to be above $0 + \varepsilon$ and some certain number that is not too big – 1000 is a good practical value. Good limitation for signal-to-noise ratios of any kind is in the range -60 dB to $+60$ dB. Suppression gains should be limited to the range 0–1. Proper limiting of the value range in the real-time execution code allows the algorithm to be more robust to unexpected input data or computational errors.

4.8.4 Minimal Gain

Zeroing some frequency bins is never a good idea. It is a harsh operation for the sound which results in musical noise and unpleasant distortions for the human ear. The background noise in the silent segments is distorted and chopped. To prevent this from happening, a minimal gain should be applied after estimating the final suppression rule:

$$H_{\text{final}}^{(n)}(k) = (1 - G_{\min})H_k^{(n)} + G_{\min}. \tag{4.89}$$

The most frequently used value for the minimum gain is 0.1, which limits the suppression to 20 dB. The level of musical noises is negligible and the background noise, while suppressed, is not distorted.

4.8.5 Overflow and Underflow

The noise suppression and all other audio processing algorithms are usually implemented as real-time processing modules. There is no time to handle all potential overflow and underflow exceptions; this is why proper measures to avoid them should be taken. Earlier in this chapter it was said that, in real-time implementations, $1/x$ becomes $1/(x + \varepsilon)$, where ε is a small positive number. The same is true for computing logarithms (log-likelihood, for example). Then $\log(x)$ becomes $\log(\max(\varepsilon, x))$ or $\log(x + \varepsilon)$. In many cases we compute exponents and large arguments can cause exceptions or undefined results. For double precision, floating-point numbers $\exp(x)$ overflows when x is slightly greater than 700. For secure execution we should use $\exp(\min(700, x))$.

4.8.6 Dealing with High Signal-to-Noise Ratios

The noise suppression is a trade-off between suppression and introduced distortions. Most of the algorithms are optimized for SNRs of 5–15 dB – the most common when capturing sound in homes, offices, and conference rooms. This means that these algorithms are suboptimal for high-quality input, when the SNR is 30–50 dB. This can happen when using a headset, for example. Then the noise-suppression algorithms just introduce distortions, decreasing the actual quality and the overall MOS results. In such cases, instead of making the algorithms more complex, it is a better idea just to turn the noise suppression off.

The noise suppressor has a signal/pause classifier anyway. For each frame we can compute the level and add it to our estimation of the signal or of the noise using a certain time constant. With the signal and noise level estimations, computing the average SNR is trivial. If it is high enough we can turn off the noise suppression. Actually the suppression rules work well under very high SNR conditions. They stay around 1 and do not harm the signal quality.

The suppression rule modifier for the uncertain presence of a speech signal (see Equation 4.54) is usually the source of increased distortions under high SNR conditions. In most cases it makes sense to turn only this feature off. This should be done with a certain hysteresis; that is, turn the feature off when the average SNR goes above a high threshold and turn it back on when the average SNR goes below a lower threshold. Typical values for these two thresholds are 20 dB and 25 dB average SNR.

4.8.7 Fast Real-time Implementation

The processing algorithms described so far require computation of many probabilities, exponents, gamma functions, and so on. While the processors used in modern personal computers have the integrated ability to work with floating-point numbers, the operations they can do beyond the four arithmetic operations are limited to exponent and tangent. This means that everything else has to be computed in a programmatic way. Not many digital signal processors have the capability to perform operations with floating-point numbers at all. This makes performance optimizations critical for the success – and even applicability – of given algorithms.

Fortunately, the majority of these computations can be performed off-line and kept as a set of tables. Most of the distributions (Gaussian, gamma, Laplace) can be tabulated with steps. In real time, a linear interpolation can be used to obtain the exact value. In many cases the nearest-neighbor algorithm provides sufficient precision. The suppression rule itself is a function of two parameters. It can be discretized as a matrix, computed off-line and used in real time. This can save computations and make the algorithm run faster.

Of course, before going to performance optimization of the execution code, the normal software engineering rules should be followed. The process starts with performance profiling, which provides the time used by the CPU to execute any of the functions in the code. Then the performance optimization starts with the functions with highest execution times. It is pointless to optimize the performance of a function or operation that takes 0.1% of the CPU time; even completely removed it will reduce the execution time only by 0.1%.

In many cases the most computationally expensive part is the conversion to the frequency domain and back – that is, FFT and iFFT functions. From this perspective, the manufacturers of most DSPs provide implementation of the FFT algorithm optimized for their processor. Using frame size, which is a power of two, decreases the execution time of these two operations as well.

4.9 Summary

This chapter has discussed algorithms and approaches for single-channel noise reduction. Clean signal estimation from the mixture of signal and noise is a gain-based process and the algorithms belong to the group of noise suppressors. The process applies a time-varying, real-valued gain to each frame in the frequency domain. The computation of this gain, called a suppression rule, is based on the statistical parameters of both the noise and clean signal. The suppression gain is a function of the a-priori and a-posteriori SNRs and is optimal in one way or another: the MMSE of the magnitudes, ML, and so on. Applying the suppression rule reduces the noise component, but introduces distortion and artifacts, called musical noise.

An important part of each noise suppressor is the voice activity detector. In its simplest form this is a two-way classifier: the current audio frame contains only noise, or it is a mixture of noise and speech. The most complex VADs provide a speech presence probability for each frequency bin. In some cases this probability is used to modify the suppression rule – the uncertain presence of speech signal approach.

Part of the noise suppression algorithm builds the noise and the speech models. Based on the VAD output, the noise model variance is updated from frame to frame. For estimation of the speech signal statistical parameters, the decision-directed approach is commonly used, which assumes high correlation of the speech signal in consequent frames and uses the previous output frame to estimate the a-priori SNR.

The goal of the noise suppressor is not to remove the noise, but to make the output sound better to humans. Therefore, optimizing the noise suppressor as a system targets maximization of the MOS results, not improvement in the SNR.

To reduce distortion and artifacts, other approaches are used such as adaptive noise cancellation (with a secondary channel for capturing just the noise) or a stationary-tones canceller (which estimates and subtracts the non-random components of the noise). A separate group are perceptually based noise suppressors, which use the masking effects in human hearing to suppress less noise – the parts we cannot hear anyway.

The noise-reduction system in modern communication devices is a real-time running complex program. It can and should be optimized for better initialization and tracking, faster adaptation, and faster execution.

In general, noise reduction is a science, an art, and a craft. It is a science because it deals with mathematical models and has reproducible results. It is an art because it is about the human perception of sounds, where not everything can be modeled with numerical models. It is a craft because there are always better implementations of the same algorithm, some "secret sauce" which makes the entire system work well. This chapter has discussed all three aspects of good noise-reduction systems.

Bibliography

[1] Wiener, N. (1949) *Extrapolation, Interpolation, and Smoothing of Stationary Time Series, with Engineering Applications*, MIT Press, Cambridge, MA.
[2] Wang, D. and Lim, J. (1982) The unimportance of phase in speech enhancement. *IEEE Transactions on Acoustics, Speech, and Signal Processing*, **30**, 679–681.
[3] Ephraim, Y. and Malah, D. (1984) Speech enhancement using a minimum mean-square error short-time spectral amplitude estimator. *IEEE Transactions on Acoustics, Speech, and Signal Processing*, **32**, 1109–1121.
[4] McAulay, R. and Malpass, M. (1980) Speech enhancement using a soft-decision noise suppression filter. *IEEE Transactions on Acoustics, Speech, and Signal, Processing*, **28**, 137–145.

[5] Boll, S. (1979) Suppression of acoustic noise in speech using spectral subtraction. *IEEE Transactions on Acoustics, Speech, and Signal Processing*, **26**, 113–120.

[6] Ephraim, Y. and Malah, D. (1985) Speech enhancement using a minimum mean-square error log-spectral amplitude estimator. *IEEE Transactions on Acoustics, Speech, and Signal Processing*, **33**, 443–445.

[7] Wolfe, P. and Godsill, S. (2001) Simple alternatives to the Ephraim and Malah suppression rule for speech enhancement. Proceedings of 11th IEEE Workshop on Statistical Signal Processing, pp. 496–499.

[8] Martin, R. (2002) Speech enhancement using MMSE short-time spectral estimation with gamma distributed speech priors. Proceedings of IEEE International Conference on Acoustics, Speech, and Signal Processing (ICASSP), Orlando, FL, pp. 253–256.

[9] Gazor, S. and Zhang, W. (2003) Speech probability distribution. *IEEE Signal Processing Letters*, **10**(7), 204–207.

[10] Van Trees, H.L. (1968) *Detection, Estimation and Modulation Theory* (Part I), MIT Press, Cambridge, MA.

[11] ITU-T (1996) *Recommendation G.729 Annex B: A Silence Compression Scheme for G.729 Optimized for Terminals Conforming to Recommendation V.70*, ITU-T, Geneva, Switzerland.

[12] Sohn, J., Kim, N. and Sung, W. (1999) A statistical model based voice activity detector. *IEEE Signal Processing Letters*, **6**(1), 1–3.

[13] Davis, A. and Nordholm, S. (2003) A low complexity statistical voice activity detector with performance comparisons to ITU-T/ETSI voice activity detectors. Information, Communications and Signal Processing, 2003 and the Fourth Pacific Rim Conference on Multimedia, Dec. 2003, vol. 1, pp. 119–123.

[14] Haykin, S. (1994) *Communication Systems*, 3rd edn, John Wiley & Sons, Chichester, England.

[15] Middleton, D. and Esposito, R. (1968) Simultaneous optimum detection and estimation of signals in noise. *IEEE Transactions on Information Theory*, **14**, 433–443.

[16] Martin, R. (2001) Noise power spectral density estimation based on optimal smoothing and minimum statistics. *IEEE Transactions on Speech and Audio Processing*, **9**, 504–512.

[17] Hendriks, R., Heusdens, R. and Jensen, J. (2005) Forward-backward decision directed approach for speech enhancement. Proceedings of International Workshop on Acoustics, Echo and Noise Control (IWAENC), Eindhoven, The Netherlands.

[18] Song, M.-S., Lee, C.-H. and Kang, H.-G. (2006) Performance of various single channel speech enhancement algorithms for automatic speech recognition. Proceedings of International Conference on Spoken Language Processing (Interspeech), Pittsburg, PA.

[19] Tashev, I., Droppo, J. and Acero, A. (2006) Suppression rule for speech recognition friendly noise suppressors. Proceedings of Eight International Conference Digital Signal Processing and Applications (DSPA), Moscow.

[20] Haykin, S. (2002) *Adaptive Filter Theory*, 4th edn, Prentice-Hall, Upper Saddle River, NJ.

[21] Zwicker, E. and Fast, H. (1999) *Psychoacoustics: Facts and Models*, Springer-Verlag, Berlin.

[22] Fletcher, H. and Munson, W. (1933) Loudness, its definition, measurement and calculation. *Journal of the Acoustical Society of America*, **5**, (2), 82–108.

[23] Robinson, D. and Dadson, R. (1956) A re-determination of the equal loudness relations for pure tones. *British Journal of Applied Physics*, **7**, (5), 166–181.

[24] Schroeder, M., Atal, B. and Hall, J. (1979) Optimizing digital speech coders by exploiting masking properties of the human ear. *Journal of the Acoustical Society of America*, **64**(S1), S139.

[25] Schraf, B. (1970) Critical bands, in *Foundations of Modern Auditory Theory*, Academic Press, New York.

[26] Malvar, H. (2004) Auditory masking in audio compression, in *Audio Anecdotes: Tools, Tips, and Techniques for Digital Audio* (eds K. Greenbaum and R. Basel), A.K. Peters, Natick, MA.

[27] Spanias, A., Painter, T. and Atti, V. (2007) *Audio Signal Processing and Coding*, John Wiley & Sons, Hoboken, NJ.

[28] Vary, P. and Martin, R. (2006) *Digital Speech Transmission: Enhancement, Coding and Error Concealment*, John Wiley & Sons, Chichester, England.

[29] Wolfe, P. and Godsil, S. (2003) A perceptually balanced loss function for short-time spectral amplitude estimator. Proceedings of IEEE ICASSP, Hong Kong, China, **5**, 425–428.

[30] Roguet, W., Martin, N. and Chehikian, A. (1996) Tracking of frequency in a time-frequency representation. Proceedings of IEEE International Symposium on TFTS, pp. 341–344.

[31] Davy, M., Leprette, B., Doncarli, C. and Martin, N. (1998) Tracking of spectral lines in ARCAP time-frequency representation. Proceedings of EUSIPCO, Rhodes Island, Greece.

[32] Andrieu, C., Davy, M. and Doucet, A. (2003) Efficient particle filtering for jump Markov systems: application to time-varying autoregressions. *IEEE Transactions on Signal Processing*, **51**, 1762–1770.

[33] Andia, B. (2006) Restoration of speech signals contaminated by stationary tones using an image perspective. Proceedings of IEEE ICASSP, Toulouse, France.

[34] Tashev, I. and Malvar, H. (2007) Stationary-tones Interference Cancellation Using Adaptive Tracking. Proceedings of IEEE ICASSP, Honolulu, HI.

[35] Drucker, H. (1968) Speech processing in a high ambient noise environment. *IEEE Transactions on Audio and Electroacoustics*, **16**, 165–168.

[36] Ephraim, Y., Cohen, I. (2006) Recent advancements in speech enhancement in *The Electrical Engineering Handbook, Circuits, Singals, and Speech and Image Processing,* Richard C. Drof (ed.), Third Edition, CRC Press, Boca Raton, FL, pp. 15-12–15-26.

[37] Benyassine, A., Shlomot, E., Su, H. *et al.* (1997) ITU-T recommendation G.729 annex B: a silence compression scheme for use with G.729 optimized for V.70 digital simultaneous voice and data applications. *IEEE Communications Magazine*, **35**(9), 64–73.

[38] Cho, Y.D., Al-Naimi, K. and Kondoz, A. (2001) Improved voice activity detector based on a smoothed statistical likelihood ratio. Proceedings of IEEE International Conference on Acoustics, Speech, and Signal Processing (ICASSP), Salt Lake City, UT, vol. 2, pp. 737–740.

[39] Fawsett, T. (2006) An introduction to ROC analysis. *Pattern Recognition Letters*, **27**, 861–874.

[40] Sohn, J. and Sung, W. (1998) A voice activity detection employing soft decision based noise spectrum adaptation. Proceedings of IEEE International Conference on Acoustics, Speech, and Signal Processing (ICASSP), Seattle, WA, pp. 365–368.

5

Sound Capture with Microphone Arrays

This chapter is dedicated to sound capture with systems of multiple microphones, called "microphone arrays." The sound propagation itself is a three-dimensional process, so that capturing it in one single point, with one microphone, is not sufficient to deal with 3D processes like ambient noise, reverberation, and multiple sound sources. Using several closely positioned microphones allows listening to the sound coming from one direction, while suppressing the noises and interference sounds coming from other directions. In addition, microphone arrays allow estimation of the direction of arrival – that is, sound source localization.

There are two major groups of microphone-array processing algorithms: time-invariant and adaptive. The first are optimal under the assumption of isotropic ambient noise, and they are fast and simple to implement for working in real time. Adaptive processing algorithms shine when we have point noise sources in low reverberant conditions. They require more CPU resources and are more complex to implement. Both approaches assume identical capturing channels and are affected by mismatch, caused mainly by the manufacturing tolerances of the microphones used. This is handled by creating a robustness to manufacturing tolerances in the algorithms, by manufacturing time-calibration procedures, or by real-time autocalibration algorithms. Multiple channels allow the creation of multichannel noise suppressors and spatial filters, which further improve the quality of the sound and increase suppression of unwanted sounds and noise.

5.1 Definitions and Types of Microphone Array

5.1.1 Transducer Arrays and their Applications

The concept of using a set of antennas for directional radio transmission and receiving has been known since World War I, but was employed in practice in radars used during

Sound Capture and Processing   Ivan J. Tashev
© 2009 John Wiley & Sons, Ltd

World War II. Currently, phased antenna arrays are part of multiple systems for early ballistic missile detection, ground- and ship-based radars, and airborne systems. These systems usually use a model in which both the signal and the interference can be modeled as plane waves impinging on the array. Similar models are used for transmitting a narrow beam in a given direction. Some complexity is added in airborne radars, looking at the ground, where additional reflections from the ground and multipaths are encountered. The theory and practice for the design of such antenna systems has been well developed through the years, as they are widely used in the military and civil aviation.

A fascinating example of a phased antenna array system is AN/FPS-115 PAVE PAWS radars. They were deployed in 1975 as part of the system for early detection of ballistic missiles. Each of the radars has two circular antennas with a diameter of 22.1 m and 3584 elements each. Without any moving parts, these radars can move the beam direction within milliseconds. A schematic view of one of these radars is shown on Figure 5.1.

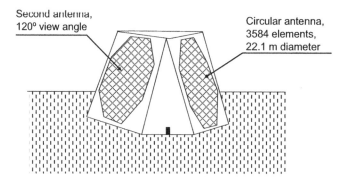

Figure 5.1   Radar with phased antenna arrays

Another interesting application of antenna array systems is radio astronomy. They are passive systems, used to detect the presence and location of celestial sources of radio waves. While the radar phased antenna arrays are usually compact, radio astronomy employs antenna arrays with a very large distance between antennas. This causes multiple additional problems with synchronization and calibration of the antennas. A good example here is the Very Large Array, one of the world's premier astronomical radio observatories. It was built by the National Radio Astronomy Observatory and consists of 27 radio antennas in a Y-shaped configuration on the Plains of San Agustin, 80 km west of Socorro in New Mexico. Each antenna is 25 m in diameter. The data from the antennas is combined electronically to give the resolution of an antenna 36 km across, with the sensitivity of a dish 130 m in diameter.

Arrays of sensors are used in sonars, which can be active or passive. While active sonars use a principle similar to radar (sending a directed beam of sound under water and listening to reflections from objects of interest), underwater sound propagation is much more complex than propagation of electromagnetic energy in the atmosphere. The complications include stronger reflections from the bottom and the surface, absorption, and much higher levels of noise. The medium parameters depend heavily on the temperature, depth, presence of air and gas bubbles, and presence of marine life. The noise can be generated by the presence of man-made objects (ships, submarines), sea life, precipitation, waves, and so on. Multiple reflections make this type of noise ambient and spread across the entire frequency spectrum. The unpleasant part of the noise is the self noise of the platform, carrying the sonar array (engines, cavitations from the propellers, etc.). The passive sonar is just a system of hydrophones (specially designed underwater microphones), listening to the noise under the water. They can be stationary or a long string of hydrophones can be dragged by a ship or submarine (Figure 5.2). The main application is detecting submarines, although some hydrophone systems are used for studying the sounds from marine life (dolphins, orcas). No submarine since World War II has been built without sophisticated sonar systems, both passive and active. Strings of hydrophones (i.e., hydrophone arrays) are dragged by military ships for detection and localization of submarines as well. From the sound-capture perspective, the theory and practice of hydrophone arrays is closest to that of microphone arrays, the subject of discussion in this chapter.

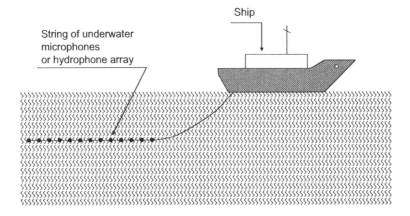

Figure 5.2   Array of hydrophones

Underground structures are studied by means of seismic experiments that involve an array of seismometers. An explosion creates a shock wave that propagates through the non-homogeneous medium underground and reflects from boundaries between layers where the characteristics of the medium change. The shock waves and the reflections are captured by an array of seismographs (Figure 5.3). Software is used to reconstruct the underground structure. Another application of the arrays of seismographs is to detect the location and the magnitude of an underground nuclear explosion – an important tool for controlling the execution of international treaties during the Cold War. The algorithms and approaches used for reconstructing the underground structure are relatively far from what is used for capturing sounds with a microphone array, but source localization and reconstruction are based on the same principles.

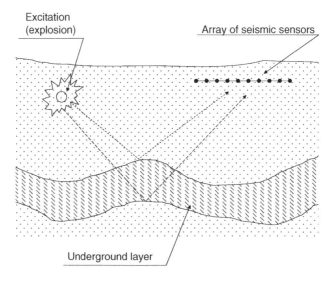

Figure 5.3   Using arrays of seismic sensors to study underground structure

Array processing is used in modern medicine too. Work on the first prototype of computer-assisted tomography was initiated by Godfrey Hounsfield, an electrical engineer working for the British company Electric and Musical Industries (EMI) in the mid-1960s, and the first human patient was examined in 1971. For this invention, in 1979 he received the Nobel Prize for Medicine together with Alan M. Cormack, a South African physicist. The idea is to send an X-ray beam through the patient's head or body and receive it from the other end with a single receiving element (fluorescent screen and photo-multiplying tube, for example), which measures the X-ray magnitude (i.e., the attenuation). This is entered into the computer using analog-to-digital conversion. Then the sensor is moved at a small angle around the head or the body of the patient in a plane perpendicular to the body axis and the process repeats. Once the transmitting area is covered, the whole system rotates a small angle and the process repeats. Then, based on the measured attenuation for all combinations of the transmitter and receiver positions, an image is computed – a perpendicular slice of the patient's head or body. Owing to the higher resolution in magnitudes and the powerful image-restoration algorithms, any abnormalities in the soft tissues, such as tumors or cysts, are visible – something impossible for the classic radiography of that time. Later, the single moving transmitter and receiver were replaced by an array of transmitters and receivers, which substantially shortened the scan time. While this revolutionary method has extremely powerful diagnostic capabilities, the algorithms used are quite different from those used in sound capture with microphone arrays.

Of course, the theory and algorithms for designing phased antenna arrays found their way to telecommunications. Today, arrays of antennas are widely used practically everywhere – from satellite communications to cellphone towers. Technologies and materials were designed to ensure robust and good directivity patterns specific to the conditions where these antennas should work.

5.1.2 Specifics of Array Processing for Audio Applications

With advancing array technologies and the available computing power, attempts were made to use arrays of microphones for capturing sounds for the needs of telecommunication and automatic speech recognition. A single microphone captures too much noise and reverberation, which limits the performance of single-channel noise-suppression algorithms, discussed in Chapter 4. The general idea of microphone array processing is to use multiple microphones and to create an analog of a single microphone with higher directivity by combining the signals from all the microphones into one signal. Such microphones with higher directivity will capture less noise and less reverberation. The output of the microphone array will have reduced noise levels, allowing the noise suppressor to perform better. The initial steps were to adapt established working algorithms from antenna arrays. With the first attempts, big differences became visible in the electromagnetic field propagation, sound wave propagation, and the conditions in which these arrays work.

One of the major differences is that the sound is an extremely wideband signal. Even the telephone-quality bandwidth from 300 to 3400 Hz has eleven times the difference in frequencies and wavelengths. For wideband audio between 200 and 7400 Hz this ratio is 38 times (not to mention 1000 times the difference for the full band of human hearing from 20 to 20 000 Hz). In the radars operating in the 10 GHz band, a 600 MHz bandwidth is considered wideband, which is just 6% of the main frequency. From this perspective, practically all antenna arrays work in the best possible mode, with elements usually placed every half wavelength. The antenna system is in resonance, allowing the creation of very narrow beams during transmission and very high directivity during receiving. This is not possible with microphone arrays for two reasons: array size and the number of elements.

Antennas are professional equipment with (usually) large size. They can be replaced with a similarly sized antenna array. The microphone array replaces a single microphone, much smaller than the wavelengths in the captured band (from 1.7 m to 4.6 cm for the band 200–7400 Hz). It is expected that the microphone array, which is just a better microphone, will be at most slightly bigger than a single microphone. In addition, many construction constraints limit the size of the microphone array – the length of the upper bezel of the laptop screen where we want to integrate the microphone array, for example.

This means that the performance of the microphone array for the lower part of the frequency band will be suboptimal owing to its limited size.

In antenna arrays the number of elements varies, but usually it is in the range of 30 to 50. There are phased antenna arrays with thousands of elements. Under price and space constraints of mass-production devices, microphone arrays usually have just two or four elements; some arrays in professional communication equipment may have six or eight. To cover the band from 200 to 7400 Hz with optimal array design (antenna array style) we should have 38 elements placed evenly every 2.3 cm, with a total array length of 85.1 cm. The computing power can be a limitation for the number of channels as well. Modern personal desktop computers can easily handle eight channels in real time, but this may not be the case for mass-production laptops or specialized DSPs, part of the communication equipment. Using fewer elements means that the array will work suboptimally in the upper part of the frequency band as well.

The noise conditions for acoustic microphone arrays are much worse owing to the wideband character of the sound. A wider bandwidth means capturing more noise, and most of the surrounding noise sources emit sound in the entire audible band as well. In offices, homes, or conference rooms, such noise sources are air-conditioning and ventilation, computers with their cooling fans and hard discs, and noises from the street. In open spaces in cities, we also have multiple people talking, buses, cars, trains, wind noise, and so on. The SNR of the captured sound from each microphone is low and this is the reason for using microphone arrays.

In radars, multipath and multiple reflections from ground objects and the terrain happen less frequently than with sound. The wavelength of the electromagnetic wave is usually smaller than the observed objects. In contrast, the sound wave reflects multiple times from walls and ceilings, creating what is known as "reverberation." Its wavelength is comparable to the objects in the room, which causes diffraction effects as well. As a result, microphones capture not only the direct path sound, but a mixture of the direct path and multiple reflected and distorted copies of the same signal. The same is true for unwanted sound sources. This mixture of coherent signals is very difficult to work with and creates processing problems specific to audio. We will discuss the reverberation effects later in this book.

Another substantial difference between antenna arrays and microphone arrays is the self noise of the microphones. Antennas do not have self noise, excluding the very low thermal noise, while a microphone's self noise is substantial and is one of the limiting factors for the performance of microphone arrays.

Microphones have higher manufacturing variations of their parameters than a passive antenna. In addition, the wavelength of the sound in the middle of the audio frequency band is around 10 cm, which corresponds to 3 GHz radio frequency, which is below the frequency band where most of the antenna arrays operate. This makes the precision for sensor placement more critical for microphone arrays – another factor to account for. The variations in the sensor parameters and positions require robust microphone-array processing algorithms.

5.1.3 Types of Microphone Arrays

A microphone array is a set of closely positioned microphones. Large arrays with hundreds of microphones are out of the scope of this book. We will assume that the microphone array is one device, containing several microphones placed in known positions. We will not discuss in this chapter algorithms for microphone arrays with unknown or variable microphone positions. The number and positions of the microphones in the array is frequently referred to as "microphone array geometry." Based on the geometry and the target scenario, we can distinguish the following types of microphone arrays.

5.1.3.1 Linear Microphone Arrays

The microphones are positioned in one line. The work area of these arrays is a half plane. They cannot distinguish sound waves arriving under the same body angle to the microphone array axis, which forms what is known as the "cone of uncertainty." This is due to the fact that under these circumstances the sound wave reaches the microphones in the same order and with the same delay.

If the desired sound sources are mostly positioned in the area perpendicular to the microphone array axis, we describe this geometry as a *broadside* linear microphone array. Typical places for such arrays are on top of a computer monitor or integrated in the upper bezel of a monitor or laptop screen. The sound source we want to capture is the voice of one or more persons positioned in front of the computer. The work area is usually in the range of $\pm 45°$ from the line perpendicular to the microphone array axis. These arrays can change the listening direction to capture the voices of several people, or to track the movements of one of them. The number of microphones varies from two to eight, but is typically two or four. Figure 5.4 shows a four-element broadside linear microphone array. The length is 195 mm – short enough to fit in the upper bezel of a monitor or laptop screen. The microphones are unidirectional, with a close to cardioid

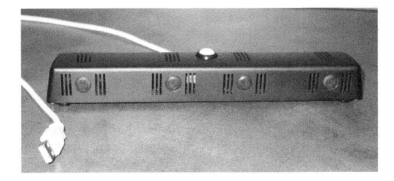

Figure 5.4 Linear four-element microphone array. Reproduced by permission of Microsoft Corporation

directivity pattern, pointing forward. In this way they partially resolve the front–back ambiguity, caused by the linear geometry.

If the sound source is positioned on the microphone array axis, we denote this as an *endfire* linear microphone array. The typical application is a highly directive, shotgun type microphone, or a small two- or three-element microphone array as part of a headset with a short boom. These arrays usually do not change the listening direction, which means that either the sound source is in a fixed position (the mouth in the headset case) or the whole microphone array is rotated to track the sound source movements (shotgun-type microphones). The number of microphones varies from two to sixteen, but usually is two or three. Figure 5.5 shows a three-element endfire microphone array for a headset, on the ear of a head-and-torso simulator. The length of the array is limited by the size of a short-boom headset and is 65 mm. This array has two unidirectional microphones, pointing towards the mouth, while the third microphone is omnidirectional.

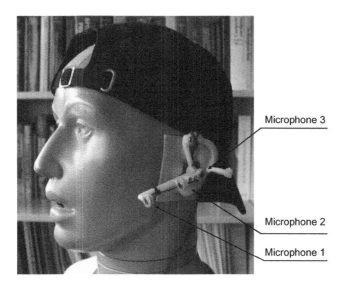

Figure 5.5   Three-element endfire microphone array for headset. Reproduced by permission of Microsoft Corporation

5.1.3.2 Circular Microphone Arrays

In this geometry the microphones are positioned in a circle. There are designs where the microphones are placed in two or more concentric circles. A typical place for such arrays is the center of a conference room table, where they work mainly in one plane and change listening direction in, or slightly above, the plane of the circle. The number of microphones varies from three to sixteen, and is typically six or eight. Figure 5.6 shows an eight-element circular microphone array with a diameter of 140 mm. The microphones are omnidirectional.

Figure 5.6 Eight-element circular microphone array. Reproduced by permission of Microsoft Corporation

5.1.3.3 Planar Microphone Arrays

The microphones of the planar array are positioned in one plane. The desired sound sources are usually towards the direction perpendicular to the microphone array plane. This is the main difference from the previous scenario, where the desired sound sources were positioned close to the array plane. Otherwise the circular array is a type of planar array as well. Such arrays can be mounted on the wall or on the ceiling of a conference room and can capture sound sources in a half sphere. They can change the listening direction both horizontally and vertically, but cannot distinguish front from back. The minimum number of microphones is obviously three (the least number of points to define a plane in a 3D space); typical geometries include three, four, or eight elements. Figure 5.7 shows an eight-element planar microphone array installed on the ceiling of a small lecture room. The microphones are unidirectional, to resolve to a certain extent front–back ambiguity. They point down, towards the lecture room. The geometry is a double square: the side of the larger square is 200 mm; the side of the small square is 50 mm and it is rotated through 45°.

5.1.3.4 Volumetric (3D) Microphone Arrays

Three-dimensional microphone arrays can capture sound sources from any direction in a 3D space; they do not have areas of confusion. The minimum number of microphones is four, not lying in one plane, and under the condition of placing them just hanging in the air. The design of practically all microphone arrays so far assumes that the array's enclosure is acoustically transparent. There are spherical array designs, where the microphones are placed on top of an acoustically non-transparent sphere. In this case

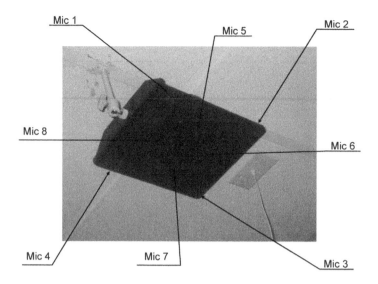

Figure 5.7 Eight-element planar microphone array for ceiling mounting. Reproduced by permission of Microsoft Corporation

the way the sound wave diffracts around the sphere is used to achieve better results. Actually the oldest microphone array that uses the different ways the sound diffracts around an oval object was created by nature and has two elements: the elements are usually called ears and the oval object is called the head. We will discuss some of these diffraction effects in the chapter on sound source localization.

5.1.3.5 Specialized Microphone Arrays

This group includes very small arrays, used in hearing-aid devices, which use a different class of algorithms. In some cases a mix of sensors is used (omni- and uni- directional microphones, sensors to capture bone vibrations, etc.), which allows better results to be achieved. All these arrays employ application-, geometry-, and sensor- specific algorithms. These approaches are not generic and we will discuss some of them towards the end of this chapter.

5.2 The Sound Capture Model and Beamforming

5.2.1 Coordinate System

In this book we will work with a three-dimensional coordinate system with the center close to the center of the microphone array (the average of the coordinates of all the microphones). We will assume that the x axis points to the most probable direction to the listener, or to the center of the listening area. Usually the xy plane will be horizontal

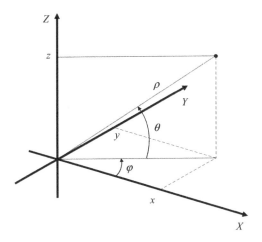

Figure 5.8   Coordinate system

and the z axis will point up. An example for an exception of this rule is the ceiling-mounted microphone array in Figure 5.7, where the x axis points down and the yz plane is horizontal. The position of each point c can be represented by three coordinates, $c = \{x, y, z\}$, or as a distance and two angles (azimuth and elevation), $c = \{\rho, \varphi, \theta\}$. The azimuth angle is in the xy plane with a positive counterclockwise direction when looking from above. The elevation angle is between the direction to the point and the xy plane. These two ways of representing the coordinates are shown in Figure 5.8. The Euclidian distance

$$d = \sqrt{(x_1 - x_2)^2 + (y_1 - y_2)^2 + (z_1 - z_2)^2} \qquad (5.1)$$

between two points with coordinates c_1 and c_2 will be denoted as $d = \|c_1 - c_2\|$. The angle ϕ between the directions to two points, where

$$\cos \phi = \cos(\varphi_1 - \varphi_2) \cdot \sin(\theta_1 - \theta_2) \qquad (5.2)$$

will be denoted as $\phi = \sphericalangle(c_1, c_2)$. In cases when we deal with directions only (i.e., ρ is unknown or does not matter), we will keep the same notation for direction as for coordinates; that is, $c = \{\rho, \varphi, \theta\}$.

While the standard unit for measuring angles is the radian, abbreviated to "rad," in this book we will also use the older and non-standard unit for angle – the degree. The full circle is 2π radians, equivalent to 360 degrees, or $360°$. The same units can be used to measure direction angles in the coordinate system and phase differences of the signals in microphone channels. To prevent confusion, we will specify the direction angles in degrees and the phase differences in radians. Note, however, that in most programming languages the mathematical functions for estimation of cosine, sine, tangent, and so on, expect the argument to be in radians. This means that inside the computer program angles for both direction and phase should be in radians.

EXERCISE

Write in the default MATLAB® directory a file named *MicArrDesc.dat*, containing the description of a microphone array in the coordinate system above. The microphone array is linear, symmetric, and has four cardioid microphones with a distance of 195 mm between the outer pair and 55 mm between the inner pair. The file has as many lines as the number of microphones. Each line contains six numbers: x, y, z of the microphone, azimuth and elevation of the microphone's MRA, and the microphone type as discussed in Chapter 3. All coordinates are in meters, all angles in degrees. The microphone's type is 0 for omnidirectional, 1 for subcardioid, 2 for cardioid, 3 for supercardioid, 4 for hypercardioid, 5 for a figure-8.

   Look at the provided *MicArrDesc4el.dat*, containing the description of the array shown in Figure 5.4.

5.2.2 Sound Propagation and Capture

5.2.2.1 Near-field Model

Let the vector $\mathbf{p} = \{p_m, m = 0, 1, \ldots, M-1\}$ denote the positions of the M microphones in the array, where $p_m = (x_m, y_m, z_m)$. This yields a set of signals that we denote by the vector $\mathbf{x}(t, \mathbf{p})$. We will consider processing in the frequency domain only, using the standard overlap–add analysis and synthesis procedure described earlier in this book. We will denote the input vector by $X_m^{(n)}(f, p_m)$. The frame index will be omitted for simplicity wherever possible. We will assume that each sensor M_m has a known directivity pattern $U_m(f, c)$, where c is the coordinates of the sound source and f denotes the signal frequency. For a sound source at location c under non-reverberant conditions, the captured signal from each microphone is

$$X_m(f, p_m) = D_m(f, c)S(f) + N_m(f) \tag{5.3}$$

where the first term on the right-hand side represents the phase rotation and the decay due to the distance to the microphone $\|c - p_m\|$, and v is the speed of sound. This can be written as:

$$D_m(f, c) = \frac{e^{-j2\pi f \frac{\|c-p_m\|}{v}}}{\|c-p_m\|} A_m(f) U_m(f, c). \tag{5.4}$$

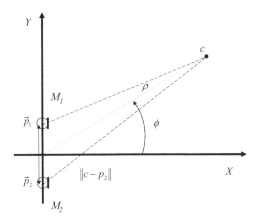

Figure 5.9   Near-field sound capture model

This is presented schematically in the xy plane only in Figure 5.9 where we have a linear two-element microphone array and a sound source. The term $A_m(f)$ is the frequency response of the system preamplifier/ADC, $S(f)$ is the source signal, and $N_m(f)$ is the noise, usually modeled as a zero-mean Gaussian random process. Later we will see that this noise consists of two major components: correlated (external) and uncorrelated (the noise of the microphones). This capturing model is geometrically precise and valid for any distance to the sound source. The model is called "near-field" because it is good for sound sources close to, and even inside, the microphone array.

5.2.2.2 Far-field Model

With increasing distance to the sound source, the difference in signal levels due to variable distances becomes smaller and smaller. For distances larger than 5–10 times the size of the microphone array, it is

$$\frac{1}{||c - p_i||} \approx \frac{1}{||c - p_j||} \quad \forall i, j. \tag{5.5}$$

Then the sound source can be described with direction ϕ and distance ρ from the center of the coordinate system, as presented in Figure 5.9, which we assumed is the center of the microphone array. For simplicity we will work in two instead of three dimensions. The sound capture (Equation 5.4) is transformed to

$$D_m(f, c) = \frac{1}{\rho} e^{-j2\pi f \frac{|p_m| \cos(\angle(p_m) - \phi)}{v}} A_m(f) U_m(f, c). \tag{5.6}$$

In this case the first term, representing the decay due to the distance, can be constant for every direction and sound source. It can be fixed to some average distance ρ_{AV}, greatly reducing the complexity of the sound capture model. The second term represents the phase difference the sound wave reaches in the corresponding microphone regardless of the center of the coordinate system. These phase shifts depend only on the direction to the sound source and the microphone array geometry. This model is called the "far-field" because it is good only for sound sources at a large distance. It is simpler and more suitable for finding optimal solutions analytically.

The near-field model is more generic and suitable for both close and far-field modeling. If the microphone array design is done by finding the optimal solution numerically, it is better to use the near-field model. There are cases when the two models can be used in parallel. If we design a microphone array to be placed close to a human mouth, we should use the near-field model for the desired sound source (human speech from the mouth) and it is possible to use the far-field model to model the ambient noise. The generic solution will use the near-field model for both the desired sound source (positioned close at given coordinates) and for the ambient noise sources (positioned on a sphere with a radius of 2 m, which is a far-field distance).

5.2.3 Spatial Aliasing and Ambiguity

Spatial aliasing occurs when, for a given frequency, we have the same signal delay in the microphones for at least two different directions. For a simple example we will take a two-element array like the one shown in Figure 5.9. Looking at the sound capture equation (Equation 5.6) for far-field capture, we notice that the direction-dependent part is $e^{-j2\pi \frac{f\|p_m\|}{v}\cos(\sphericalangle(p)-\phi)}$. Here we will ignore for a moment the directivity pattern of the capturing microphones, $U_m(f, c)$, and the frequency response of the preamplifier $A_m(f)$. If the microphones are placed symmetrically, we can say that the distance between them is $d = 2\|p\|$. In the same geometry, $p_m = \pm 90°$. Then the phase rotation is $e^{-j\pi \frac{d}{\lambda}\sin(\phi)}$, where $\lambda = v/f$ is the sound's wavelength. This corresponds to a phase shift of $\pi(d/\lambda)\sin(\phi)$. To have ambiguity for certain frequencies (or wavelengths), this phase shift should be the same for at least two different direction angles. The phase shift is the same if the difference in the delay is the integer number of periods, so we are looking for conditions when

$$\pi \frac{d}{\lambda}\sin(\phi_1) = \pi \frac{d}{\lambda}\sin(\phi_2) \pm n2\pi, \quad \phi_1 \neq \phi_2. \tag{5.7}$$

can be possible. As $-1 \leq \sin(\phi) \leq 1$ we can have more than one solution only when $d < \lambda/2$. This means that, to prevent spatial aliasing and direction ambiguity, we have to have a distance between the two microphones smaller than half a wavelength of the highest frequency in the work band.

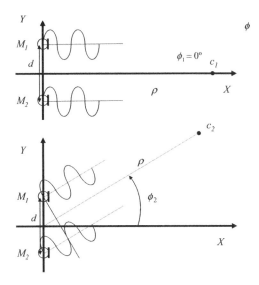

Figure 5.10   Spatial aliasing for two-element microphone array

The spatial aliasing is shown in Figure 5.10. In the first case we have a sound source right in front of the microphone pair M_1M_2. The direction angle is $\phi_1 = 0°$ and the sound arrives with the same delay to each of the microphones. Then the phase difference between the two signals, captured by the microphones, will be zero. In the second case the sound source is positioned at direction ϕ_2. The far-field difference in the sound path is $d \sin(\phi_2)$, which happens to be exactly the wavelength of the sound. Then the phase difference between the two signals is 2π, which cannot be distinguished from 0; that is, the two signals have the same phase difference as when the sound source was right in front of the microphone pair.

It is easy to note that at a distance $d = \lambda/2$ between the pair of microphones the phase difference range of $\pm\pi$ is fully used, which means that the spatial resolution is best. The wavelength of a sound wave with a frequency of 3400 Hz is $\lambda = c/f = 0.101$ meters. If we have a two-element microphone array with a distance of 5 cm (half a wavelength) between the microphones, then the phase difference for a sound with frequency 3400 Hz coming from a direction of 90° will be $-\pi$ radians, for 0° it will be 0 radians, and for $-90°$ it will be π radians. The same microphone pair for a sound with a frequency of 300 Hz will provide phase differences of -0.22 rad, 0 rad, and $+0.22$ rad. The phase difference in the second case is almost 14 times smaller owing to the larger wavelength. Considering the noise, which will smear the phase differences, the ability of the microphone array to distinguish the direction the sound comes from will be substantially lower for this microphone array and this signal frequency. The microphone pair performs best when the distance between the microphones is one half the wavelength of the signal frequency.

As the sound is a wideband signal, the solution is to use more than two microphones in a way to cover the entire frequency band. Considering sound capture with telephone quality (300–3400 Hz), we have to have a microphone pair with a distance between the microphones of 5 cm (see above) to cover 3400 Hz. To cover 300 Hz, we have to have a distance of 57 cm between the microphones. Assuming a linear microphone array with equal distances between the microphones, we have to have $57/5.0 \approx 11$ microphones. This microphone array will have $M(M-1)/2$ potential microphone pairs which will cover relatively evenly the entire frequency band. Using uneven and asymmetrical spacing, the number of microphones can be reduced to eight or even six with minimal hit on the performance. Such an approach will complicate the construction of the array and will make the analytical design of the array more difficult.

Besides being aware of the problems of spatial aliasing, the microphone array geometry by itself can cause spatial ambiguity. In Figure 5.11 we have two sound sources and one microphone pair. The directions to the sound sources are ϕ_1 and $\phi_2 = 180° - \phi_1$, respectively. In the far-field the differences in the sound paths will be $d_1 = d \sin(\phi_1)$ and $d_2 = d \sin(\phi_2)$. It is easy to see that $d_1 = d_2$, which means that for the microphone pair these two sound sources will have the same phase difference; that is, they cannot be distinguished as coming from different directions. The microphone array pair works in half a plane and cannot distinguish sound sources with a direction symmetric to the microphone pair.

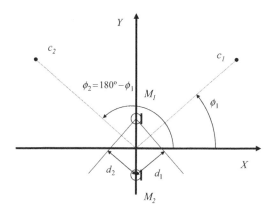

Figure 5.11   Front–back ambiguity for linear microphone array

The statement above is under the assumption of far-field propagation. In a near-field things are more complex. Every given difference between the signals, captured by the two microphones, defines a parabola in the xy plane. Every point from the parabola has the same difference in the signal paths to the two microphones; that is, it cannot be distinguished as coming from a different location from the microphone pair. In three dimensions this looks even worse as the ambiguous zone has the shape of a parabolic cone. To make the microphone array work for sound sources in the entire plane, we have to have at least three microphones that are not in one line. It is easy to see that three microphones in one line will have the same problem in distinguishing front from back. A triangular microphone array will have problems distinguishing sound sources symmetric to the plane, defined by the three points, for the same reasons. For a 3D space we have to have at least four microphones, not lying in one plane. With the wideband audio signal in the example above, the number of microphones increases to an unacceptably large number. Such a volumetric microphone array will fit in a sphere with a diameter of half a meter – and this is just for telephone-quality sound capture!

Generalizing further, we sample a multidimensional process (the sound wave propagation) in a fixed number of points. To have a fully unambiguous sampling, we have to have a number of points larger than the number of dimensions plus one, and the sampling points should be placed closer than half the wavelength of the highest frequency. This is nothing more than spatial extension of the sampling theorem discussed in Chapter 2.

Design algorithms for such large microphone arrays with a number of microphones are interesting from the research perspective, but they are not quite applicable in real practice. The microphone array substitutes for the microphone – a relatively small device. It is expected that the better quality replacement should have comparable size. The number of microphones is typically two or four and in rare cases reaches eight or sixteen. Real designs put restrictions to the size of the array as well – it has to be shorter than the upper bezel of the laptop screen, fit on the top of a monitor, and so on. These restrictions lead to the fact that the microphone array will not be optimal for substantial parts of the frequency band.

Additional measures can be taken for spatial disambiguation – using directional microphones in the linear array, pointing towards the working half plane, for example. Then a symmetric sound source from the back side will be at least partially suppressed owing to the directivity of the microphones used. Additional acoustic measures can be taken towards spatial disambiguation by placing the microphone on an acoustically opaque casing. Humans have only two ears and can easily distinguish the direction of the sound source in three dimensions. Such solutions are usually computationally expensive – just to remind readers that humans have 100 billion neurons between their two ears, a computing power we will not have for microphone array processing in the near future. The art of designing microphone arrays under such practical restrictions is to make them as efficient as possible in the most common scenarios in which they will be used.

5.2.4 Spatial Correlation of the Microphone Signals

If we look at the noise component $N_m(f)$ in the sound capture (Equation 5.3) we will find that it has two distinct components. The first component is the self noise of the microphones. This noise is generated inside each microphone and is uncorrelated across the microphone channels. We will see later that this is one of the limiting factors for the spatial selectivity of microphone arrays. A typical electret microphone provides 60 dB signal-to-noise ratio with a reference level of 96 dB SPL. Considering that a normal human voice has levels of 60–65 dB SPL, and that the cables, preamplifiers, and ADCs will add some noise, we can say that the uncorrelated noise is typically 30 dB below the speech level; that is, even in complete silence such a microphone system will provide 30 dB SNR.

The second component of the noise is the external ambient noise, captured by the microphones. The main difference is that it is the same noise, slightly delayed and decayed across all microphones, and this component is highly correlated across the channels. Of course this will happen in a perfect microphone array, which samples the sound field without violating the sampling theorem. In real practice this is not valid in the upper part of the frequency band and the captured noise is not as correlated. This simply means that the ambient noise itself can be modeled as having highly correlated and uncorrelated components. The ratio of these components is frequency-dependent and varies with the microphone array geometry and the microphones used. The levels of ambient noise vary as well, but we can say that the microphones typically will provide an SNR of 5–20 dB. This means that the ambient noise is 10–25 dB above the level of the uncorrelated self noise from the microphones.

It is understandable that the speech signal from the sound source is highly correlated across the microphone channels. This is especially true when the sound source is very close, or when the noise level is low. With higher reverberation (i.e., larger distance), the correlation of the captured sound source decreases. This is due to the fact that the reflections interfere with the direct-path signal. The resulting signal for each frequency can still be modeled as arriving from one direction, but this direction will be different for different frequencies, which leads to smearing of the correlation peaks. Still the desired signal in most practical cases is the strongest and has the highest correlation across the microphone channels.

5.2.5 Delay-and-Sum Beamformer

Considering an array of M microphones and a far-field sound source at direction ϕ, each of the microphones captures the same signal slightly delayed owing to the different times when the sound reaches the corresponding microphone. The sound capture is according to Equations 5.3 and 5.6. The most intuitive approach to estimate the source signal is to delay each of the microphone signals in a way that the sound from the source signal are in phase, and to sum them. Then we will take the source signal (the same in all channels after the delay) to have an amplitude M times bigger than in each of the channels. This is usually compensated for by dividing the sum by the number of channels:

$$Y(f) = \frac{1}{M} \sum_{m=0}^{M-1} e^{j2\pi f \frac{\|p_m\|\cos(\mathfrak{I}(p_m)-\phi_0)}{v}} X_m(f) \tag{5.8}$$

where ϕ_0 is the listening direction. Combining Equations 5.8, 5.3, and 5.4 under the assumption of omnidirectional microphones $(U_m(f, c) \equiv 1)$ and perfect pre-amplifier/ ADC system $(A_m(f) \equiv 1)$ for the working band, we have

$$Y(f) = S_0(f) + \frac{1}{M}\sum_{m=0}^{M-1} e^{j2\pi f \frac{\|p_m\|\cos(\sphericalangle(p_m)-\phi_0)}{\nu}} N_m(f). \tag{5.9}$$

If we assume for a moment that the captured noise is not correlated across the microphones and is modeled as a zero-mean Gaussian process $N_m(f) = \mathbb{N}(0, \lambda(f))$, then the output signal becomes

$$Y(f) = S_0(f) + \mathbb{N}\left(0, \frac{\lambda(f)}{M}\right). \tag{5.10}$$

This means that the deviation of the uncorrelated noise is decreased $M^{1/2}$ times. This is a 3 dB suppression for a two-element microphone array, 6 dB for four, and so on. Alternative sound sources or just ambient noise will not be in phase and they will partially cancel each other out, or will be with decreased magnitude – that is, suppressed. Let us assume we have a sound source $S(f)$ at direction $\phi \neq \phi_0$. For simplicity we will assume that there is no noise at all. Then the output signal will be

$$Y(f) = S(f)\frac{1}{M}\sum_{m=0}^{M-1} e^{j2\pi f \frac{\|p_m\|}{\nu}(\cos(\sphericalangle(p_m)-\phi_0)-\cos(\sphericalangle(p_m)-\phi_1))} \tag{5.11}$$

It is easy to show that the multiplicand after $S(f)$ has values smaller than or equal to 1. This means that in the majority of the cases a sound source coming from a direction different from the listening direction will be suppressed to a certain degree.

In real practice it is much more convenient to use the center of the microphone array as a reference point; that is, under no-noise and no-reverberation conditions to have an output signal the same as the ideal microphone in the center of the array. Then our sound source estimator converts to the following:

$$Y(f) = \sum_{m=0}^{M-1} \left\{ \frac{F(f)\|p_m-c\|\exp\left[j\frac{2\pi f}{\nu}(\|p_m-c\|-\|p_c-c\|)\right]}{M\|p_c-c\|U_m(f,c)\cdot A(f)} \right\} X_m(f). \tag{5.12}$$

Here

$$p_c = \frac{1}{M}\sum_{m=0}^{M-1} p_m$$

is the microphone array center. In acoustics $2\pi f/\nu$ is referred to as the "wave number."

In Equation 5.12, $F(f)$ is the desired frequency response of the microphone array. Usually we want to have a zero response below a certain frequency f_{min} and above a certain frequency f_{max}. Zeroing the beginning is to remove DC components and low-frequency noise, usually from the power net. Removing the signals above a certain frequency is to mitigate the imperfect anti-aliasing filter that precedes the ADC. We want a flat frequency response in the middle of the frequency band between the frequencies f_{beg} and f_{end}. Usually for the transitions $f_{min} - f_{beg}$ and $f_{end} - f_{max}$ a smooth function like cosine is used. A typical analytic representation of $F(f)$ is

$$F(f) = \begin{vmatrix} 0 & f < f_{min} \\ \dfrac{1}{2}\left[1 + \cos\left(\dfrac{2\pi(f - f_{beg})}{(f_{beg} - f_{min})}\right)\right] & f_{min} \leq f \leq f_{beg} \\ 1 & f_{beg} < f < f_{end} \\ \dfrac{1}{2}\left[1 + \cos\left(\dfrac{2\pi(f - f_{end})}{(f_{max} - f_{end})}\right)\right] & f_{end} \leq f \leq f_{max} \\ 0 & f > f_{max}. \end{vmatrix} \quad (5.13)$$

For a sampling rate of 16 kHz, the typical values are $f_{min} = 100$ Hz, $f_{beg} = 200$ Hz, $f_{end} = 7000$ Hz, and $f_{max} = 7500$ Hz. The shape of the desired frequency response computed according to Equation 5.13 with these values is shown in Figure 5.12.

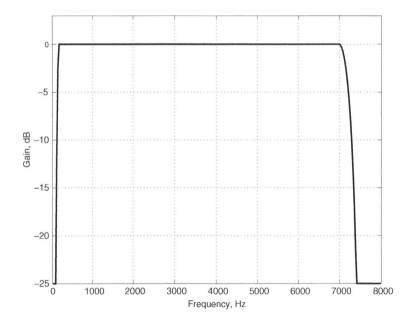

Figure 5.12 Desired frequency response

On substituting Equations 5.3 and 5.4 in 5.12 and ignoring the noise component, we obtain

$$Y(f) = F(f) \frac{1}{||p_c - c||} \exp\left[-j \frac{2\pi f}{\nu} ||p_c - c||\right] S(f) \qquad (5.14)$$

which is exactly what a perfect omnidirectional microphone will capture if placed at the center of the microphone array in the absence of noise and reverberation.

Equation 5.12 is precise and valid for near-field sound sources. For a far-field model $\frac{||p_m - c||}{||p_c - c||} \approx 1$ and can be ignored. The difference $||p_m - c|| - ||p_c - c||$ in the exponent can be replaced with the simpler $||p_m||\cos(\sphericalangle(p_m) - \phi_0)$.

In summary, the effect of this processing is that we will capture the sounds coming from the desired direction, while noises and sounds coming from other directions will be suppressed; that is, a *beam of sensitivity* will be formed. This most intuitive approach is called a "delay-and-sum beamformer."

The directivity pattern of a microphone array is a complex function showing the relationship between the microphone array output and a perfect capturing device in the center of the array as a function of the frequency and incident angle. Traditionally, complex numbers are represented as magnitude and phase. A magnitude close to 1 means perfect capture, close to 0 means perfect suppression. A phase close to 0 means no phase distortions. From this perspective, we want a magnitude of 1 and a phase of 0 for all frequencies in the desired frequency band towards the listening direction, with magnitudes close to 0 for all other directions. The directivity pattern of a delay-and-sum beamformer of the four-element linear microphone array from Figure 5.4 is shown in Figure 5.13 (magnitude in decibels, phase in degrees). The listening direction ϕ_0 is at 0 degrees. The magnitude is 1 and the phase is 0 for the working band at the listening direction, which means that we will capture the desired sound without any frequency distortions. The magnitude of the directivity pattern for angles outside the listening direction decreases (i.e., we have suppression) and then increases again owing to spatial aliasing above 3000 Hz due to the distance between the microphones. In addition it is clear that the directivity is lower for the frequencies in the lower part of the frequency band; that is, our microphone array counts only on the directivity of the unidirectional microphones to suppress sound sources coming from other directions. A better look at the directivity pattern magnitude is shown in Figure 5.14 for frequencies of 500 Hz and 5000 Hz. The symmetry due to the front–back ambiguity is visible, the suppression of signals coming from the back side is only due to the directional microphones. For 5000 Hz we have areas with increased sensitivity towards unwanted directions owing to spatial aliasing. These directions are called "side lobes."

EXERCISE

Look at the provided MATLAB script *WriteWeights.m* which writes to a file with a given format the sampling rate, the number of bins, the number of beams, the weights

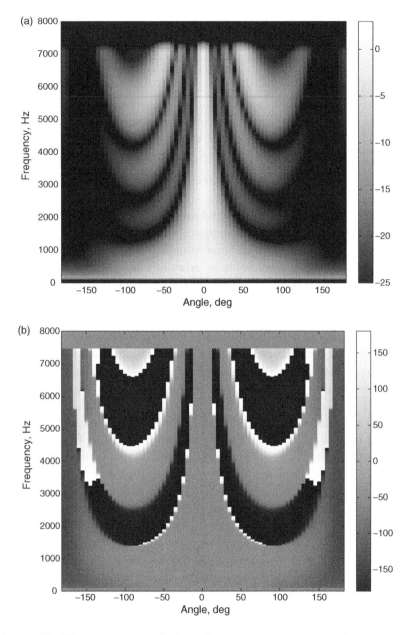

Figure 5.13 Directivity pattern of delay-and-sum beamformer: (a) magnitude; (b) phase

matrix, and the array with beam directions. Write a MATLAB program that reads the microphone array geometry from *MicArrDesc.dat* and computes and writes the weights using the delay-and-sum algorithm.

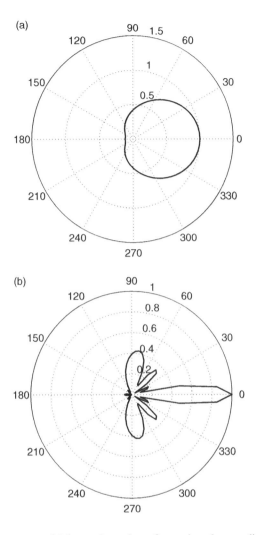

Figure 5.14   Directivity pattern of delay-and-sum beamformer in polar coordinates: (a) for 500 Hz; (b) for 5000 Hz

Compare the results with the provided *MASythesis.m*. Note that this script uses noise models and some scripts from Chapter 2 and microphone models and scripts from Chapter 3.

5.2.6 Generalized Filter-and-Sum Beamformer

Equation 5.12 can easily be generalized to

$$Y(f) = \sum_{m=0}^{M-1} W_m(f) \cdot X_m(f) \qquad (5.15)$$

and presented in matrix form as

$$\mathbf{Y}(f) = \mathbf{W}(f)\mathbf{X}(f). \tag{5.16}$$

The spectrum of each input audio frame $\mathbf{X}$ is an $M \times K$ complex matrix, where M is the number of microphones and K is the frame size; that is, the number of frequency bins. $\mathbf{X}(f)$ is a $M \times 1$ complex vector. The weights matrix $\mathbf{W}$ has a size of $K \times M$, and $\mathbf{W}(f)$ is the $1 \times M$ complex vector. The frame indices are omitted for simplicity. The equations above are known as a "filter-and-sum beamformer" because the weights matrix $\mathbf{W}(m)$ acts as a filter for each channel. That is, first we filter each of the channels separately and then we sum them. In the delay-and-sum beamformer we have the same magnitude for all filter coefficients and linear phase response, corresponding to a constant delay. The filter-and-sum beamformer has more degrees of freedom and allows us to design better microphone arrays. Practically all microphone-array processing algorithms for any given moment (i.e., frame) can be derived in the form of a filter-and-sum beamformer.

EXERCISE

Look at the provided MATLAB script *ReadWeights.m* which reads from a file with a given format the sampling rate, the number of bins, the number of beams, the weights matrix, and the beam directions array. Write a MATLAB program that reads the microphone array geometry from *MicArrDesc.dat*, the weights using the script above, and a multichannel WAV file, containing the recorded microphone channels. The script should process the input multichannel file using the weights and write the output as a WAV file.

Compare with the provided *MicArrProcess.m*, which uses the scripts for conversion to the frequency domain and back from Chapter 2.

Process the provided *Record1.WAV* file with the weights, designed in the previous exercise. Measure and compare the input and output SNRs using the MATLAB script from Chapter 3. Listen to the output.

5.3 Terminology and Parameter Definitions

5.3.1 Terminology

The process of computing the output signal as a linear combination of the signals from the microphones of the array, equivalent to the output of highly directional micro-phones, is called *beamforming*. It was discussed in the previous section. The name comes from the beam-like shape of the directivity pattern.

The listening direction is called the *main response axis* (MRA). This is the listening direction where usually the microphone array sensitivity is highest. It presumably points towards the sound source we want to capture. The sensitivity of the microphone array decreases with increasing incident angle between the sound source and MRA, suppressing sound sources and noises coming from other directions.

By changing the filter coefficients, we can change the listening direction without moving any of the microphones of the array. This process of electronically changing of the listening direction is called *beamsteering*. Most microphone-array geometries are capable of steering the beam in a certain range of angles.

We can make the microphone array completely suppress sounds coming from a given direction. Placing an explicit null in the microphone-array sensitivity is called *nullforming*. The process can be combined with beamforming; that is, we can have a microphone array listening to a given direction and completely suppressing the signals coming from another direction. This is very convenient when there are two sound sources: the one we want to capture and the one that is unwanted. This scheme will perform well in low reverberant conditions. In a normal room the nullformer will suppress only the direct path, but the microphone array will capture the portion of the reverberations of the unwanted sound source.

In the same manner as with beamsteering, by changing the filter coefficients we can electronically move the direction of the null sensitivity. This process is called *nullsteering*.

Based on the multiple signals captured from the microphones in different locations, the array can be used for *sound source localization* (SSL). This is the process of estimating the *direction of arrival* (DOA) of the sound source. One or multiple sound sources can be located and tracked, the sound source localization can be performed for the entire frequency band (i.e., per object), or even per frequency bin, which provides cues which help separate multiple simultaneously talking speakers. The main application of sound source localization is to know where to point the beam to capture the desired sound source, or to place a null for the unwanted sound source. The source localization results can be used for detecting and tracking several sound sources (participants in a meeting) as a powerful additional cue in sound source separation, and so on. The algorithms and approaches for sound source localization will be discussed in Chapter 6.

Chapter 4 covers the algorithms and approaches for single-channel noise suppression. The speech and noise signals are distinguished by their frequency and timing characteristics. Capturing sound from multiple positions allows additional dimensions. In single-channel noise suppression, the noise model was the vector of the noise variance as a function of the frequency $\lambda_d(f)$. Using the multiple channels, we can build a noise model thath implicitly or explicitly has direction $\lambda_d(f, c)$. This group of algorithms is called *multichannel noise suppressors*.

There are some operations that should be performed on each channel separately, before the beamformer, to combine them into one signal. These operations are generally called *microphone array pre-processors*. A typical example here is the compensation for the manufacturing tolerances, which may be different for each microphone.

Based on information obtained from the processing of all input channels, there is additional processing we can do on the single channel output of the beamformer. These algorithms are usually called *microphone array post-processors*. Multichannel noise suppression can be such a post-processor.

Spatial filtering is another post-processing approach. Based on the phase differences between the channels, we can estimate the direction of arrival for each frequency bin, which is itself a DOA estimation per bin, and apply a real gain to the beamformer output based on the desired or non-desired direction of arrival.

5.3.2 Directivity Pattern and Directivity Index

The directivity pattern of a microphone array is a complex function showing the relationship between the microphone array output and a perfect capturing device in the center of the array. It represents the spatial selectivity of the microphone array and is a function of the direction, elevation, and distance of the coordinates of the sound source. The directivity pattern of a microphone array is defined as

$$\mathbf{B}(f, c) = \mathbf{W}(f)\bar{\mathbf{D}}(f, c) \tag{5.17}$$

where $\mathbf{W}(f)$ is the weights vector and $\bar{\mathbf{D}}(f, c)$ is the capture vector, with each term derived from Equation 5.4 as

$$\bar{D}_m(f, c) = \frac{\|p_c - c\|}{\|p_m - c\|} \exp\left[-j\frac{2\pi f}{\nu}(\|p_c - c\| - \|p_m - c\|)\right] U_m(f, c). \tag{5.18}$$

Frequently it is presented only as a magnitude; that is, $|\mathbf{B}(f, c)|$ is used instead. In many cases it is completely acceptable to omit the distance and to use a far-field model. Then the directivity pattern becomes a function of the direction and elevation angles and can be presented as a three-dimensional plot like the one shown in Figure 5.15. This figure presents the 3D directivity pattern of the four-element microphone array for 5000 Hz. The distance between the center of the coordinate system and the surface is the microphone array sensitivity. The array listens to the direction perpendicular to its axis. In addition we have two side-lobes on each side of the main lobe with the shape of the parabolic cone of ambiguity discussed earlier in this chapter. When the microphone array has a symmetric directivity pattern, or we are interested in what happens in the main plane only, or we cannot change what happens outside the main plane (which is the linear array case), the directivity pattern can be presented as polar plots – see Figure 5.14. If we want to combine the directivity pattern for the entire frequency band, 2D plots (like the one in Figure 5.13a) can be used.

The directivity index (DI) is measured in decibels and represents how directive the microphone array is. It is computed according to the definition given in Section 3.4.5, Equation 3.21. Just in the microphone array case, $P(f, \varphi, \theta) = |B(f, c)|^2$, $\rho = \rho_0 =$ constant. The directivity index is a function of the frequency and is in general the noise suppression we will have in the case of isotropic ambient noise. The DI of the four-element microphone array we analyzed above as a function of the frequency is

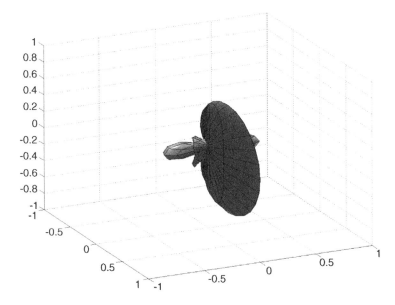

Figure 5.15   Directivity pattern of delay-and-sum beamformer for 5000 Hz in three dimensions

presented in Figure 5.16. In a similar manner we can compute the total DI using Equation 3.22. For this four-element array with a directional microphone and delay-and-sum beamformer, the average directivity index is $DI_{tot} = 10.21$ dB. It is much higher than a single ideal cardioid microphone with its DI_{tot} of 4.8 dB.

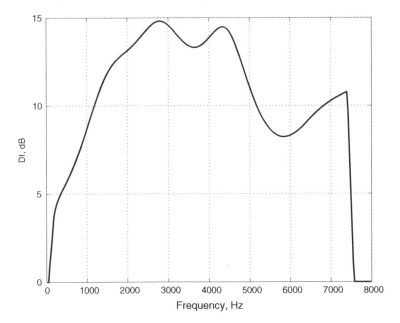

Figure 5.16   Directivity index of delay-and-sum beamformer as a function of frequency

5.3.3 Beam Width

The beam width is defined as the difference between the angles where the directivity pattern goes below $-3\,dB$ off the MRA. For most microphone arrays it varies as a function of the frequency. The directivity pattern of the previously discussed two-element microphone array for three frequencies is shown in Figure 5.17. The figure

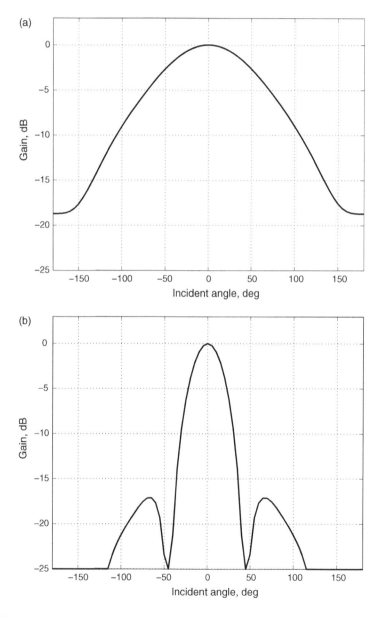

Figure 5.17 Directivity patterns in decibels for (a) 500 Hz, (b) 2000 Hz, and (c) 5000 Hz with various beam widths

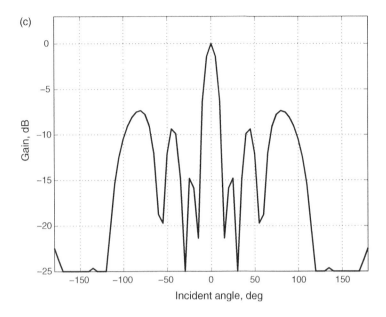

Figure 5.17 *(Continued)*

shows another way to represent the directivity – as a linear plot of the incident angle and the microphone array gain. The plots for 500 and 5000 Hz are equivalent to the polar plots in Figure 5.14. It is easy to measure the angles where the directivity goes below -3 dB for the three frequencies. The results are shown in Table 5.1. Note that there might be side-lobes reaching levels above -3 dB. Measuring the beam width is for the main lobe only.

Table 5.1   Beamwidth of two-element microphone array

Frequency, Hz	Beamwidth, deg
500	120
2000	50
5000	30

The main use of this parameter is to guarantee that we point the beam to the sound source with a precision better than the beam width, otherwise we will capture the desired sound source with certain frequency distortions.

5.3.4 Array Gain

The array gain is defined as the ratio between the beamformer output and the signal, captured by a single omnidirectional microphone, placed in the center of the micro-phone array, under conditions of no noise and no reverberation. The array gain is a function of the incident angle (the direction and elevation to the sound source) and of the frequency (we can have different array gains for different frequencies). The array gain is *de facto* the directivity pattern as it was defined in Equations 5.17 and 5.18, and shown in Figures 5.14 and 5.17.

5.3.5 Uncorrelated Noise Gain

The uncorrelated noise gain is the portion of the uncorrelated noise power we have on the beamformer output. The main source of uncorrelated noise is the microphones of the array. Assume that each of the microphones has a self noise with variance $\lambda_{ml}(f) = |N_{ml}(f)|^2$, where m is the microphone index and $|N_{ml}(f)|$ is the average magnitude. Then the uncorrelated noise gain is given by

$$G_{I}(f) = \frac{\sum\limits_{m=0}^{M-1} \left(|W(f,m)| \cdot |N_{ml}(f)|\right)^2}{\sum\limits_{m=0}^{M-1} \left(|N_{ml}(f)|\right)^2}. \tag{5.19}$$

If we have an array with identical microphones, the uncorrelated noise gain looks simpler:

$$G_{I}(f) = \sum\limits_{m=0}^{M-1} \left(|W(f,m)|\right)^2 \tag{5.20}$$

or in matrix form:

$$\mathbf{G_{I}}(f) = \mathbf{W}(f)\mathbf{W}^{\mathrm{T}}(f). \tag{5.21}$$

The uncorrelated noise gain is a function of the frequency and it is usually represented in decibels.

5.3.6 Ambient Noise Gain

The ambient noise gain represents the ratio of the isotropic (i.e., equality spread spatially) ambient noise power on the output of the beamformer, compared to what an omnidirectional microphone in the center of the microphone array would capture. The ambient noise gain $G_A(f)$ is simply the directivity index:

$$G_{A}(f) = \frac{|B(f, \varphi_{\mathrm{T}}, \theta_{\mathrm{T}})|^2}{\frac{1}{4\pi}\int\limits_{0}^{\pi} d\theta \int\limits_{0}^{2\pi} d\varphi \cdot |B(f, \varphi, \theta)|^2} \tag{5.22}$$

and it is function of the frequency. Here $(\varphi_{\mathrm{T}}, \theta_{\mathrm{T}})$ is the direction and elevation of the MRA. For this case we assume far-field conditions; that is, $\rho = \rho_0 = $ constant.

5.3.7 Total Noise Gain

The total noise gain is the ratio of the noise in the output of the beamformer and the noise captured by an omnidirectional microphone, placed in the center of the microphone array:

$$G(f) = \frac{G_A(f) \cdot |N_A(f)|^2 + G_I(f) \cdot |N_I(f)|^2}{|N_A(f)|^2 + |N_{oI}(f)|^2}. \tag{5.23}$$

Here we assume the magnitude spectrum of the isotropic ambient noise $N_A(f)$, self-noise spectrum of the microphones in the array $N_I(f)$, and self noise of the omnidirectional microphone $N_{oI}(f)$. For an ideal microphone, $N_{oI}(f) \equiv 0$, but for more realistic conditions it can be assumed equal to the quantization noise; that is:

$$N_{oI}(f) \equiv \frac{1}{12.2^n}$$

where n is the number of bits of the ADC (see Section 2.4.2 for more details). Of course $N_{oI}(f)$ can be replaced by a small number to prevent division by zero in Equation 5.23.

The total noise gain is one of the most important parameters of microphone arrays because in most cases they are designed and used for better sound capture – reducing the captured ambient noise. In some literature sources this parameter is referred to as "array gain."

EXERCISE

Using the provided MATLAB script *ReadWeights.m*, write a MATLAB program that reads the microphone-array geometry from *MicArrDesc.dat*, and the weights using the script above. Compute and plot the magnitude and phase responses; correlated, uncorrelated, and total noise gains; and directivity index as a function of the frequency.

Compare the results with the provided *MAAnalysis.m*. Note that this script uses noise models and some scripts from Chapter 2, plus microphone models and scripts from Chapter 3.

5.3.8 IDOA Space Definition

Microphone arrays can improve sound capture due to the fact that they capture the sound at multiple points, in most of the cases with *a priori* known positions. For the far-field sound capture model, this means that the sounds will be captured in slightly different moments, leading to phase differences between the signals from different microphones. These phase differences can be used for spatial selection and sound source localization. We can find the *instantaneous direction of arrival* (IDOA) for each frequency bin based on these phase differences. For each frequency bin, the IDOA space has the phase differences between the unique non-repetitive microphone pairs as coordinates; that is, 1–2, 1–3, 1–4 for a four-element array. For M microphones, these phase differences form an $(M-1)$-dimensional space, spanning all potential IDOAs. If we define an IDOA vector in this space as

$$\Delta(f) \triangleq [\delta_1(f), \delta_2(f), \ldots, \delta_{m-1}(f)] \qquad (5.24)$$

where

$$\delta_l(f) = \arg(X_0(f)) - \arg(X_l(f)) \quad l = \{1, 2, \ldots, M-1\} \qquad (5.25)$$

then we can see that the uncorrelated noise will be equally spread in this space, while the signal and ambient noise (correlated components) will lie in a hypervolume that represents all potential positions of a sound source in the real three-dimensional space. For far-field sound capture this is a $(M-1)$-dimensional hypersurface, as the distance is presumed constant or approaching infinity. Linear microphone arrays, a common case, can distinguish only one dimension – the incident angle – and the real space is represented by a $(M-1)$-dimensional hyperline.

Figure 5.18 shows the distribution of 1000 audio frames in the IDOA space for the frequency bin at 750 Hz for a four-element linear array. The solid line is a set of

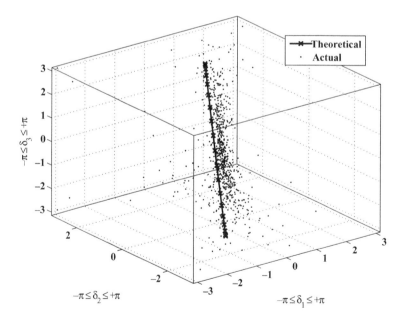

Figure 5.18   IDOA space for four-element microphone array

theoretical positions for sound sources in the range $-90°$ to $+90°$. The markers are placed every 10 degrees. The actual distribution is a cloud around the theoretical line owing to the presence of a uncorrelated component and reverberation.

Note that for each point in the real space we have a corresponding point in the IDOA space but it may not be unique. The converse is not true; there are points in the IDOA space without a corresponding point in the real space. A point there is either uncorrelated noise, or a sum from a real sound source and uncorrelated noise, or a sum of two (or more) real sound sources, or a sum of all three. This is an important feature – it gives us an estimation of the disturbance, caused by the uncorrelated noise, and important cues for the presence of multiple sound sources. From this perspective, the IDOA space is not quite so useful for two-element arrays as there is no redundant information. We have one phase difference; hence the IDOA space is one-dimensional. On the other hand, the two-element array works in one dimension – the direction of arrival. Linear arrays with three or more microphones, planar arrays with four or more microphones, and volumetric arrays with five or more microphones, utilize the IDOA space better.

We will use the IDOA space for spatial filtering later in this chapter and for sound source localization in the next chapter.

5.3.9 Beamformer Design Goal and Constraints

With M number of microphones, each with directivity pattern $U_m(f, c)$ and placed in a given position $p_m = (x_m, y_m, z_m)$, we want to capture a sound source at a known location c_T. Besides the prior knowledge of the microphone-array geometry (the microphone positions) and the desired sound source location c, we know the input from each microphone – the audio frame $X_m^{(n)}(f)$. For real-time implementations we do not have knowledge about the frames in the future, since we only know the current and previous audio frames.

Based on the information above, the goal of beamformer design is for each frame to find a weights matrix $\mathbf{W}^{(n)}$ optimal according to one criterion or another. Among the most commonly used criteria is minimizing the noise gain. This means that we want to capture the desired sound source with the highest possible SNR. As we cannot influence the desired sound source, the only thing we can do is minimize the noise in the beamformer output. Then the design goal is to find weights such as

$$\mathbf{W} = \arg\min_{\mathbf{W}}(\mathbf{G}). \tag{5.26}$$

Here $\mathbf{G}$ is the total noise gain vector, defined in Equation 5.23. One obvious solution is to have all the weights equal to zero, which obviously will completely suppress the ambient noise, but will also remove the desired sound source. To prevent our design process from going towards that undesired direction we should add design constraints. The most natural constraint is to require unit gain and zero phase shift towards the listening direction; that is, $B(f, \varphi_T, \theta_T) \equiv 1$ for $f \in [f_{min}, f_{max}]$. With this constraint we want to minimize the output noise keeping the captured signal from the desired sound source undistorted. This excludes the all-zeros solution.

5.4 Time-invariant Beamformers

In time-invariant beamformers the weight matrix is designed in advance for each listening direction and does not change until the microphone array listens to that direction. This means that we cannot use the input signals to design the beamformer and the only assumption we can make is about isotropic noise (same noise spectrum in every point of the space) and spherical noise (uncorrelated noise sources from all directions).

Time-invariant beamformers typically operate with a set of pre-computed beams, spread in the listening space. For example, if we have a linear four-element array that has to listen in the range between $-50°$ and $+50°$, we can compute off-line the weights for 11 beamformers, listening to every $10°$ in the listening space; that is, for $-50°$, $-40°$, $-30°$, and so on, to $+50°$. Then, in real time, the processing code uses these pre-computed weight matrices; switching the beam direction is just changing the table of coefficients. This approach allows the creation of fast real-time beamformers. However, there is an unrealized potential in using the microphone array for better suppression of point noise sources by creating nulls in real time. Later in this chapter we will discuss adaptive beamforming algorithms that can do that.

5.4.1 MVDR Beamformer

This is one of the first beamforming algorithms. The microphone arrays inherited it from the design of antenna arrays. MVDR stands for *minimum-variance distortionless response*. The idea for constrained optimal beamformer design was derived in the work of Frost [1] for use with antenna arrays. While in the original paper the optimal beamformer is derived for work in the time domain, here we will provide a derivation for the frequency domain, as with most algorithms in this book. The optimal weights derivation below follows the derivation of the MVDR beamformer in [2].

The input vector $\mathbf{X}(f)$ with size $M \times 1$, where M is the number of microphones, is formed as

$$\mathbf{X}(f) = \mathbf{D}_c(f)S_c(f) + \mathbf{N}(f). \tag{5.27}$$

Here $S_c(f)$ is the desired sound source at position c, $\mathbf{D}_c(f)$ is the capturing vector from Equation 5.4 for position c with size $M \times 1$, and $\mathbf{N}(f)$ is the noise vector. The frame index is omitted for simplicity. The beamformer output is

$$Y(f) = \mathbf{W}_c(f)\mathbf{X}(f) \tag{5.28}$$

where $\mathbf{W}_c(f)$ is the weights vector with size $1 \times M$. We want to capture the desired sound source undistorted; that is

$$Y(f) = S(f) \tag{5.29}$$

under no-noise conditions, which leads to

$$\mathbf{W}_c(f)\mathbf{D}_c(f) = 1. \tag{5.30}$$

In addition we want to find weights that minimize the noise variance in the beamformer output:

$$Y(f) = \mathbf{W}_c(f)\mathbf{D}_c(f)S(f) + \mathbf{W}_c(f)\mathbf{N}(f) = S(f) + \mathbf{W}_c(f)\mathbf{N}(f) = S(f) + Y_n(f) \tag{5.31}$$

which is

$$Q_{\text{constr}} = E\left[|Y_n|^2\right] = \mathbf{W}_c(f)\mathbf{\Phi}_{NN}(f)\mathbf{W}_c^{\mathrm{H}}(f). \tag{5.32}$$

Here Q_{constr} is the constrained optimization criterion, and $\mathbf{\Phi}_{NN}(f)$ is the noise cross-power spectral matrix:

$$\mathbf{\Phi}_{NN}(f) = \mathbf{N}(f)\mathbf{N}^{\mathrm{H}}(f) = \begin{pmatrix} \Phi_{11}(f) & \Phi_{12}(f) & \cdots & \Phi_{1M}(f) \\ \Phi_{21}(f) & \Phi_{22}(f) & \cdots & \Phi_{2M}(f) \\ \vdots & \vdots & \ddots & \vdots \\ \Phi_{M1}(f) & \Phi_{11}(f) & \cdots & \Phi_{MM}(f) \end{pmatrix} \tag{5.33}$$

where $\Phi_{ij}(f) = N_i(f)N_j^*(f)$, * denotes the complex conjugate, and $N^{\mathrm{H}}(f)$ denotes conjugate transpose. The constrained minimization problem

$$\mathbf{W}_c(f) = \underset{\mathbf{W}_c(f)}{\arg\min}\, \mathbf{W}_c(f)\mathbf{N}(f)\mathbf{W}_c^{\mathrm{H}}(f) \tag{5.34}$$
$$\text{subject to } \mathbf{W}_c(f)\mathbf{D}_c(f) = 1$$

can be solved analytically by using Lagrange multipliers. The modified, already non-constrained, optimization criterion is

$$Q = \mathbf{W}_c(f)\mathbf{N}(f)\mathbf{W}_c^{\mathrm{H}}(f) + \lambda(f)[\mathbf{W}_c(f)\mathbf{D}_c(f) - 1] + \lambda^*(f)[\mathbf{D}_c^{\mathrm{H}}(f)\mathbf{W}_c^{\mathrm{H}}(f) - 1]. \tag{5.35}$$

Taking the complex gradient of Q with respect to $\mathbf{W}$ and solving it gives the optimal weights:

$$\mathbf{W}_{c-\text{opt}}(f) = -\lambda(f)\mathbf{D}_c^{\mathrm{H}}(f)\mathbf{\Phi}_{NN}^{-1}(f) \tag{5.36}$$

which when compared with Equation 5.30 gives the following evaluation for $\lambda(f)$:

$$\lambda(f) = -[\mathbf{D}_c^{\mathrm{H}}(f)\mathbf{\Phi}_{NN}^{-1}(f)\mathbf{D}_c(f)]^{-1}. \tag{5.37}$$

Then the constrained optimal weights vector is

$$\mathbf{W}_{c-\mathrm{opt}}(f) = \frac{\mathbf{D}_c^{\mathrm{H}}(f)\mathbf{\Phi}_{NN}^{-1}(f)}{\mathbf{D}_c^{\mathrm{H}}(f)\mathbf{\Phi}_{NN}^{-1}(f)\mathbf{D}_c(f)}. \tag{5.38}$$

The mathematical derivation so far provided us with the ability to estimate the optimal weights given knowledge of the noise cross-power spectral matrix. As we design a time-invariant beamformer, our best assumption for the noise field is spatially homogeneous (the correlation function is not dependent on the absolute position of the sensors), and also spherically isotropic (uncorrelated noise sources from all directions). The original correlation for two omnidirectional microphones spaced at distance d_{ij} was published in 1955 by Cook *et al.* [3]. The cross-spectral density for an isotropic noise field is the average cross-spectral density for all spherical directions φ and θ:

$$\Phi_{ij}(f) = \frac{N_O(f)}{4\pi} \int_0^\pi \int_0^{2\pi} e^{-jkd_{ij}\cos\varphi}\sin(\varphi)\mathrm{d}\varphi\mathrm{d}\theta =$$

$$= \frac{N_O(f)\sin\left(\dfrac{2\pi f d_{ij}}{\nu}\right)}{\dfrac{2\pi f d_{ij}}{\nu}} = \tag{5.39}$$

$$= N_O(f)\mathrm{sinc}\left(\frac{2\pi f d_{ij}}{\nu}\right).$$

Here $N_o(f)$ is the noise spectrum captured by an omnidirectional microphone, and ν is the speed of sound. Under the assumption for isotropic noise it is the same at each location. Elko [4] generalized this further for unidirectional microphones with given directivity patterns $U_i(\varphi, \theta, f)$ and $U_j(\varphi, \theta, f)$:

$$\Phi_{ij}(f) = N_O(f)\frac{N_{ij}(f)}{\sqrt{D_i D_j}} \tag{5.40}$$

where

$$N_{ij}(f) = \int\limits_{0}^{\pi}\int\limits_{0}^{2\pi} \widehat{U}_i(\varphi,\theta,f)\widehat{U}_j^*(\varphi,\theta,f)e^{-j\frac{2\pi}{v}d_{ij}\cos\varphi} \sin\varphi\,\mathrm{d}\varphi\,\mathrm{d}\theta$$

$$D_i(f) = \int\limits_{0}^{\pi}\int\limits_{0}^{2\pi} \left|\widehat{U}_i(\varphi,\theta,f)\right|^2 \sin\varphi\,\mathrm{d}\varphi\,\mathrm{d}\theta \qquad (5.41)$$

$$D_j(f) = \int\limits_{0}^{\pi}\int\limits_{0}^{2\pi} \left|\widehat{U}_j(\varphi,\theta,f)\right|^2 \sin\varphi\,\mathrm{d}\varphi\,\mathrm{d}\theta.$$

Note that $D_i(f)$ and $D_j(f)$ are simply the inversed directivity indices of the microphones for this frequency. While $U_i(\varphi,\theta,f)$ and $U_j(\varphi,\theta,f)$ are the directivity patterns of the microphones with MRA pointing towards the x axis, $\widehat{U}_i(\varphi,\theta,f)$ and $\widehat{U}_j(\varphi,\theta,f)$ are the directivity patterns of the microphones as they are oriented in the microphone array.

At this point, in order to compute the optimal weights of the MVDR beamformer listening towards location c, all we have to know is the ambient noise spectrum $N_o(f)$, the directivity pattern of each microphone $U_m(\varphi, \theta, f)$, and their positions which are necessary to compute the capturing vector $\mathbf{D}_c(f)$.

5.4.2 More Realistic Design – Adding the Microphone Self Noise

Using the methodology described above to compute the weights that are optimal in this sense leads to high instrumental noise gain, especially in the lower part of the frequency band. This practically removes the effect of the suppressed ambient noise, which is replaced by the magnified self noise of the microphones. In certain designs, the output SNR can be even worse than the input SNR in each of the microphone channels. The instrumental noise is uncorrelated across the channels and shows only in the diagonal of the noise cross-power spectral matrix [5]:

$$\mathbf{\Phi}_{N'N'}(f) = \mathbf{\Phi}_{NN}(f) + \mathbf{\Phi}_{II}(f) \qquad (5.42)$$

where $\mathbf{\Phi}_{II}(f)$ is either $N_I^2(f)\mathbf{I}$ (if we assume the same instrumental noise for all microphones and $\mathbf{I}$ is the identity matrix), or it is a diagonal matrix with members $\Phi_{ii}(f) = N_{ii}^2(f)$. Here $N_{ii}(f)$ is the noise magnitude of the instrumental noise in the i-th microphone. Using the modified cross-power matrix gives lower directivity, but better performance of the microphone array, optimal in the total noise suppression sense.

It is interesting to see what happens if the cross-power spectral matrix is diagonal; that is, the noise in the microphone channels is not correlated. This can happen if the microphone array works under very low ambient noise conditions, where the instrumental noise of the microphones dominates. If we assume that all microphones have the same noise variance, we will have

$$\Phi_{N'N'}(f) = N_I^2(f)\mathbf{I}. \tag{5.43}$$

Substituting this in Equation 5.38 leads to

$$\mathbf{W}_{c-\text{opt}}(f) = \frac{1}{M\mathbf{D}_c(f)} \tag{5.44}$$

which is exactly the delay-and-sum beamformer from Equation 5.12 excluding the desired frequency response. This means that the delay-and-sum beamformer is an MVDR beamformer optimal for uncorrelated noise across the channels.

EXERCISE

Modify the microphone array weights synthesis program to design an MVDR beamformer using the equations above. Generate the weights for the four-element microphone arrays we used as an example.

Use the microphone array analysis program to visualize the parameters of the beamformer above. Compare the results with the previously designed delay-and-sum beamformer.

5.4.3 Other Criteria for Optimality

While the requirements for minimizing the output noise variance under the constraints for unit gain and zero phase shift of the desired signal sound reasonable, and the MVDR beamformer is one of the most frequently designed and used, there can be other criteria for optimality.

One potential beamformer is the *maximum-likelihood* (ML) estimator. If we add the assumption that $\mathbf{N}(f)$ is a circular complex Gaussian random vector, then the likelihood function is

$$l_R(f) = [\mathbf{X}^H(f) - S^*(f)\mathbf{D}^H(f)]\Phi_{NN}^{-1}(f)[\mathbf{X}(f) - S(f)\mathbf{D}(f)]. \tag{5.45}$$

Here, all constant multipliers and additives terms that do not depend on $S(f)$ have been dropped. Taking the complex gradient and solving it with respect to $S^*(f)$, setting the results to zero, and solving the resulting equation gives

$$\hat{S}(f)|_{\text{ML}} = \frac{\mathbf{D}_c^H(f)\Phi_{NN}^{-1}(f)}{\mathbf{D}_c^H(f)\Phi_{NN}^{-1}(f)\mathbf{D}_c(f)}\mathbf{X}(f) \tag{5.46}$$

which corresponds to the solution for MVDR beamformer weights. Without going into further detail, the conclusion is that the MVDR beamformer provides the ML estimate of the desired signal with known position.

Another potential criterion of optimality is maximizing the SNR of the output signal. Without deriving the weights optimal in this sense, it is logical that they will lead to the same MVDR beamformer, which keeps the desired signal constant and minimizes the noise. For more details, see [2].

In general for a wide class of criteria, the optimum processor is the MVDR beamformer. Later in this chapter we will talk about the minimum-power distortionless response (MPDR) and mean-square-error estimators.

5.4.4 Beam Pattern Synthesis

A typical problem when designing wideband beamformers is that the beamformer usually cannot cover the entire frequency band in an optimal way. With a reasonable number of microphones (two or four, maximum eight) the microphone array will not work well in the lower and upper parts of the frequency band. The weights design algorithm will provide the best possible results, but the overall efficiency will decrease. In addition, the beam width in the upper part of the frequency band tends to decrease; that is, beside the side-lobes we have a very narrow beam towards the listening direction. This increases the necessary precision with which we need to know where the desired sound source is. For time-invariant beamformers, the narrow beam in the upper part of the frequency band means that we have to pre-compute the beams in smaller intervals, for example every 5° instead of every 10°, which requires more memory to store the beam weights. Considering the fact that in the upper part of the frequency band the noise energy is usually very low, and that the reverberation decays much faster there, we can afford to design a suboptimal beam. Now we add to the design criteria requirements for the beam shape, aiming to keep the central lobe wider than a given constant.

5.4.4.1 Beam Pattern Synthesis with the Cosine Function

One of the most straightforward approaches is to do minimum mean-square-error pattern design. It can be done analytically, using the approach described above (derive the gradient and solve to estimate the weights), or numerically. The latter is easier to program and more robust to derivation errors.

Assume a discrete set of L directions $C = \{c_1, c_2, \ldots, c_L\}$, usually uniformly spread in the work listening area. To this set corresponds a set of propagation vectors, represented by an $M \times L$ matrix $\mathbf{D}(f)$, each member of which is the propagation vector between the Mth microphone and the Lth direction according to Equation 5.4. The direction set is usually lying on a full or half circle around the microphone array center. The radius of this circle is typically 1.5 m, which practically corresponds to far-field sound propagation. In special cases for near-field sound capture it can be different. For linear array geometry with omnidirectional microphones they are spread $\pm 90°$, as the linear array cannot distinguish front from back. For microphone arrays with planar geometry, or linear arrays with unidirectional microphones, the points should be spread full circle; for volumetric arrays they should cover a relatively even hemisphere or the entire sphere. Usually spreading the points every $5°$ body angle is sufficient. For this set of directions we define the values of the desired beam shape. Various functions are suitable. We can define one as a function of one parameter – the beam width δ:

$$\Delta(\phi) = \begin{vmatrix} 0 & \phi < -2\delta \\ 0.5 + 0.5\cos(\pi\phi/2\delta) & \phi \in [-2\delta, 2\delta] \\ 0 & \phi > 2\delta. \end{vmatrix} \qquad (5.47)$$

Here, ϕ is the body angle between the listening direction and the corresponding direction from the discrete set. Figure 5.19 shows the targeted beam shape for $\delta = 40°$. Given the desired beam shape Δ (vector of length L), and assuming $L \gg M$, we have an over-determined system of L equations with M unknowns (the weights vector $\mathbf{W}$):

$$\mathbf{WD} = \boldsymbol{\Delta}. \qquad (5.48)$$

On solving it for $\mathbf{W}$ we receive a solution that satisfies the MMSE criterion for the difference between the desired beam shape Δ and the actual beam shape $\hat{\boldsymbol{\Delta}} = \mathbf{WD}$. As

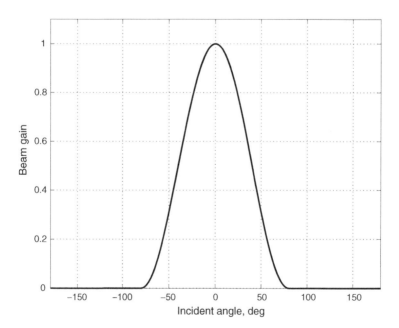

Figure 5.19 Target beam shape approximation with cosine function

the only criterion is to minimize the MMSE between desired and actual beam shapes, the estimated weights are not normalized towards the listening direction and they have to be scaled with the gain at the listening direction; that is

$$\mathbf{W}_{\text{beam}} = \frac{\hat{\mathbf{W}}}{\hat{A}(\phi_c)}.$$ (5.49)

No constraints for noise suppression are considered, but this technique allows us to control the beam shape width.

5.4.4.2 Beam Pattern Synthesis with Dolph–Chebyshev Polynomials

A more advanced technique for satisfying requirements to beam shape is described in [2]. It uses Dolph–Chebyshev polynomials for the target beam shape. The n-th degree Chebyshev polynomial is defined as

$$T_n(x) = \begin{vmatrix} \cos(n\cos^{-1}x) & |x| \leq 1 \\ \cosh(n\cosh^{-1}x) & x > 1 \\ (-1)^n\cosh(n\cosh^{-1}x) & x < -1. \end{vmatrix}$$ (5.50)

The target beam shape can be defined as

$$\Delta_l(\phi) = \frac{1}{R}T_M\left(x_0\cos\left(\frac{\phi}{2}\right)\right)$$ (5.51)

where R is the desired ratio between the main lobe maximum $T_{M-1}(x_0)$ and the maximum side-lobe level. Then x_0 can be computed as

$$x_0 = \cosh\left(\frac{1}{M}\cosh^{-1}R\right).$$ (5.52)

The targeted beam shape from Equation 5.51 will have the main lobe with a maximum of 1, and once the gain goes below $1/R$ it will be always be below that level. With this approach, we set the desired side-lobe maxima, but have no control over the main lobe width. Another design approach is to satisfy the requirement for the main lobe width at the -3 dB δ and first compute x_0:

$$x_0 = \frac{1}{\cos\delta}$$ (5.53)

and then the suppression ratio

$$R = \cosh\left(\frac{1}{M}\cosh^{-1}x_0\right).$$ (5.54)

In this case we will have the desired beam width, but no control over the side-lobe maxima.

Figure 5.20 shows the shape of the target beam shape, computed with Dolph–Chebyshev polynomials. The target beam width is set to $\delta = 40°$ and the maxima of the side lobes happen to be at $R = -20.5$ dB. The number of microphones is $M = 4$. Note that the vertical scale is in decibels when comparing this to Figure 5.19. The actual noise suppression of a beam with such shape is the integral of the beam shape curve.

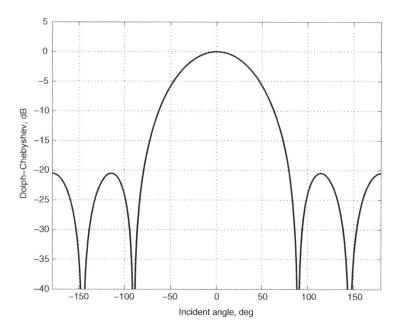

Figure 5.20 Target beam shape approximation with Dolph–Chebyshev polynomial

Figure 5.21 shows the noise suppression of a Dolph–Chebyshev beamformer as a function of the desired main lobe width. It has a well-defined minimum around 40°. Making the main lobe narrower increases the side-lobe levels; making it wider decreases the side-lobe levels further, but the main lobe is too wide and captures a lot of ambient noise.

Once we have the desired beam shape, computed using the Dolph–Chebyshev polynomial, we can use the MMSE approach above to compute the weights, or to compute them directly. The desired target beam has exactly M zeros at

$$\phi_{0p} = \frac{(2p-1)}{M}\pi, p = 1, 2, \ldots, M. \tag{5.55}$$

Using the zeros and the propagation vector $\mathbf{D}(f)$ for these directions allows direct computation of the weights.

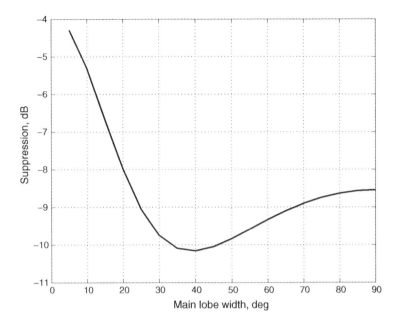

Figure 5.21   Noise suppression as a function of the desired beam width of the Dolph–Chebyshev polynomial

The Dolph–Chebyshev beamformer was initially used to design antenna arrays. It works at its best for linear, equidistant arrays with omnidirectional elements, and a relatively larger number of elements. As with the previous beam pattern synthesis algorithm, this design does not account for the instrumental noise component.

5.4.4.3 Practical Use of Beam Pattern Synthesis

The practical application of beam pattern synthesis is to limit the beam width. With increasing frequency, both delay-and-sum and MVDR beamformers tend to become very narrow. Above a certain frequency it is desirable to switch from these two algorithms to one of the beam pattern synthesis algorithms above to guarantee minimal width of the beam. The solution will be suboptimal from the noise-suppression perspective, but in the upper part of the frequency band there is not much noise and reverberation anyway. On the other hand, reducing the number of pre-calculated beams saves memory for storing the beamformer coefficients.

5.4.5 Beam Width Optimization

The beam pattern synthesis approach, described in the previous subsection, can be extended for the entire bandwidth. Of course, in the lower part of the frequency band we will not be able to achieve a narrow beam. In addition, an even narrower beam shape will come at the cost of high instrumental noise gain owing to the high magnitude of the computed weights. In the end, what we want from the microphone array is the best noise suppression with unit gain and zero phase shift towards the capturing direction, which is exactly the MVDR criterion.

Varying the desired beam width δ in Equation 5.47 with consequent beam pattern syntheses, we can note two controversial trends. When the beam is wider the microphone array picks up more ambient noise; a narrower beam means less ambient noise in the beamformer output. On the other hand, achieving a narrower beam comes at the cost of increased instrumental noise gain; that is, more instrumental noise will be in the beamformer output. Widening the beam usually reduces the instrumental noise gain. With these two controversial trends for a given ambient and instrumental noise spectra, there exists an optimal beam width that provides minimal noise level in the beamformer output. This allows us to compute the beamformer weights using single-dimensional optimization for each frequency bin:

1. For a given beam width δ, compute the desired beam shape according to Equation 5.47 or 5.51.
2. Then compute the beamformer weights giving an MMSE solution for this directivity pattern according to Equation 5.48: $\mathbf{W} = \mathbf{\Delta D}^{-1}$.
3. Normalize the weights according to Equation 5.49.
4. Compute the total noise gain of the solution according to Equation 5.23.
5. Repeat the previous steps, varying the beam width δ to minimize the noise gain using some of the one-dimensional optimization algorithms. Any single-dimensional algorithm for one-dimensional optimization can be used. See [6] for some of them.

Figure 5.22 shows the noise suppression as a function of the beam width and frequency for the target beam shape using Equation 5.47. The optimal beam width is marked with a black line. In Figure 5.23, the directivity pattern of this type of beamformer is shown. Comparing it with the directivity pattern of the delay-and-sum beamformer, shown in Figure 5.13a, it is easy to see that the beam is substantially narrower in the lower part of the frequency band where most of the noise energy is.

This method for the time-invariant beamformer design has the same optimization goal as the MVDR beamformer. Owing to the indirect optimization, theoretically it finds a suboptimal solution, but in practice it is as good as the MVDR beamformer. The advantage is that it is simpler to implement and debug, and thus more robust to programming errors. It is not as computationally expensive as it uses computationally efficient single-dimensional optimization. In addition, the computational performance can be optimized further. A fully blown one-dimensional optimization can be done only for the first frequency bin. For the next bins, single-dimensional gradient descent with a variable step and starting point of the optimal beam width from the previous bin can be used. Assuming correctly that the next bin will have an optimal beam width close to the previous, we can say that the optimal solution will be found with few estimations of the optimization criterion. This beamformer design approach is described in detail in [7].

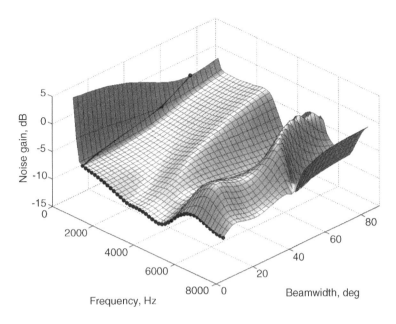

Figure 5.22 Noise gain as a function of the target beam width and frequency. The black line with the dots shows the optimal beam width

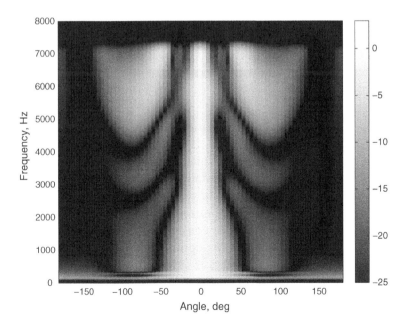

Figure 5.23 Directivity pattern of a beamformer designed using beam pattern synthesis

5.4.6 Beamformer with Direct Optimization

In real design practice, the requirements are more than the simple MVDR criterion we minimized and derived the weights estimation formula for in Equation 5.38. The algorithm from the previous subsection can be extended to a fully blown optimization of the weights with an optimization criterion that represents the actual requirements. In general we want to minimize the noise gain, so the optimal weights can be found using constrained optimization:

$$
\mathbf{W}_c(f) = \arg\min_{\mathbf{W}_c(f)} \mathbf{W}_c(f)\mathbf{N}(f)\mathbf{W}_c^{\mathrm{H}}(f)
$$

$$
\text{subject to} \left| \begin{array}{c} \mathbf{W}_c(f)\mathbf{D}_c(f) = 1 \\ \delta > \delta_{\min} \\ \text{other requirements.} \end{array} \right. \tag{5.56}
$$

Equation 5.56 is just a generalization of Equations 5.26 and 5.34 by adding more constraints.

Where we do not have analytical expressions for some of the constraints (beam width at -3 dB, for example), or they are very complex, we cannot solve Equation 5.56 analytically. This means using a numerical solution with some of the algorithms for multidimensional mathematical optimization.

The optimization criterion should be converted to non-constrained by using penalty functions:

$$
\mathbf{W}_{nc} = \arg\min_{W_c} \left[\mathbf{W}_c\mathbf{N}\mathbf{W}_c^{\mathrm{H}} + c_{\mathrm{G}}|\mathbf{W}_c\mathbf{D}_c - 1|^2 + c_\delta|\min(0, \delta - \delta_{\min})|^2 + \sum_C c_i P_i \right]. \tag{5.57}
$$

Frequency dependency is omitted for clarity, but the optimization still has to be performed for each frequency bin. Here, C is the number of additional constraints, c_i and P_i are the weight and the penalty function for the corresponding constraint, respectively. Note that the two known constraints for unit gain and minimal beam width are weighted with c_{G} and c_δ respectively. In general, the constraint weight normalizes the different magnitudes of the main optimization criterion and the penalty function and sets how much we want this particular constraint to be considered compared to the others. In many cases it is normal for the optimization procedure to find a solution slightly outside the permitted values.

The optimization parameters are actually twice the number of microphones. Most of the optimization algorithms assume real-valued optimization parameters. This means that each microphone weight provides two parameters for optimization – the real and the imaginary parts.

As an optimization starting point, we can use the weights computed using the delay-and-sum algorithm, or the weights computed by solving Equation 5.38, or even by setting a weight of 1 to the closest microphone and zeros to the others. A good approach is to use the optimal weights computed from the optimization in the previous frequency bin. In general, the closer the starting point is to the optimal point, the faster the optimization algorithm will find the optimal solution.

For the optimization algorithm, most of the well-known types [6] can be used, such as gradient descent with variable step, steepest gradient descent, simplex method (low efficiency, but it works), and some of the more sophisticated optimization algorithms included in packages like MATLAB (look at `fminunc`).

Potential drawbacks are the same as with any multidimensional optimization: an inefficient optimization algorithm owing to the specifics of the optimization criterion, and finding a local minimum instead of the global one. This can be mitigated by adding constraints and corresponding penalty functions that lead to the desired global optimum, verified by starting the optimization from different points.

Figure 5.24a shows the directivity pattern of a beam pointing at zero degrees for the four-element microphone array from Figure 5.4. Part (b) shows the noise gains (instrumental, ambient, and total). Part (c) shows the directivity index of the designed microphone array as a function of the frequency, compared to the directivity index of the delay-and-sum beamformer. The directivity index of the optimal design is substantially higher up to 3000 Hz, providing better noise suppression. This is at the price of increased instrumental noise, as is visible in Figure 5.24b.

Table 5.2 compares the directivity indices and weighted noise suppression of a single unidirectional microphone, the delay-and-sum beamformer, the two algorithms based on beam pattern synthesis, and the optimal beamformer. While the directivity indices of the four design approaches are practically the same, the weighted noise gain of the delay-and-sum beamformer is almost 3 dBA below the noise gain of the interpolation and optimized beamformers. The beamformer designed with beam pattern synthesis and interpolation with the cosine () function is very close to the optimized beamformer.

In general, the proposed algorithm for computing the beamformer weights by direct optimization gives more freedom in microphone array design than any other approach. The price for this is longer computation time for the beamformer design. Considering the fact that the design is offline, this is not a severe limitation. We will see additional reasons for beamformer design using this approach later in this chapter.

EXERCISE

Modify the microphone-array weights synthesis program to design a beamformer using direct optimization.

Use the microphone-array analysis program to visualize the parameters of the beamformer above. Compare the results with the previously designed MVDR and delay-and-sum beamformers.

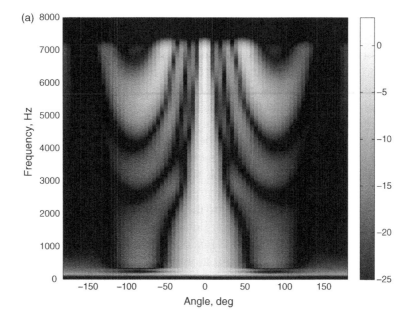

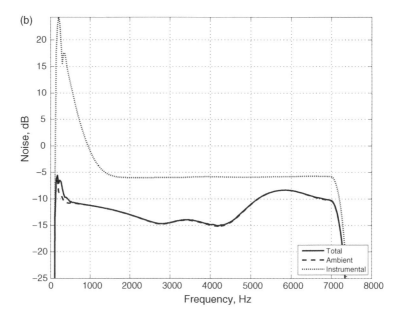

Figure 5.24 Directivity pattern (a), noise gains (b), and directivity index (c) of a beamformer designed with direct optimization

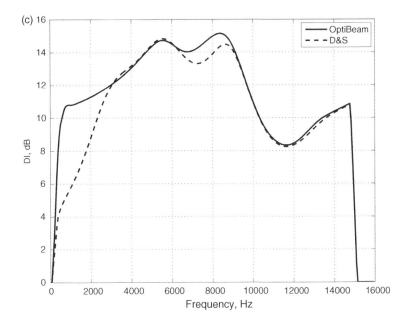

Figure 5.24 (*Continued*)

Table 5.2 Comparison between various techniques for beamformer design

Parameter	Unidir Mic	Delay-and-sum	Interpolation cos()	Interpolation Dolph-Chebyshev	Optimal Beam
DI total, dB	4.34	10.21	10.67	9.75	10.90
Total noise gain, dBC	−9.19	−9.59	−11.27	−10.84	−11.53
Weighted noise gain, dBA	−4.96	−6.79	−9.56	−9.45	−9.63
Weighted ambient noise gain, dBA	−4.96	−6.80	−10.45	−10.60	−10.39
Weighted instrumental noise gain, dBA	−0.37	−6.37	14.78	15.84	14.02

5.5 Channel Mismatch and Handling

5.5.1 Reasons for Channel Mismatch

The beamformer design algorithms and procedures so far have assumed perfect channel matching – identical microphones, preamplifiers, and ADCs. These algorithms rely on small differences in the magnitudes and the time of the arrival, measured as the phase difference between the signals in different channels. Many microphone arrays used for beamforming or sound source localization do not provide the estimated shape of the beam, estimated noise suppression, or estimated precision for sound source localization. One of the reasons for this is the difference in the signal

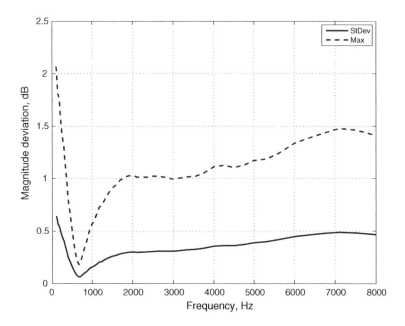

Figure 5.25 Microphone manufacturing tolerances in magnitude

paths – that is, impulse responses of the systems of microphone–preamplifier–ADC for different channels.

The source of these differences is mostly the manufacturing tolerances of the microphones, even those of the same type. Manufacturing tolerances affect the sensitivity, magnitude, and phase–frequency responses towards MRA, the overall shape of the directivity pattern. The typical sensitivity for widely used electret microphones is 55 ± 4 dB, where 0 dB is 1 Pa/V. This means that in one microphone array we can have microphones with a difference in sensitivity up to 8 dB.

Figure 5.25 shows the standard deviation and the maximum deviation of the sensitivities of 30 cardioid electret microphones from the same model and manufacturer. The deviations in the sensitivity are frequency-dependent. The manufacturing tolerances affect the phase response as well. Figure 5.26 shows the frequency-dependent standard deviation and maximum deviation in the phase response for the same 30 microphones. There can be 10 or more degrees of deviation in the phase response due to manufacturing tolerances.

Tolerances of the elements of a preamplifier introduce gain and phase errors as well. In addition, microphones and preamplifier parameters depend on external factors such as temperature, atmospheric pressure, power supply, and so on. In general these have less impact – we can use capacitors and resistors with tolerances below 1%, and temperature and humidity have even less impact.

An additional disturbance to the expected directivity pattern, magnitude, and phase response is the array's enclosure. The design algorithms described so far assume

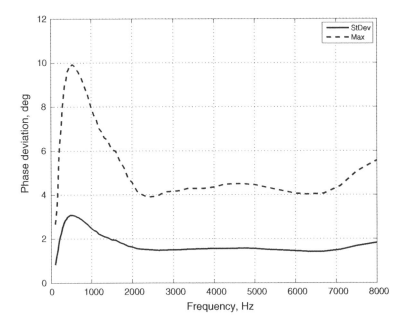

Figure 5.26 Microphone manufacturing tolerances in phase

microphones hanging in the air, or in acoustically completely transparent enclosures. The real enclosure changes the directivity pattern, as well as having its own manufacturing tolerances that affect the microphone positions. In real microphone arrays, higher differences in the overall directivity patterns than those in the previous two figures can be expected, where the microphones were measured hanging in the air in an anechoic chamber.

5.5.2 How Manufacturing Tolerances Affect the Beamformer

We can model the manufacturing tolerances as a random, frequency-dependent variable added to the microphone average directivity pattern $\bar{U}(f, c)$:

$$U(f, c) = \bar{U}(f, c)M(f)\exp(-j\varphi(f)). \tag{5.58}$$

Here, $M(f)$ represents the effects of the variation in sensitivity, and $\varphi(f)$ of the phase. The first can be modeled with a cut to a $\pm 2.5\sigma$ random Gaussian variable with mean 1 and standard deviation $\sigma_M(f)$. We assume that the quality control in the microphone manufacturing plant rejected all microphones with substantial differences in their parameters. In the same way we can model the phase tolerances as a zero-mean random Gaussian process with standard deviation $\sigma_\varphi(f)$. We can assume that both are known or measured.

To evaluate how the manufacturing tolerances affect the beamformer parameters, we shall use the same four-element linear microphone array as an example. Assume that the

Table 5.3 Sensitivity results for magnitude deviation (in dB): noise gain and DI in dB

StDev	NG	NG Min	NG Max	DI	DI Min	DI Max
0.00	−11.55	−11.55	−11.55	9.28	9.28	9.28
0.06	−11.54	−11.65	−11.37	9.28	9.27	9.28
0.13	−11.51	−11.84	−11.22	9.28	9.27	9.28
0.25	−11.49	−12.25	−10.91	9.27	9.23	9.28
0.50	−11.27	−12.98	−9.84	9.25	9.12	9.28
1.00	−10.79	−14.18	−7.59	9.18	8.78	9.28
2.00	−9.45	−14.65	−3.76	9.01	7.99	9.28
4.00	−6.55	−14.71	3.83	8.43	6.40	9.27
8.00	−0.08	−13.67	14.04	7.14	3.91	9.27

four cardioid electret microphones have manufacturing tolerances $\sigma_M(f)$ and $\sigma_\varphi(f)$ shown in the previous two figures. The beamformer weights, computed using the beam pattern approach with optimization of the beam width as described in [7], provide near-optimal noise suppression. Evaluation parameters were used in the total noise gain from Equation 5.23 and the beam directivity index – Equation 3.21 and Section 5.3.2. The effects of variations in the magnitude and phase were evaluated separately.

For a set of standard deviations of the magnitude, we generated 100 random instances of the four-element microphone array, evaluated them for noise gain and DI, averaged, and found the minimum and maximum values. The results are shown in Table 5.3 and the noise gains are graphically represented in Figure 5.27. The same

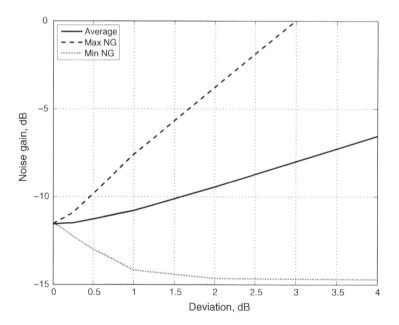

Figure 5.27 Effect of magnitude tolerances on the noise gain

Table 5.4 Sensitivity results for phase deviation (in degrees): noise gain and DI in dB

StDev	NG	NGMin	NGMax	DI	DIMin	DIMax
0.00	−11.55	−11.55	−11.55	9.28	9.28	9.28
5.00	−11.56	−12.49	−10.30	9.27	9.25	9.28
10.00	−11.25	−12.68	−8.11	9.25	9.20	9.28
15.00	−10.97	−12.61	−6.69	9.24	9.17	9.28
20.00	−10.76	−12.68	−5.69	9.23	9.11	9.28
25.00	−10.44	−12.73	−4.80	9.21	9.06	9.26
30.00	−9.91	−12.67	−4.37	9.19	9.09	9.27

approach was used to evaluate the effect of the phase variations; the results are shown in Table 5.4 and Figure 5.28. The evaluation results show higher sensitivity of the beamformer parameters to variations in magnitude than to variations in phase. This is somewhat counterintuitive, as the beamformer exploits the fact that the sound reaches the microphones with different delays – that is, with certain phase differences. The explanation is that we explored a range of variations for typical manufacturing tolerances. It is narrower for the phase response.

If we assume a level of acceptable degradation of the noise gain to be 3 dB in the worst case, for this particular microphone array we have to have channels matching with standard deviations better than 0.5 dB in magnitude and 10 degrees in phase. These requirements exceed what the industry provides for electret microphones in mass production. This means that, for guaranteed microphone-array parameters, channel

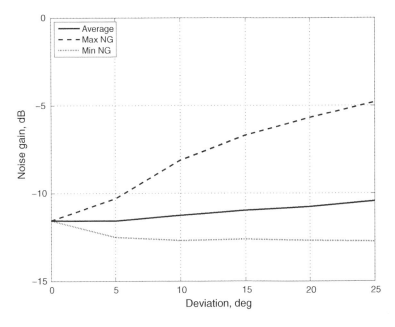

Figure 5.28   Effect of phase tolerances on the noise gain

matching should be ensured by selecting sets of microphones with similar parameters, or by using a calibration procedure. More details from this study can be found in [8].

Another approach is to have a beamformer design robust to manufacturing tolerances. For building good-quality microphone arrays in large series, most probably a combination of the approaches described above should be employed.

5.5.3 Calibration and Self-calibration Algorithms

5.5.3.1 Classification of Calibration Algorithms

Difficulties with the calibration of microphones and microphone arrays are well known. It can be an expensive and difficult task, particularly for broadband arrays. There are several groups of approaches to calibrate the microphones in the array.

Calibration of each microphone channel can be done separately by comparison with a reference microphone in a specialized environment: acoustic tube, standing-wave tube, or anechoic sound chamber [9]. This can be done pre-design and post-design. The pre-design calibration includes measurement of the directivity pattern of each microphone channel and using it in the beamformer design. This approach is very expensive as it requires manual calibration and design for each microphone and specialized equipment, but provides the best results. Another approach is to use an averaged directivity pattern for the design and to measure and compensate for the differences in the microphone responses only. The obtained impulse responses for each microphone channel towards the MRA can be used to design individual filters (i.e., gain and phase shift versus frequency), which need to be applied to each channel before the beamformer. While less expensive, this approach requires individual calibration of each device after manufacture.

The next group of calibration methods is post-installation. It employs calibration signals (speech, sinusoidal, white noise, acoustic pulses, swiping sinusoidal) sent from the speaker(s) or other source with a known location or from a live speaker [10]. In [11], far-field white noise is used to calibrate a microphone array with two microphones; the filter parameters are calculated using an NLMS algorithm. Other works use optimization methods to find the microphone array parameters. The algorithms in this group require manual calibration after installation of the microphone array and specialized equipment to generate test sounds. The advantages are that the calibration of the array does not consume CPU time during normal exploitation, and it can be combined with the calibration of other parts of the whole audio system. Calibration results are used for compensation in real time. They do not reflect changes in the equipment during exploitation time due to temperature or pressure.

The next group of algorithms apply self-calibration. The general approach is described in [2]: find the direction of arrival (DOA) of a sound source assuming that the microphone array parameters are correct, use a DOA to estimate the microphone array parameters, and repeat the iteration until the estimates converge. Various papers discuss the estimation of many parameters of the microphone array: sensor positions, gains, phase shifts. Different techniques are used, from simple normalized mean-square-error minimization to complex matrix methods and high-order statistical parameter estimation [11]. The main downside of these algorithms is that they require a certain amount of CPU time during normal working of the microphone array. Many of them are not suitable for practical real-time implementation.

In many cases the calibration algorithms are just a measurement of the directivity pattern or frequency response with well-studied techniques. Self-calibration algorithms are more interesting and attractive for arrays manufactured in large quantities. We will present two self-calibration algorithms that are computationally inexpensive and suitable for integration in real-time systems.

5.5.3.2 Gain Self-calibration Algorithms

One of the simplest, but quite effective, approaches is to compute the compensation gains during silent frames. The assumption of a compact array with omnidirectional microphones and far-field sound propagation means that the sound level from all microphone channels should be approximately the same. Then for each audio frame we can compute the signal level as RMS in each channel, $L_m^{(n)}$, and to average it to obtain $\bar{L}^{(n)}$. Then the individual gain for each channel should be

$$G_m = \frac{\bar{L}^{(n)}}{L_m^{(n)}} \tag{5.59}$$

which should be smoothed to obtain the individual gains:

$$G_m^{(n)} = \left(1 - \frac{T}{\tau_G}\right) G_m^{(n-1)} + \frac{T}{\tau_G} G_m. \tag{5.60}$$

Here, T is the frame duration and τ_G is the adaptation time constant. The latter can be chosen quite high as we do not expect the microphones' sensitivities to change rapidly. Then the individual gains are applied to each microphone channel before the beamformer. Here the microphone variance model assumes that the directivity patterns are the same and only the sensitivity changes. It is trivial to implement this algorithm using several (5–10) frequency bands, logarithmically spread in the working band, to reflect eventual differences in the frequency responses of the individual microphones.

The major disadvantage of this algorithm is the assumption of omnidirectional microphones. While for a linear microphone array with unidirectional microphones pointing in the same direction the assumption of the same signal level still partially holds, for circular microphone arrays with unidirectional microphones pointing

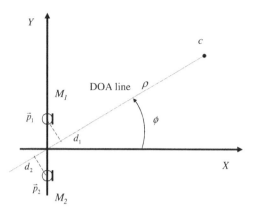

Figure 5.29   Projecting the microphone positions on the DOA line

outwards this is may not be not true. To mitigate these issues we can use a binary voice activity detector (VAD) and do estimation of the gains in the silent frames only, keeping the assumption of isotropic ambient noise. In a noisy environment this can work well, but in a quieter environment (office, conference room) the instrumental noise will reduce the precision, as it will be a substantial part of the signal level. Going in the opposite direction and estimating the gains when the external signal is dominant would be a better solution.

Most microphone arrays have an integrated DOA estimator – we have to know where the sound source is to point the beam towards it. Then we can estimate the gains only for frames where the DOA estimator gives direction $\{\varphi_{\mathrm{DOA}}, \theta_{\mathrm{DOA}}\}$ above a certain confidence level. With a known DOA for the sound source, we can project the microphone positions on the DOA direction and interpolate the signal levels with a straight line. The first part is shown in Figure 5.29 for a two-element array in two dimensions. The DOA line goes through the coordinate system center, which is the origin of the single-line coordinate system on the DOA line as well. A microphone with coordinates $c_m = \{\rho_m, \varphi_m, \theta_m\}$ in the radial coordinate system can be projected at coordinate:

$$d_m = \rho_m \cos(\varphi_{\mathrm{DOA}} - \varphi_m)\cos(\theta_{\mathrm{DOA}} - \theta_m). \tag{5.61}$$

The level interpolation with a straight line, as shown in Figure 5.30 for a four-element array, models the effects of signal attenuation arising from the different distances (i.e., we go beyond the far-field assumption), and microphone directivity for a cardioid and below (i.e., no longer assuming omnidirectional microphones). A requirement is symmetric geometry of the array and orientation of the microphones. This is valid for linear, circular, and spherical arrays, which are among the most common geometries

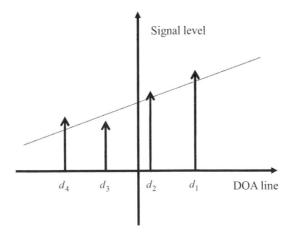

Figure 5.30   Interpolation of the signal levels with a straight line

used in practice. It is valid for planar microphones facing the major direction to the sound source with directional microphones pointing in that direction. Assume that the expected level can be interpolated as a straight line towards the direction of arrival:

$$\tilde{L}^{(n)}(d) = a_1^{(n)}d + a_0^{(n)}. \tag{5.62}$$

Then we can find parameters a_1 and a_0 to satisfy the criterion

$$\min\left(\sum_{m=0}^{M-1}(\tilde{L}(d_m)-L_m)^2\right). \tag{5.63}$$

For two microphones we can obtain an exact solution. For three or more, this leads to an overloaded system of equations, which can be solved to find the MMSE solution. Once the parameters a_1 and a_0 are estimated for the current frame, we can compute the expected levels at the microphone projection points using Equation 5.62 and then compute the channel gains following the procedure in Equations 5.59 and 5.60. The errors, introduced by the perpendicular projection of the microphones to the DOA and linear interpolation in the signal decay, instead of the inverse distance law, are minimal. For reasonable working distances they are below 0.4% each.

   This algorithm can be implemented for several sub-bands as well. It can be combined with in-factory calibration for better initial values of the compensation gains. During normal working of the microphone array, the gain calibration procedure has to track changes in the microphone gains only due to differences in the temperature, pressure, humidity, and aging. All of these changes are slow, so the self-calibration procedure will have enough time to average and track the gains. In Section 5.5.2 it was stated that, in order to keep the quality degradation below a certain threshold, we have to have a standard deviation of the gains below 0.5 dB. This self-calibration procedure can achieve the required precision for gain calibration. More details for this algorithm can be found in [12].

5.5.3.3 Phase Self-calibration Algorithm

As was mentioned earlier, the normal phase tolerances for most microphones on the market have a lower impact on the degradation of beamformer parameters. If there is a suspicion that the microphones phase tolerances will harm sound capture quality and the beamformer parameters, one potential solution is to apply a phase self-calibration procedure. This is based on the IDOA space, described in Section 5.3.8.

The basic idea is that in the IDOA space we have a set of possible values – the image of the real three-dimensional space in the $(M - 1)$-dimensional IDOA space. For far-field propagation it is an $(M - 1)$ hyperline, as shown in the earlier figure. Then for each frame and frequency bin we can compute the point in the IDOA space. Theoretically this point should lie on the hyperline. It can be shifted from there for several reasons: uncorrelated (instrumental) noise, multiple coherent sound sources (reverberation, for example), and phase mismatch. Assuming the first two average to the hyperline, the phase mismatch will generate a constant shift that can be measured and compensated for. Here is the proposed algorithm for each frequency bin and frames marked as "signal" by the VAD:

- Compute the IDOA coordinates for this frequency bin for the current frame.
- Find the closest point on the $(M - 1)$-dimensional hyperline, which marks the instantaneous direction of arrival for this frequency bin and frame.
- Compute the differences in phase for the microphones from 1 to $M - 1$, which will bring the current frame point on the hyperline.
- Average the estimated corrections with the current values using a relatively large time constant.

This procedure is not very expensive computationally and compensates for the differences between the first microphone, which is used as a reference, and the rest of the microphones in the array.

5.5.3.4 Self-calibration Algorithms – Practical Use

Note that both magnitude and phase self-calibration algorithms require persistent memory for the autocalibration correction filters. The next time the microphone array is turned on, the calibration starts from the point at whch it stopped the last time the array was turned off. This allows use of larger time constants and slows down the adaptation speed to achieve a better estimation for the gain and phase corrections. Another advantage of remembering the correction filters is that the self-calibration procedure can be run in the background, saving some CPU time.

Additional improvements can be applied to these basic algorithms. The speech presence probability from the VAD for the corresponding frequency bin can be used for to weight the adaptation speed (higher probability – more weight of the current correction). To save on additional CPU time, a threshold can be applied – for example, do not process frames with speech presence probability under a certain threshold.

5.5.4 Designs Robust to Manufacturing Tolerances

Another way to handle microphone manufacturing tolerances and the consequent channel mismatching is to design the beamformer weights to be robust to these tolerances. Frequently with a minimal decrease in the beamformer performance a much higher robustness can be achieved.

5.5.4.1 Tolerances as Uncorrelated Noise

The manufacturing tolerances can be represented as an additive random variable, which is a different form of Equation 5.58:

$$U(f,c) = \bar{U}(f,c) + \mathbb{N}(0,\sigma^2(f)) \tag{5.64}$$

where $\bar{U}(f,c)$ is the average directivity pattern of the microphones and $\sigma^2(f)$ is the deviation variance. It is not direction-dependent and the model includes magnitude and phase manufacturing tolerances, position errors, enclosure tolerances, and so on. On substituting further the microphone model in Equations 5.3, 5.4, and 5.15, we obtain the expression for the microphone array output:

$$Y(f) = \sum_{m=0}^{M-1} W_m(f) \cdot \left(\frac{e^{-j2\pi f\frac{\|c-p_m\|}{v}}}{\|c-p_m\|} A_m(f)(\bar{U}(f,c) + \mathbb{N}(0,\sigma^2(f)))S(f) + N_m(f) \right). \tag{5.65}$$

After splitting the beamformer sum:

$$\begin{aligned} Y(f) = & \sum_{m=0}^{M-1} W_m(f).(D_m(f)\bar{U}(f,c)S(f) + N_m(f)) \\ & + \sum_{m=0}^{M-1} W_m(f).\mathbb{N}(0,|D_m(f)|\sigma^2(f)).S(f) \end{aligned} \tag{5.66}$$

or

$$Y(f) = Y_{ND}(f) + \sum_{m=0}^{M-1} W_m(f) \cdot \mathbb{N}(0, |D_m(f)|\sigma^2(f)) \cdot S(f). \qquad (5.67)$$

Assuming that manufacturing tolerances across the channels are statistically independent, we can say that the effect of the channel mismatch can be modeled as additional uncorrelated noise. Improving the robustness to manufacturing tolerances in general can be achieved by reducing the uncorrelated noise gain. This makes the delay-and-sum beamformer the most robust to manufacturing tolerances. The question is: How much better beamformer parameters can be achieved given a model of the manufacturing tolerances?

5.5.4.2 Cost Functions and Optimization Goals

One of the first studies of the effects of the channel mismatch is that of Er [13]. The author considers a narrowband array for a specific directivity pattern, uniform PDF and least-squares cost function. Minimizing the uncorrelated noise gain leads to better robustness. In a further generalization of this by Doclo and Moonen [14–17], the authors use the microphone array directivity pattern and the deviations caused by the channel mismatch. The microphone model is close to that of Equation 5.58, and they split the further derivation on magnitude and phase variations. For the design of the beamformer they consider three cost functions:

- *weighted least-squares*, minimizing the weighted LS error between the desired and actual directivity patterns;
- the *total least-squares error* which leads to the generalized eigenvalue problem;
- the *difference between the magnitudes* of the desired and actual directivity patterns.

Finding an analytical solution for each of these cost functions is complex and the attempt to finalize the derivation usually leads to assuming very simple cases and specific designs (linear symmetric microphone array with omnidirectional microphones). When the channel mismatch is involved there can be considered two potential optimum solutions:

- the *mean performance* – that is, the weighted sum of the cost functions for all feasible microphone characteristics, using the probability of the microphone characteristics as weights;
- the *worst-case performance* – that is, the maximum cost function for all feasible microphone characteristics, leading to a minimax criterion.

To simplify further, the authors consider a uniform PDF of the microphone characteristics in a given interval. In general, the derived solutions are under numerous assumptions and contain a relatively large number of parameters to estimate. While this is good from a theoretical standpoint, it would be difficult to use the results from this work in a real microphone array design.

An interesting approach for a minimax-based robust design has been proposed by Chen *et al.* [18]. After analytically deriving the lowest error bounds, the authors design a robust beamformer for a seven-element array using direct optimization. Evaluation against deviations in magnitude, phase, position, and speed of sound is provided in the paper.

In general, mismatch beamformer design robust to channels leads to either partial cases in analytical form, or to numerical optimization.

5.5.4.3 MVDR Beamformer Robust to Manufacturing Tolerances

An easy and computationally inexpensive way to increase the robustness of the beamformer design is to incorporate manufacturing tolerances into instrumental noise. Then we can use the MVDR beamfromer design described earlier. Considering Equation 5.67, a very rough estimation of the uncorrelated noise is

$$\widehat{N}_{\mathrm{I}}^{2}(f) = N_{\mathrm{I}}^{2}(f) + \sigma^{2}(f)N_{\mathrm{A}}^{2}(f) \qquad (5.68)$$

where $N_{\mathrm{I}}^{2}(f)$ is the instrumental noise variance, $N_{\mathrm{A}}^{2}(f)$ is the ambient noise variance, and $\sigma^{2}(f)$ is the relative manufacturing tolerance variance $\widehat{N}_{\mathrm{I}}^{2}(f)$ is the noise variance to substitute in Equation 5.43. After this, the MVDR beamformer design goes through the procedure already described.

5.5.4.4 Beamformer with Direct Optimization Robust to Manufacturing Tolerances

Easy to implement, but computationally very expensive, is robust beamformer design using direct optimization, as described earlier. Instead of optimizing one microphone array, computation of the optimization criterion (Equation 5.57) uses just a large number of microphone array instances, generated using the microphone model in Equation 5.58. Assuming known variances in magnitude and phase, a large number of instances (1000, for example) of the designed microphone array can be created before the optimization. In each of these instances the microphone parameters are randomly generated using the known deviations and distributions. Then the optimization can go either by minimizing the mean, or by minimizing the worst-case scenario – that is, a minimax design. Here we will try to achieve the same results with much less computation than this brute-force approach.

With typical unit gain and zero phase towards the listening direction, the MVDR beamformer actually minimizes the noise by maximizing the microphone-array directivity pattern. In the most common case we distinguish the listening area, usually centered in the listening direction, from the area where we want to suppress all signals. Then we define the microphone array directivity ratio as the proportion of the average energy from the listening area and the average energy from the whole working area. If we assume L probe points are evenly spread in the work area S (which depends on the microphone array geometry – from a semicircle to a whole sphere), then L_{pass} belongs to the listening area A. The directivity ratio is then

$$\mathbb{R} = \frac{\frac{1}{L_{pass}} \sum_{s \in A} |\mathbf{W}(\mathbf{D}_s \circ \mathbf{U}_s)|^2}{\frac{1}{L} \sum_{s} |\mathbf{W}(\mathbf{D}_s \circ \mathbf{U}_s)|^2}. \tag{5.69}$$

To achieve maximum directivity without accounting for channel mismatches, we should maximize

$$C_{MD} = \max_{\mathbf{W}}(\mathbb{R}), \quad \text{subject to } \mathbf{W} \circ \mathbf{D} = 1. \tag{5.70}$$

Solving this leads to unusable weights owing to high uncorrelated noise gain. Adding penalty functions for large weight magnitude partially solves the problem and leads to a constrained maximum directivity optimization criterion, as described in the previous sections:

$$C_{MDC} = \max_{\mathbf{W}} \left(\mathbb{R} + \sum_{M} (\max(1, |W_i|) - 1)^2 \right), \quad \text{subject to } \mathbf{W} \circ \mathbf{D} = 1. \tag{5.71}$$

On adding the channel mismatch model (Equation 5.64) in the expression for directivity (Equation 5.69)

$$\frac{L}{L_{pass}} \frac{\sum_{s \in A} |\mathbf{W}(\mathbf{D}_s \circ \mathbf{U}_s)|^2}{\sum_{s} |\mathbf{W}(\mathbf{D}_s \circ \mathbf{U}_s)|^2} = \frac{L}{L_{pass}} \frac{\sum_{s \in A} \left| \sum_{i=1...M} W_i D_{si} (\bar{U}_{si} + \mathbb{N}_i(0, \sigma^2)) \right|^2}{\sum_{s} \left| \sum_{i=1...M} W_i D_{si} (\bar{U}_{si} + \mathbb{N}_i(0, \sigma^2)) \right|^2} \tag{5.72}$$

we can take the random component out of the absolute value. Then it becomes real Gaussian noise instead of complex:

$$\mathbb{R} = \frac{L}{L_{pass}} \frac{\sum_{s \in A} \left(\left| \sum_{i=1...M} W_i D_{si} \bar{U}_{si} \right| + \mathbf{N}\left(0, \frac{\sigma^2}{2} \sqrt{\sum_{i=1...M} |D_{is} W_i|^2} \right) \right)^2}{\sum_{s} \left(\left| \sum_{i=1...M} W_i D_{si} \bar{U}_{si} \right| + \mathbf{N}\left(0, \frac{\sigma^2}{2} \sqrt{\sum_{i=1...M} |D_{is} W_i|^2} \right) \right)^2}. \tag{5.73}$$

The next logical step is to expand the square, denoting the new deviations by σ_A and σ_S:

$$\mathbb{R} = \frac{L}{L_{pass}} \frac{\sum\limits_{s\in A} |\sum\limits_{i=1...M} W_i D_{si} \bar{U}_{si}|^2 + 2N(0,\sigma_A^2) \sum\limits_{s\in A} |\sum\limits_{i=1...M} W_i D_{si} \bar{U}_{si}| + L_{pass}\sigma_A^2 \mathbf{Q_1}}{\sum\limits_{s} |\sum\limits_{i=1...M} W_i D_{si} \bar{U}_{si}|^2 + 2N(0,\sigma_S^2) \sum\limits_{s} |\sum\limits_{i=1...M} W_i D_{si} \bar{U}_{si}| + L\sigma_S^2 \mathbf{Q_1}} \quad (5.74)$$

Here, $\mathbf{Q_1}$ is chi-squared distributed random variable with one degree of freedom. Denoting the average power in the listening area and in the whole volume by $\bar{P}_A$ and $\bar{P}_S$, respectively, gives us the next simplification:

$$\mathbb{R} = \frac{|\bar{P}_A| + \frac{2}{L_{pass}} N(0,\sigma_A^2) \sum\limits_{s\in A} |\sum\limits_{i=1...M} W_i D_{si} \bar{U}_{si}| + \sigma_A^2 \mathbf{Q_1}}{|\bar{P}_S| + \frac{2}{L} N(0,\sigma_S^2) \sum\limits_{s} |\sum\limits_{i=1...M} W_i D_{si} \bar{U}_{si}| + \sigma_S^2 \mathbf{Q_1}} \quad (5.75)$$

Then we expand the standard deviations and take the powers in front:

$$\mathbb{R} = \bar{\mathbb{R}} \frac{1}{1 + 2N(0,\sigma_S^2)\sqrt{\bar{P}_S} + \sigma_S^2 \mathbf{Q_1}} + \frac{2N(0,\sigma_A^2)\sqrt{\bar{P}_A} + \sigma_A^2 \mathbf{Q_1}}{1 + 2N(0,\sigma_S^2)\sqrt{\bar{P}_S} + \sigma_S^2 \mathbf{Q_1}} \quad (5.76)$$

Here $\bar{\mathbb{R}}$ is the directivity assuming channel matching. We expand the standard deviations, replace the random variables with their means and derive the average directivity:

$$\tilde{\mathbb{R}} = \bar{\mathbb{R}} \frac{1}{1 + \frac{\sigma^2}{2\bar{R}_S}\sqrt{\sum\limits_{i=1...M} |W_i|^2}} + \frac{\frac{\sigma^2}{2\bar{R}_A}\sqrt{\sum\limits_{i=1...M} |W_i|^2}}{1 + \frac{\sigma^2}{2\bar{R}_S}\sqrt{\sum\limits_{i=1...M} |W_i|^2}} \quad (5.77)$$

where $\bar{R}_A$ and $\bar{R}_S$ are the average distances in the listening area and in the overall volume, respectively. As $\bar{\mathbb{R}} \geq 1$ we will always have degradation in directivity if $\sigma > 0$, which is clearly visible in Figures 5.27 and 5.28.

Then the robust beamformer design can maximize the modified directivity:

$$C_{RobMD} = \max_{\mathbf{W}}(\tilde{\mathbb{R}}), \quad \text{subject to } \mathbf{W} \circ \mathbf{D} = 1. \quad (5.78)$$

If we assume that the Gaussian distribution of the manufacturing tolerances is pruned to 3.0σ – that is, if the quality control at the manufacturer rejected microphones that differed too much from the specification – we can find the limiting estimate for the worst directivity:

$$\tilde{\mathbb{R}}_{min} = \bar{\mathbb{R}} \frac{1}{1 + \frac{(3.0\sigma)^2}{2\bar{R}_S}\sqrt{\sum\limits_{i=1...M} |W_i|^2}} - \frac{\frac{(3.0\sigma)^2}{2\bar{R}_A}\sqrt{\sum\limits_{i=1...M} |W_i|^2}}{1 + \frac{(3.0\sigma)^2}{2\bar{R}_S}\sqrt{\sum\limits_{i=1...M} |W_i|^2}} \quad (5.79)$$

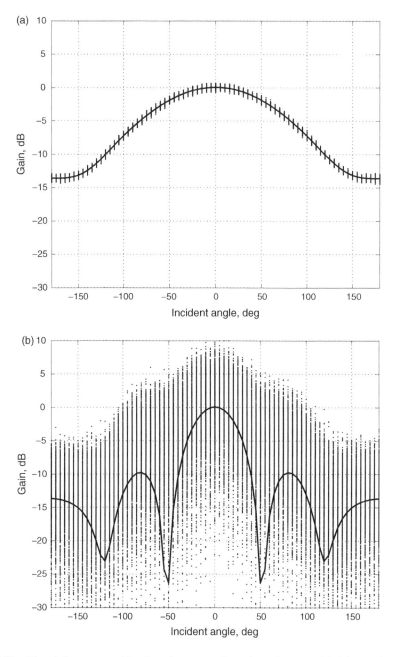

Figure 5.31 Directivity pattern with microphone manufacturing tolerances: (a) delay-and-sum beamformer; (b) optimal beamformer without accounting for the instrumental noise; (c) robust minimax optimized beamformer

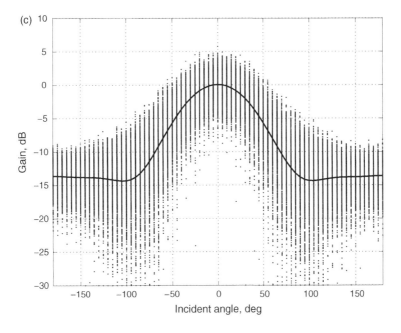

Figure 5.31 (*Continued*)

and we can maximize the worst case:

$$C_{\mathrm{RobMaxMinD}} = \max_{\mathbf{W}}(\tilde{\mathbb{R}}_{\min}), \quad \text{subject to } \mathbf{W} \circ \mathbf{D} = 1. \tag{5.80}$$

To illustrate the design process robust to manufacturing tolerances, we generated 1000 instances of the four-element linear array with a variation in the microphone parameters according to Equation 5.58 and gain and phase deviations, presented in Figures 5.25 and 5.26. We selected the noise gain at 300 Hz as a parameter for comparison. This is in the lower part of the frequency band, where there are higher deviations of the microphone parameters and higher noise levels. The overall noise-suppression capabilities of the microphone array heavily depend on the noise suppression in this area.

Table 5.5 presents the results for best microphone, delay-and-sum beamformer, optimal beamformer according to Equation 5.57 without accounting for the instrumental noise, optimal beamformer according to the same equation with instrumental noise in the model, and the beamformer designed with robust minimax optimization according to Equation 5.80. The table shows the noise gain under channel-matching conditions, the standard deviation of the noise gain in these 1000 realizations of the array, and the minimum and maximum noise gain. Under matched conditions, the best noise suppression comes from the optimal beamformer without accounting for the

Table 5.5   Noise gain parameters for 300 Hz and 1000 realization of the microphone array

Noise Gain for 300 Hz, dB	Matched	Deviation	Average	Minimal	Maximal
Best microphone	−5.03	0.24	−5.02	−6.22	−3.65
Delay and sum	−5.31	0.11	−5.31	−5.91	−4.65
Opt. beam, no noise	−10.30	3.89	−4.67	−14.92	17.42
Optimal beam	−8.78	1.17	−7.50	−13.33	4.72
Robust MiniMax optimization	−8.08	0.87	−7.38	−12.33	0.36

instrumental noise. Its performance has the highest deviation in real conditions; the average noise gain is worse even than a single microphone, and sometimes this beamformer boosts the noise up to 17 dB. The delay-and-sum beamformer has the lowest deviation of the noise gain, as expected, but the average noise suppression is better only than the single microphone. The minimax-optimized robust beamformer design has a lower difference than the optimal beamformer and a lower difference between the minimal and maximal noise gains. The optimal beamformer, which is practically an MVDR beamformer, actually has pretty good robustness for manufacturing tolerances – this is due to the fact that the instrumental noise in the cross-power matrix acts as regularization. Figure 5.31 shows the beam pattern for these 1000 instances of the four-element microphone array at 300 Hz.

5.5.4.5 Balanced Design for Handling the Manufacturing Tolerances

In conclusion, it is clear that designs that are robust to manufacturing tolerances reduce the overall performance of the microphone array. Using in-factory calibration is expensive and not suitable for mass production. The autocalibration procedure requires additional resources (CPU time and memory) during real-time processing, but can substantially reduce the residual channel mismatch. From this perspective, a way to achieve good practical design is to combine calibration or autocalibration measures with the robust design. If the calibration or autocalibration can reduce the microphone tolerances from ±4 dB to ±0.5 dB, then the overall performance of the designed robust beamformer will be better.

EXERCISE

Modify the microphone-array weights synthesis program to design beamformers that are robust to manufacturing tolerances using direct optimization and the equations above.

Modify the microphone-array analysis program to visualize the parameters of the beamformers accounting for the microphone manufacturing tolerances.

Compare the results with the previously designed delay-and-sum beamformer.

5.6 Adaptive Beamformers

Time-invariant beamformers assume isotropic ambient noise, as they are designed in advance and do not depend on, and change with, the input signals. This leaves unrealized opportunities to achieve better noise suppression and to provide a higher quality signal in the beamformer output. Adaptive beamformers, in contrast, estimate the weights in real time, based on the microphone-array geometry and the input signals. They are computationally more expensive and consume more memory in real-time implementations.

5.6.1 MVDR and MPDR Adaptive Beamformers

We have already derived the optimal weights in the section on time-invariant beamformers:

$$\mathbf{W}_{\text{MVDR}}(f) = \frac{\mathbf{D}_c^{\text{H}}(f)\mathbf{\Phi}_{NN}^{-1}(f)}{\mathbf{D}_c^{\text{H}}(f)\mathbf{\Phi}_{NN}^{-1}(f)\mathbf{D}_c(f)}. \tag{5.81}$$

If there is enough computational power available, the cross-power matrix can be updated and inverted in real time. Then we will have optimal weights for every frame, guaranteeing the best possible suppression. Such beamformers will adapt to the noise distribution and will modify the beamformer directivity pattern, and – with a point noise source – will place a null towards its direction. The only inconvenience is the estimation of the noise cross-power matrix during the pauses in the speech signal. If we use the cross-power matrix of the input signals instead, that is

$$\mathbf{W}_{\text{MPDR}}(f) = \frac{\mathbf{D}_c^{\text{H}}(f)\mathbf{\Phi}_{XX}^{-1}(f)}{\mathbf{D}_c^{\text{H}}(f)\mathbf{\Phi}_{XX}^{-1}(f)\mathbf{D}_c(f)} \tag{5.82}$$

then we will have what is known as a *minimum-power distortionless-response* (MPDR) beamformer. It can be shown that while an MVDR beamformer was derived to minimize the noise magnitude in the output, under the distortionless constraints the MPDR beamformer minimizes the noise power. The advantage of this is that it does not rely on a VAD to compute and update the noise cross-power matrix. Its behavior towards the main response axis is the same as that of the MVDR beamformer, but the beam's shape is slightly different in the side-lobes. In some literature sources this beamformer is called "sample matrix inversion" [2].

5.6.2 LMS Adaptive Beamformers

The adaptive beamformers of this group rely on the least-minimum-square (LMS) error algorithm for adaptive filtering. Instead of estimating the optimal weights for each frame entirely, they will be estimated using the recursive LMS algorithm.

5.6.2.1 Widrow Beamformer

One of the first published LMS adaptive algorithms is the Widrow beamformer [19]:

$$\mathbf{W}_k^{(n+1)} = \mathbf{W}_k^{(n)} - \mu \tilde{\nabla}_k^{(n)}. \tag{5.83}$$

Here, $\tilde{\nabla}_k^{(n)}$ is the estimated gradient of the error vector with respect to $\mathbf{W}_k^{(n)}$:

$$\tilde{\nabla}_k^{(n)} = -2\varepsilon_k^{(n)} \mathbf{X}_k^{(n)} \tag{5.84}$$

which leads to

$$\mathbf{W}_k^{(n+1)} = \mathbf{W}_k^{(n)} - 2\mu \varepsilon_k^{(n)} \mathbf{X}_k^{(n)}. \tag{5.85}$$

The only problem left is estimation of the error in the beamformer output. When the desired signal $S_k^{(n)}$ is known, estimation of the error is trivial:

$$\varepsilon_k^{(n)} = S_k^{(n)} - \mathbf{W}_k^{(n)} \mathbf{X}_k^{(n)}. \tag{5.86}$$

The original algorithm was designed for antenna arrays. The desired signal consists of a carrier, or pilot signal, which is modulated with the information signal. In the original paper there is an internal pilot signal generator and the adaptation has two modes. The first is when the inputs of the beamformer is a properly delayed pilot signal from the internal generator. The second mode is when the inputs of the beamformer are connected to the corresponding antenna and the desired signal is zero; that is, maximum noise suppression. It is easy to see that the second mode minimizes the captured energy, while the first mode imposes the constraint for correct reception of the desired signal and prevents the adaptive filter from finding the all-zeros solution. Apparently this algorithm is not applicable directly to audio microphone arrays, while some software equivalent exists. We will see later in this chapter how this approach can be used in audio processing. The Widrow beamformer is an example of an unconstrained adaptive beamformer, where the constraint is enforced externally.

5.6.2.2 Frost Beamformer

The first published adaptive algorithm with integrated constraint was due to Frost [1]. In that paper, the optimum set of weights was derived for an array of radio antennas. The derivation is close to the derivation of the MVDR beamformer already described, and leads to the optimal weights in Equation 5.38. Computing the noise cross-power spectral matrix and inverting it in real time is computationally expensive and can happen when we have only the noise sources active. Frost derived an adaptive LMS estimator of the optimal weights. The derivation is in the time domain, but when it is converted for processing in the frequency domain it looks like this:

$$
\begin{aligned}
\mathbf{W}_k^{(0)} &= \mathbf{F} \\
\mathbf{W}_k^{(n+1)} &= \mathbf{P}(\mathbf{W}_k^{(n)} - \mu \mathbf{\Phi}_{XX}^{(n)}(k)\mathbf{W}_k^{(n)}) + \mathbf{F}.
\end{aligned}
\tag{5.87}
$$

Here, $\mathbf{\Phi}_{XX}^{(n)}(k)$ is the cross-power spectral matrix of the input signal, and the constraints are

$$
\begin{aligned}
\mathbf{F} &\triangleq \frac{\mathbf{D}_k^H}{\mathbf{D}_k\mathbf{D}_k^H} \\
\mathbf{P} &\triangleq \mathbf{I} - \frac{\mathbf{D}_k^H\mathbf{D}_k}{\mathbf{D}_k\mathbf{D}_k^H}.
\end{aligned}
\tag{5.88}
$$

It is easy to transform Equation 5.87 into

$$
\begin{aligned}
\mathbf{W}_k^{(0)} &= \mathbf{F} \\
\mathbf{W}_k^{(n+1)} &= \mathbf{P}(\mathbf{W}_k^{(n)} - \mu Y_k^{(n)}\mathbf{X}_k^{(n)}) + \mathbf{F}
\end{aligned}
\tag{5.89}
$$

which requires less computational resources. This LMS adaptive algorithm is known as a "Frost beamformer." It is the most computationally inexpensive among the adaptive beamformers. The proposed LMS adaptive filter is relatively easy to convert to NLMS for increased stability and better convergence.

5.6.3 Generalized Side-lobe Canceller

5.6.3.1 Griffiths–Jim Beamformer

The adaptive beamformers so far use a set of LMS or NLMS adaptive filters to minimize the output noise, imposing directly or indirectly the constraint for unit gain and zero phase shift towards the listening direction. The constraint actually causes complications in the real-time algorithm – compare the simplicity of the Widrow algorithm with the Frost algorithm. A second observation is that, while the beam is usually flat towards the listening direction and it is difficult to make it narrow, the nulls in the beamformer directivity pattern are sharp and almost completely suppress the signal coming from that direction. The Griffiths and Jim [20] generalized side-lobe canceller (GSC) beamformer utilizes the second observation and imposes the constraint implicitly.

The proposed alternative architecture is shown in Figure 5.32. It consists of a conventional time-invariant beamformer, blocking matrix, and adaptive interference

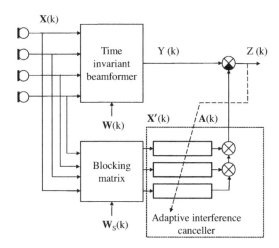

Figure 5.32   Generalized side-lobe canceller

canceller. In the original paper all the processing is performed in the time domain, while we will provide the equations converted for processing in the frequency domain.

The conventional time-invariant beamformer **W** works according Equation 5.15 and can have weights, designed using any of the already mentioned algorithms for time-invariant beamforming. Its output $Y(k,n) = \mathbf{W}(k)\mathbf{X}(k,n)$ contains the desired signal undistorted and with ambient noise suppressed to a certain degree. The constraint is explicitly integrated into the beamformer weights.

The blocking matrix has M input channels and $M-1$ output channels. The purpose of this matrix preprocessor $\mathbf{W}_s$ is to block the desired signal S_k. Since it is common to each path, we can have a maximum of $M-1$ output signals:

$$\mathbf{X}'(k,n) = \mathbf{W}_S(k)\mathbf{X}(k,n). \tag{5.90}$$

This can be implemented as a set of two-element time-invariant beamformers, each of which has a null towards the listening direction. One of the ways to group the signals into pairs is to take one of them as a reference and then combine it with the others. Each of the outputs of the blocking matrix contains some portion of the ambient noise and does not contain the desired signal (or at least the direct path).

The adaptive interference canceller $\mathbf{A}(k)$ tries to minimize the output of the GSC, but as the blocking matrix outputs do not contain the desired signal, it will minimize only the noise:

$$Z(k,n) = Y(k,n) - \mathbf{A}(k)\mathbf{X}'(k,n). \tag{5.91}$$

The filter $\mathbf{A}(k)$ is a complex vector of size $M-1$ adapted using either the LMS algorithm

$$\mathbf{A}(k,n+1) = \mathbf{A}(k,n) - \mu Y(k,n)\mathbf{X}'(k,n) \qquad (5.92)$$

or the NLMS algorithm if we normalize μ with the total power of the audio frame:

$$\mu = \frac{\alpha}{\displaystyle\sum_{m=1}^{M}\sum_{k=1}^{K}|X_m(k,n)|^2}. \qquad (5.93)$$

In this case, $0 < \alpha < 1$ guarantees the convergence.

GSC is computationally efficient, as it contains only unconstrained adaptive filters. A potential problem with microphone arrays using GSC is the reverberation. In the blocking matrix outputs we will have portions of the reverberated desired signal. This issue can be mitigated by stopping the adaptation, or at least reducing the learning step α, when we have a presence of the desired signal. A simple binary VAD can be used to provide this adapt/not-adapt flag.

5.6.3.2 Robust Generalized Side-lobe Canceller

The GSC beamformer relies on the fact that the nulls after the blocking matrix are sharp and remove only the desired signal. The sharp nulls impose increased requirements for the precision and we have to know the position of the desired sound source. Owing to the sharp nulls, even small errors in the localization of the desired sound source will cause leakage of it to the outputs of the blocking matrix, which will substantially reduce the efficiency of the GSC beamformer.

To address this issue, Hoshuyama et al. [21] propose a structure robust to such leakage as shown in Figure 5.33. The adaptive blocking matrix ABM, which is a set of NLMS adaptive filters that try to remove the leaked desired signal from the blocking matrix outputs, is added to the time-invariant blocking matrix. The time-invariant beamformer output is used as the desired signal. As it contains ambient noise, the ABM filter coefficients are constrained to suppress only the signals originating from the area of interest, which is defined as a certain zone around the estimated direction to the desired sound source. This zone is converted to delays and overrides of the NLMS adaptation procedure, keeping the ABM's null varying in the area of interest.

In [22], this robust adaptive filter is converted for processing in the frequency domain. The paper shows that this is computationally more efficient, regardless of the fact that to impose constraints in the ABM the authors convert the filter to the time domain and use the conditions designed in the original paper.

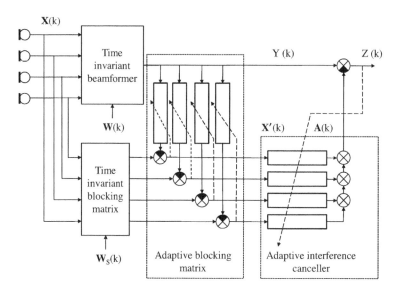

Figure 5.33 Robust generalized side-lobe canceller

5.6.4 *Adaptive Algorithms for Microphone Arrays – Summary*

Adaptive array algorithms were initially designed for antenna arrays. There they are considered more efficient and suppress more noise due to the fact that they place nulls towards the undesired signal sources. In audio, the efficiency of the adaptive microphone array algorithms is reduced mostly owing to the reverberation. In reverberant conditions there are no point sources – desired and undesired sound sources are smeared. At a critical distance, where the received energy directly from the sound source equals the reverberant energy, even if we place a perfect null towards the unwanted sound source the best we can achieve is a 6 dB attenuation – the direct path. The reverberation is practically isotropic ambient noise. For most rooms and offices the critical distance is close to the work distance for hands-free sound capture – around 1.5 m. From this perspective, an isotropic ambient noise model is closer to reality than a point noise source model. The time-invariant beamformers were optimal exactly for isotropic ambient noise. Under these conditions, a time-invariant beamformer with perfect channel matching is as efficient as an adaptive beamformer; that is, the adaptive beamformer replaces the autocalibration procedure. Still, the adaptive beamformers provide slightly better results in a reverberant environment, with the price of increased CPU and memory use.

5.7 Microphone-array Post-processors

Both time-invariant and adaptive beamformers are linear processors and compute the output signal as a linear combination of the input signals. Even if the weights of different microphones change, which is the case with adaptive beamformers, they do it much slower than the audio frame rate. Microphone-array post-processors apply real-valued gain which varies from frame to frame in the same way as the static noise suppressors do. The difference is that the gain estimation is based on the additional information about the positions of the desired and undesired sound source, which we have from the multiple channels and eventually the microphone positions.

5.7.1 Multimicrophone MMSE Estimator

Assume that the source signal $S_c(f)$ in Equation 5.27 has variance $\lambda_c(f)$. The noise contains correlated and uncorrelated components, with spectral matrix presented in Equation 5.42:

$$\mathbf{\Phi}_{N'N'}(f) = \mathbf{\Phi}_{NN}(f) + \lambda_{NC}(f)\mathbf{I} \tag{5.94}$$

where $\lambda_{NC}(f)$ is the variance of the uncorrelated noise and $\mathbf{\Phi}_{NN}(f)$ is the spectral matrix of the correlated (spatial) noise. Then the spectral matrix of the input signals is

$$\mathbf{\Phi}_{XX}(f) = \lambda_c(f)\mathbf{D}_c(f)\mathbf{D}_c^{H}(f) + \mathbf{\Phi}_{N'N'}(f). \tag{5.95}$$

Let the estimation of the desired signal be provided by a matrix processor $\mathbf{H}(f)$, which is an M-element complex vector. The derivation provided here follows [2]. Then the mean square error is

$$\begin{aligned} \varepsilon &= \mathrm{E}\{|S_c(f)-\mathbf{H}(f)\mathbf{X}(f)|^2\} \\ &= \mathrm{E}\{(S_c(f)-\mathbf{H}(f)\mathbf{X}(f))(S_c^*(f)-\mathbf{X}^H(f)\mathbf{H}^H(f))\}. \end{aligned} \tag{5.96}$$

Taking the complex gradient with respect to $\mathbf{H}^H(f)$ and setting the result equal to zero gives

$$\begin{aligned} \mathrm{E}[S_c(f)\mathbf{X}^H(f)]-\mathbf{H}(f)\mathrm{E}[\mathbf{X}(f)\mathbf{X}^H(f)] &= 0 \\ \text{or} \quad \mathbf{\Phi}_{dX^H}(f) &= \mathbf{H}_0(f)\mathbf{\Phi}_{XX}(f). \end{aligned} \tag{5.97}$$

From Equation 5.27 and the assumption for uncorrelated signal and noise components:

$$\mathbf{\Phi}_{dX^H}(f) = \lambda_c(f)\mathbf{D}_c^{H}(f). \tag{5.98}$$

This leads to the solution for the optimal MMSE estimator:

$$\mathbf{H}_0(f) = \lambda_c(f)\mathbf{D}_c^{H}(f)\mathbf{\Phi}_{XX}^{-1}(f). \tag{5.99}$$

On inverting Equation 5.95 using the matrix inversion formula, we obtain

$$\mathbf{\Phi}_{xx}^{-1} = \mathbf{\Phi}_{N'N'}^{-1} - \mathbf{\Phi}_{N'N'}^{-1}\lambda_c\mathbf{D}_c(1 + \mathbf{D}_c^H\mathbf{\Phi}_{N'N'}^{-1}\lambda_c\mathbf{D}_c)^{-1}\mathbf{D}_c^H\mathbf{\Phi}_{N'N'}^{-1} \qquad (5.100)$$

where the frequency indices are suppressed for simplicity. After defining

$$\mathbf{\Lambda}^{-1}(f) = \mathbf{D}_c^H(f)\mathbf{\Phi}_{N'N'}^{-1}(f)\mathbf{D}_c(f) \qquad (5.101)$$

and on substituting (5.99) in (5.100) we obtain the optimal solution:

$$\mathbf{H}_0(f) = \frac{\lambda_c(f)}{\lambda_c(f) + \mathbf{\Lambda}(f)}\mathbf{\Lambda}(f)\mathbf{D}_c^H\mathbf{\Phi}_{NN}^{-1}(f). \qquad (5.102)$$

This MMSE processor is practically a multichannel Wiener filter. Its block diagram is shown in Figure 5.34. Taking a closer look, it is easy to see that it consists of an MVDR beamformer – compare the right part of Equation 5.102 with Equation 5.38, and something close to the single-channel Wiener filter, described in Chapter 4. The noise variance $\lambda_d(f)$ is computed by $\mathbf{\Lambda}(f)$, defined in Equation 5.101. In this chapter we ignore the beamformer and focus on the post-processor only.

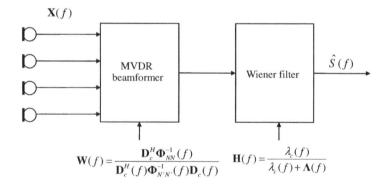

Figure 5.34   Multimicrophone MMSE estimator

5.7.2 Post-processor Based on Power Densities Estimation

Equation 5.102 provides the optimal solution under the assumption of the known desired signal variation $\lambda_c(f)$ and the noise cross-power matrix $\mathbf{\Phi}_{XX}(f)$. Estimation of these is not trivial in real conditions, where all we know is the input signals and the microphone-array geometry. One of the first practical applications of post-processor for microphone arrays was published by Zelinski [23]. Calculating the auto- and cross-spectral densities of the aligned (i.e., properly delayed) channels i and j leads to (all frequency indices omitted for simplicity)

$$\phi_{X_iX_i} = \phi_{SS} + \phi_{N_iN_i} + 2\Re\{\phi_{SN_i}\} \tag{5.103}$$

$$\phi_{X_iX_j} = \phi_{SS} + \phi_{N_iN_j} + \phi_{SN_j} + \phi_{N_iS}. \tag{5.104}$$

Zelinski makes the following assumptions:

- The signal and noise are uncorrelated ($\phi_{N_iS} = 0, \forall i$), which is in general true unless we do not consider the early reverberation of the desired signal as noise.
- The noise power spectrum is the same on all sensors, ($\phi_{N_iN_i} = \phi_{NN}, \forall i$), which restricts the algorithm to microphone arrays with the same type of sensors.
- The noise is uncorrelated between sensors, ($\phi_{N_iN_j} = 0, \forall i \neq j$), which again excludes the early reverberation.

Under these assumptions, Equations 5.103 and 5.104 are reduced to

$$\phi_{X_iX_i} = \phi_{SS} + \phi_{N_iN_i} \tag{5.105}$$

$$\phi_{X_iX_j} = \phi_{SS}. \tag{5.106}$$

They can be estimated using a standard smoothing in time:

$$\hat{\phi}_{X_iX_j}^{(n)} = (1-\alpha)\hat{\phi}_{X_iX_j}^{(n-1)} + \alpha X_i X_j^* \tag{5.107}$$

where $\alpha = T/\tau$, T is the frame direction, and τ is the update time constant. Then the numerator and denominator in the post-processor part of Equation 5.102 can be estimated more robustly by averaging the spectral densities over all the possible channel combinations, resulting in the post-filter estimator (frame indices omitted for simplicity):

$$\hat{H}_{PF} = \frac{\frac{2}{M(M-1)} \sum_{i=1}^{M-1} \sum_{j=i+1}^{M} \Re\{\hat{\phi}_{X_iX_j}\}}{\frac{1}{M} \sum_{i=1}^{M} \hat{\phi}_{X_iX_i}}. \tag{5.108}$$

The real operator $\Re\{\cdot\}$ is used because the term, estimated in the nominator, is required to be a real.

The Zelinksi post-processor is a good approximation and works well for the lower part of the frequency band, where the assumptions above hold better, but it is less efficient in the upper part of the band. The author tries to compensate for this by using a "coherence detector" which decides whether there is a coherent signal (i.e., desired speech signal) or pure noise. It is a soft decision control of the attenuation of the post-processor, estimated for the whole frame, and it is based on comparison of the amplitudes of positive and negative values of the cross-spectral density:

$$\hat{C}(k) = \sum_{i=1}^{M-1} \sum_{j=i+1}^{M} \hat{\phi}_{X_i X_j}(k)$$

$$\beta = \frac{\sum_{k'} \hat{C}(k')|_{\hat{C}(k')>0}}{\sum_{k''} \hat{C}(k'')|_{\hat{C}(k'')<0}}. \qquad (5.109)$$

If the input signals are pure noise, β is close to 1; and if there is a speech signal, the value of β is considerably larger than 1. Using this parameter, additional attenuation gain during the silence periods can be computed and applied.

5.7.3 Post-processor Based on Noise-field Coherence

A step further in the generalization leads to the algorithm described by McCowan and Bourlard [24]. A common measure used to characterize noise fields is the complex coherence function. The coherence between two signals at points p_i and p_j is defined as

$$\Gamma_{ij} = \frac{\phi_{ij}}{\sqrt{\phi_{ii}\phi_{jj}}}. \qquad (5.110)$$

The coherence has the range $|\Gamma_{ij}| \leq 1$ and is essentially a normalized measure of the correlation that exists between the signals at two discrete points in a noise field. The initial correlation for two omnidirectional microphones spaced at distance d_{ij} was published in 1955 by Cook et al. [3]. We have already used it in Equation 5.39. It can be shown that the coherence of a diffused noise field is real-valued and given by

$$\Gamma_{ij} = \mathrm{sinc}\left(\frac{2\pi f d_{ij}}{\nu}\right). \qquad (5.111)$$

With the assumption of aligned signals on all sensors and zero correlation between the desired signal and the noise, we can write the following four equations:

$$\begin{aligned}
\phi_{X_i X_i} &= \phi_{SS} + \phi_{N_i N_i} \\
\phi_{X_j X_j} &= \phi_{SS} + \phi_{N_i N_i} \\
\phi_{X_i X_j} &= \phi_{SS} + \phi_{N_i N_j} \\
\Gamma_{N_i N_j} &= \frac{\phi_{N_i N_j}}{\sqrt{\phi_{N_i N_i}\phi_{N_i N_i}}}.
\end{aligned} \qquad (5.112)$$

Then we have four equations and four unknowns, which allows estimation of ϕ_{SS}. Assuming isotropic ambient noise and identical sensors, $\phi_{N_i N_i} = \phi_{NN}$ for all i, then $\Gamma_{N_i N_j} = \phi_{N_i}/\phi_{NN}$ and the system of equations above can be simplified to

$$\begin{aligned}
\phi_{X_i X_i} &= \phi_{SS} + \phi_{NN} \\
\phi_{X_j X_j} &= \phi_{SS} + \phi_{NN} \\
\phi_{X_i X_j} &= \phi_{SS} + \Gamma_{N_i N_j} \phi_{NN}.
\end{aligned} \tag{5.113}$$

This allows analytical estimation of the cross-power spectrum of the desired signal using the channel pair ij:

$$\hat{\phi}_{SS}^{(ij)} = \frac{\mathfrak{R}\left\{\hat{\phi}_{X_i X_j}\right\} - \frac{1}{2}\mathfrak{R}\left\{\hat{\Gamma}_{N_i N_j}\right\}\left(\hat{\phi}_{X_i X_i} + \hat{\phi}_{X_j X_j}\right)}{\left(1 - \mathfrak{R}\left\{\hat{\Gamma}_{N_i N_j}\right\}\right)} \tag{5.114}$$

where the average of $\phi_{X_i X_i}$ and $\phi_{X_j X_j}$ is taken to improve robustness. Using the same approach for estimation of the denominator and for averaging across the unique sensor pairs as in the Zelinski post-processor (see Equation 5.108), the suppression gain can be estimated as

$$\hat{H}_{PF} = \frac{\frac{2}{M(M-1)} \sum_{i=1}^{M-1} \sum_{j=i+1}^{M} \hat{\phi}_{SS}^{(ij)}}{\frac{1}{M} \sum_{i=1}^{M} \hat{\phi}_{X_i X_i}}. \tag{5.115}$$

A problem arises if $\hat{\Gamma}_{N_i N_j} = 1$, $i \neq j$, as this leads to an indeterminate solution in Equation 5.113. In practice this issue can be mitigated by applying a maximum value of the estimated coherence that is slightly smaller than 1.

This microphone-array post-processor provides good results and suppresses more noise than the Zelinski post-processor. In addition, the authors evaluated their algorithm using a speech recognition test and proved that, for SNRs below 10 dB, they achieved reduction in the word error rate. The coherence-based post-processor is computationally more expensive than the previous one. In addition, the computational expenses increase with the square of the number of microphones.

5.7.4 Spatial Suppression and Filtering in the IDOA Space

The instantaneous direction of arrival (IDOA) space defined in Equations 5.24 and 5.25 provides an opportunity to spatially estimate the input signal for each frequency bin and audio frame separately. This opens the door to several techniques, based on spatial models and filters.

5.7.4.1 Spatial Noise Suppression

The post-processors based on power-density estimation and noise-coherence estimation rely only on the input signals and use a spatial equivalent of a Wiener-type suppression rule. In the chapter on stationary noise suppressors we saw how much the decision-directed approach and newer suppression rules improved noise suppression and output sound quality. To apply the same technique here we have to have spatial noise and signal models:

$$\begin{aligned}
\lambda_S(f|\Delta) &\triangleq \mathrm{E}[|S(f|\Delta)|^2] \\
\lambda_D(f|\Delta) &\triangleq \mathrm{E}[|D(f|\Delta)|^2].
\end{aligned} \tag{5.116}$$

Let $Y(f|\Delta)$ be the beamformer output, with coordinates Δ in the IDOA space, estimated according to Equations 5.24 and 5.25. The spatial noise and signal variances are not vectors, as in the single-channel case, but are functions of M parameters – the frequency, and the $(M-1)$-dimensional IDOA space. Then the a-priori and a-posteriori SNRs can be defined as

$$\begin{aligned}
\xi(f|\Delta) &\triangleq \frac{\lambda_S(f|\Delta)}{\lambda_D(f|\Delta)} \\
\gamma(f|\Delta) &\triangleq \frac{|Y(f|\Delta)|^2}{\lambda_D(f|\Delta)}.
\end{aligned} \tag{5.117}$$

Based on these definitions, any of the suppression rules derived in Chapter 4 can be used. For estimation of the a-priori SNR, the decision-directed approach can be applied:

$$\xi^{(n)}(f|\Delta) \triangleq \alpha \frac{\hat{S}^{(n-1)}(f)}{\lambda_D(f|\Delta)} + (1-\alpha)\max\left[0, \gamma^{(n)}(f|\Delta)-1\right]. \tag{5.118}$$

This technique is used for a small three-element microphone array for a headset by Tashev *et al.* [25]. The design is under the constraints of the low-powered CPU in the headset, which prohibits using more sophisticated and more CPU-expensive algorithms, discussed earlier. In this particular design, is used one of the computationally efficient alternatives of the Ephraim and Malah suppression rules [26]. Authors report additional noise suppression in the range 5–6 dB. The approach is computationally inexpensive, has a low memory footprint, and can be implemented for the CPU in a small headset.

On the other hand, this algorithm does not scale well when increasing the number of microphones. For the three-microphone case the noise models are two-dimensional matrices for each frequency bin. Let assume that we discretized the IDOA space to ten regions. This means that the size of the noise model matrix will be 10×10, or a total of 100 values to store and update for each frequency bin. Increasing the number of microphones to four leads to a $10 \times 10 \times 10$ matrix, or 1000 values to store and update. It is obvious that such a large number of values, with only one updated for each audio frame, it is not possible to keep up to date and follow the noise-field changes. The output signal estimation has no spatial property when the DDA is applied to estimate the a-priori SNR in Equation 5.118. From this perspective, this spatial noise suppressor has a better noise model estimate and the output will sound better in a non-isotropic noise field, but the spatial position of the desired sound source is not accounted for.

In Selzer *et al.* [27], the authors convert the values from the IDOA space back to a three-, or two-, or one-dimensional space, based on the microphone-array geometry. With known geometry for a point with coordinates Δ in the IDOA space, it is trivial to find the closest point with an image in the real space. For a linear microphone array, which distinguishes only one dimension, the noise and signal models will be function of the incident angle θ:

$$\begin{aligned} \lambda_Y(f|\theta) &\triangleq \mathrm{E}[|Y(f|\theta)|^2] \\ \lambda_D(f|\theta) &\triangleq \mathrm{E}[|D(f|\theta)|^2]. \end{aligned} \qquad (5.119)$$

The listening space of a four-element linear microphone array is split on 10 non-overlapping sectors in the range $[-90°, +90°]$. The discrete set of directions $[\theta_1, \theta_2, \ldots, \theta_L]$ contains the listening direction θ_c. For each audio frame, and frequency bin one of the models based on the decision from the VAD and the estimated sector is updated. For the SNR and suppression rule estimation, the noise model variance is computed while accounting for the undesired sound sources:

$$\hat{\lambda}_D(f|\theta_l) = \begin{vmatrix} \lambda_D(f|\theta_l) & \theta_l = \theta_c \\ \lambda_D(f|\theta_l) + \lambda_Y(f|\theta_l) & \text{otherwise} \end{vmatrix} \qquad (5.120)$$

which means that we treat the signals in non-listening sectors as noise. In this particular implementation, the authors in [27] use a machine-learning approach to learn the transfer function $\Delta \Rightarrow \theta$ and an EM algorithm to estimate the signal presence probability, which is applied as suppression gain. In other implementations a numerical approach can be used to find the closest point in the IDOA space, which has an image in the real space. Any of the algorithms already discussed for computing the suppression gain can be applied.

The second spatial noise-suppression algorithm is best for suppressing stationary (spatially and in time) unwanted sound sources. Building two models for each sector allows quick switching of the listening direction without adaptation time to build the separate noise and signal statistical models. If an unwanted sound source changes its position the system will need some time to update the models.

5.7.4.2 Spatial Filtering

The IDOA approach allows sound source localization per frequency bin; that is, we can estimate the source location of the signal in each frequency bin. Based on this estimation we can apply a spatial filter – real-valued gain that is close to 0 for regions we are not interested in and close to 1 if the signal comes from the listening direction. There are two factors that limit our ability to do spatial filtering: the uncorrelated (instrumental) noise and the reverberation. They both cause smearing of the signal in the IDOA space around the image of the real space and smearing of the point noise sources around their position. Taking a close look at Figure 5.18, we an see how the measured IDOA points for a sequence of frames in this frequency bin are smeared around the computed image of the real space – a line for far-field sound capture and linear microphone array. A second visible effect is that smearing is higher around $\pm 90°$, where the signal level is lower owing to the unidirectional microphones, and the instrumental noise has a higher influence.

We can present the noise in each microphone channel as a sum of two components: acoustic noise with variance $\lambda_N(f)$ and uncorrelated noise with variance $\lambda_{NC}(f)$. The acoustic noise can be modeled as one noise source in a different position for each frame. The probability distribution of this random noise source position determines the noise type. Then the noise in each microphone channel can be presented as

$$N_m^{(n)}(f) = D_m(f, c_n)\mathbb{N}(0, \lambda_N(f, c_n)) + \mathbb{N}(0, \lambda_{NC}(f)) \qquad (5.121)$$

where $D_m(c_n)$ is the capturing equation for this microphone, according to Equation 5.6, and c_n is random for each frame position with a given distribution. Note that in this equation the acoustic noise variance is position-dependent. For isotropic noise, c_n is uniformly distributed in the space and $\lambda_N(f, c_n)$ is constant for all c_n.

For simplicity and without reducing the generality further, we will consider a linear microphone array, sensitive only to the incident angle θ – the direction of arrival in one dimension. Let $\Psi_k(\theta)$ denote the function that generates the vector Δ for an angle θ and frequency bin k according to Equations 5.6, 5.24, and 5.25. In each frame the k-th bin is represented by one point Δ_k in the IDOA space. Let us have a sound source at θ_S with an image in the IDOA at $\Delta_S(k) = \Psi_k(\theta_S)$. With added noise, the resultant point in the IDOA space will be spread around $\Delta_S(k)$.

We can convert the distance to the theoretical hyperline in the IDOA space to the distance in the incident angle space (real world, one-dimensional in this case) by

$$\Upsilon_k(\theta) = \frac{\|\Delta_k - \Psi_k(\theta)\|}{\left\|\frac{d\Psi_k(\theta)}{d\theta}\right\|} \qquad (5.122)$$

where $\|\Delta_k - \Psi_k(\theta)\|$ is the Euclidean distance between Δ_k and $\Psi_k(\theta)$ in the IDOA space, $d\Psi_k(\theta)/d\theta$ is the derivative, and $\Upsilon_k(\theta)$ is the distance from the observed IDOA point to the points in the real-world one-dimensional incident angle in our consideration. Note that the dimensions of the axes in the IDOA space are measured in radians as a phase difference, while $\Upsilon_k(\theta)$ is measured in radians as the units of the incident angle.

To estimate the variance of the sound source position fluctuation, an adaptation model with smoothing in the spatial and frequency directions is used. In real implementations $\Upsilon(\theta)$ is a $K \times L$ matrix, where K is the number of frequency bins and L is the number of discrete values of the direction angle θ. Variation estimation goes through two stages. During the first stage we build a rough variation estimation matrix $\breve{\lambda}_k(\theta)$. If $\theta_{min}^{(n)}(k)$ is the angle that minimizes $\Upsilon_k^{(n)}(\theta_l)$ in the n-th frame, only corresponding values of the spatial model are updated in each frame:

$$\breve{\lambda}_k^{(n)}(\theta_{min}) = (1-\alpha)\breve{\lambda}_k^{(n-1)}(\theta_{min}) + \alpha\Upsilon_k^{(n)}(\theta_{min})^2 \tag{5.123}$$

where $\Upsilon_k^{(n)}(\theta)$ is estimated for the current frame according to Equation 5.122, and $\alpha = T/\tau_A$ (τ_A is the adaptation time constant, T is the frame duration). As the data per frequency bin is quite noisy, during the second stage a direction–frequency smoothing filter $H_k(\theta)$ is applied after each update to estimate the spatial variation matrix $\lambda_k(\theta) = H_k(\theta)*\breve{\lambda}_k(\theta)$. Here we assume a Gaussian distribution of the uncorrelated component, which allows us to assume the same deviation in the real space towards θ. The spatial variance represents the smearing in space of a point noise source owing to uncorrelated noise and reverberation.

With known spatial variation $\lambda_k(\theta_l)$ and distance $\Upsilon_k^{(n)}(\theta_l)$, the probability density for frequency bin k to come from direction θ_l, assuming Gaussian distribution, is given by

$$p_k^{(n)}(\theta_l) = \frac{1}{\sqrt{2\pi\lambda_k^{(n)}(\theta_l)}} \exp\left(-\frac{\Upsilon_k^{(n)}(\theta_l)^2}{2\lambda_k^{(n)}(\theta_l)}\right) \tag{5.124}$$

which allows us to estimate the probability that the signal comes from the listening direction θ_S:

$$P_k(\theta_S) = \frac{p_k^{(n)}(\theta_S)}{\sum_{l=1}^{L} p_k^{(n)}(\theta_l)}. \tag{5.125}$$

This probability estimation is based only on the information in the current bin of the current frame. Naturally this probability estimation is quite noisy. To utilize the temporal properties of the speech signal, we can apply a smoothing filter towards the time axis. We can use the same HMM-based smoothing technique, described in

Chapter 4:

$$P_k^{(n)}(\theta_S) = \frac{a_{01} + a_{11}P_k^{(n-1)}(\theta_S)}{a_{00} + a_{10}P_k^{(n-1)}(\theta_S)} P_k(\theta_S) \tag{5.126}$$

where a_{00}, a_{11}, a_{01}, and a_{10} are the probabilities to stay in the same state (speech or non-speech), or to change it; and $P_k(\theta_s)$ is estimated by Equation 5.125 while $P_k^{(n)}(\theta_S)$ is the probability to have signal at look-up direction θ_S for the n-th frame. The suppression gain then can be computed as

$$H_k^{(n)} = \frac{P_k^{(n)}(\theta_S)}{\sqrt{2\pi\lambda_k^{(n)}(\theta_{\min})}} \tag{5.127}$$

where the denominator is just a scaling parameter to ensure that if a point is lying on the IDOA hyperline we will have a suppression gain of 1. The spatio-temporal filter to compute the post-processor output $Z_k^{(n)}$ from the beamformer output $Y_k^{(n)}$ is

$$Z_k^{(n)} = P_k^{(n)}(\theta_S) \cdot Y_k^{(n)}; \tag{5.128}$$

that is, we use the signal presence probability as a suppression rule which is an MMSE solution.

The implementation of this spatial filter for a linear four-element microphone array is discussed in [28]. Applying the spatial filter improves the directivity pattern of the beamformer, as shown in Figure 5.35 for 1000 Hz. The overall directivity index is

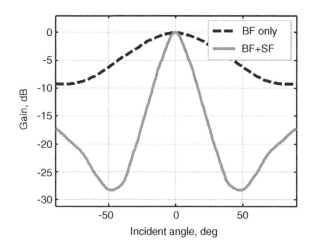

Figure 5.35 Spatial filter directivity measured at 1000 Hz

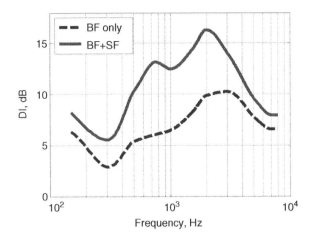

Figure 5.36   Directivity index of a microphone array with spatial filter

increased 3–8 dB in the frequency band 500–8000 Hz, as shown in Figure 5.36. Note that the plots on these two figures are based on actual measurements in an anechoic chamber; they are not simulation results. This IDOA-based spatial filter substantially improves the selectivity capabilities of the microphone array. It is not computationally expensive and can be implemented to work in real time. On the other hand, it is non-linear processing with the specific distortions and artifacts. For the best results it should be tuned carefully using the approaches described in Chapter 4.

5.7.4.3 Spatial Filter in Side-lobe Canceller Scheme

The generalized side-lobe canceller architecture is discussed earlier in this chapter. One of the problems was increasing the robustness of the blocking matrix to DOA estimation errors. Regardless of the deep null we can create with the nullformers in the blocking matrix, these DOA estimation errors result in leakage of the desired signal after the blocking matrix. In [21], a set of constraints are introduced to the adaptive filters in the blocking matrix to allow moving of the null in a certain interval $[\theta_S - \Delta\theta, \theta_S + \Delta\theta]$. A spatial filter can be used to do the same in a more elegant way [29].

We change the generalized side-lobe canceller scheme as shown in Figure 5.37. The spatial filter is used to let through only the signals in the desired interval and employs the suppression gain modified from Equation 5.125:

$$P_k(\theta_S) = \frac{\displaystyle\sum_{l=S-\Delta L}^{S+\Delta L} p_k^{(n)}(\theta_l)}{\displaystyle\sum_{l=1}^{L} p_k^{(n)}(\theta_l)} \tag{5.129}$$

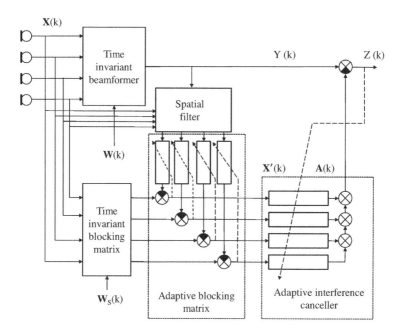

Figure 5.37   Enhanced generalized side-lobe canceller

passing everything in the range $[\theta_S - \Delta\theta, \theta_S + \Delta\theta]$, corresponding to discrete values $[S - \Delta L, S + \Delta L]$ to the blocking matrix. The smoothing in Equation 5.126 should be applied. With this new scheme, constraints around DOA estimation are applied in a simpler and efficient manner, and the sound quality is better as we do not have the spatial filter output directly in the output signal.

5.7.4.4 Combination with LMS Adaptive Filter

It is trivial to show that the generalized side-lobe canceller, regardless of its complex structure, can be normalized to the generalized filter-and-sum beamformer form in Equation 5.15. Technically, if we can find a good criterion for updating a generic adaptive filter-and-sum beamformer we can achieve the same results. In addition we have a spatial filter algorithm, which has very good filtering capabilities, but is non-linear and introduces distortion and artifacts. The general idea in [30] is to use the spatial filter output as a reference and to change the adaptive filters in a way to achieve an output as close as possible to the adaptive beamformer. The block diagram is shown in Figure 5.38. To prevent the system from shutting itself down during pauses, a VAD should be used to block the adaptation during the silent segments.

This adaptive beamformer is close to the Widrow beamformer, but instead of generating a reference signal the output of the spatial filter is used. The output of this beamformer is a linear combination of the input signals and does not contain

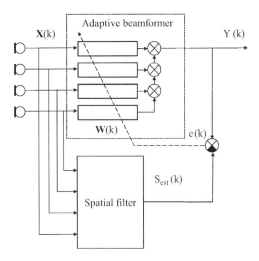

Figure 5.38   Adaptive beamformer combined with spatial filter

distortions and musical noise. In the converged state it should achieve the maximum efficiency of a linear beamformer. Note that the constraint for unity gain and zero phase shift is implicitly enforced by the spatial filter.

As evaluation criteria, besides improvement in SNR, the authors of this paper use perceptual sound quality, measured using the PESQ algorithm. Another interesting point in the paper is that a new adaptive filter algorithm is derived which minimizes the log-MMSE error, contrary to the classic NLMS adaptive filter which minimizes the MMSE. Let the output of the spatial filter be $\hat{S}_k^{(n)}$ and the beamformer output be $Y_k^{(n)} = \mathbf{W}_k^{(n)} \mathbf{X}_k^{(n)}$. We want to find a weights vector such that

$$W = \arg\min_{W} \mathrm{E}\left[(\log(|\hat{S}|^2) - \log(|Y|^2))^2\right]. \tag{5.130}$$

For online adaptations we can replace the estimator with an instantaneous error

$$\varepsilon_k^{(n)} = \frac{1}{2}\left[\log(|\hat{S}_k^{(n)}|^2) - \log(|Y_k^{(n)}|^2)\right]^2. \tag{5.131}$$

Taking the derivative with respect to the weights leads to

$$\begin{aligned}
\frac{\partial \varepsilon}{\partial \mathbf{W}} &= -\frac{\left[\log(|\hat{S}_k^{(n)}|^2) - \log(|Y_k^{(n)}|^2)\right]}{|Y_k^{(n)}|^2} \mathbf{X}_k^{(n)} (\mathbf{X}_k^{(n)})^{\mathrm{H}} \mathbf{W} \\
&= -\left[\log(|\hat{S}_k^{(n)}|^2) - \log(|Y_k^{(n)}|^2)\right] \frac{\mathbf{X}_k^{(n)}}{Y_k^{(n)}}
\end{aligned} \tag{5.132}$$

which can be used to update the log-LMS adaptive filter coefficients:

$$\mathbf{W}_k^{(n+1)} = \mathbf{W}_k^{(n)} - \mu \left[\log(|\hat{S}_k^{(n)}|^2) - \log(|Y_k^{(n)}|^2) \right] \frac{\mathbf{X}_k^{(n)}}{Y_k^{(n)}}. \tag{5.133}$$

A variable adaption step converts this to a log-NLMS adaptive filter:

$$\mu = \tilde{\mu} \frac{|Y|^2}{X^H X}, 0 < \tilde{\mu} < 1. \tag{5.134}$$

The paper provides, for comparison with this approach, the Frost algorithm and the delay-and-sum as reference. Table 5.6 summarizes the results. The adaptive beam-former can use as reference a close-talk microphone channel (an experiment to show the maximum this method can achieve) or the spatial filter output (realistic design). An interesting conclusion is that, while the classic NLMS filter gives better results using the close-talk microphone as a reference channel, in the realistic experiment the log-NLMS filter provides better results owing to increased robustness to the distortions in the spatial filter output.

Table 5.6 Output SNR and perceptual sound quality

Algorithm	SNR	MOS
Delay and sum	12.73	2.20
Frost	13.64	2.18
Spatial filter	21.20	2.06
Log-NLMS	18.15	2.41

Overall, this adaptive beamformer provides a well audible improvement of 0.21 MOS points in the perceptual sound quality, compared to the conventional delay-and-sum beamformer.

5.8 Specific Algorithms for Small Microphone Arrays

To be an effective spatial filter of the linear type (a beamformer) the microphone array should have a size comparable to the wavelength of the sound wave. The most commonly used linear arrays have a length of 10–20 cm and are effective down to 500–300 Hz. These arrays cannot be integrated into small form-factor devices such as cellphones, PDAs, camcorders, and so on. On the other hand, mobile devices are increasingly being used in situations that require hands-free communication. With advancement of the 3G and 4G wireless technologies, high-speed connection to the Internet will soon become standard in calling plans. This opens the door to mobile video telephony. Often, a multimodal user interface is used for quick access to information on the Internet, combining speech recognition and a graphic screen. In these cases, moving the phone close to the mouth to speak and then returning it to roughly an arm's length away to see the screen can be quite awkward. In the camcorder mode the device is supposed to capture sound sources from distances of 1–3 m. In these usage modes the sound source is located at some distance from the microphone.

Unfortunately, these devices historically have only one microphone which, in most cases, is omnidirectional. This leads to the picking up of too much ambient noise and reverberation, making these devices useless under higher noise conditions. The most trivial way to improve the sound capture quality is to use a unidirectional microphone. This increases the SNR with about 4.3 dB, but worsens the audio quality during video recording, as usually the video camera is on the opposite side of the phone. Still, even this small advantage in SNR helps the stationary noise suppressor to do a better job, as it is less efficient with input SNRs below 5–10 dB.

The next logical improvement is to add one more microphone and to use one of the standard beamforming technologies to improve the SNR and sound quality. Owing to the small size of the device and the need to keep the microphones as far as possible from the loudspeaker, the distance between the two microphones is limited to 30–50 mm. The efficiency of the classic beamforming techniques with small-base microphone arrays is quite low.

The majority of the algorithms discussed so far (linear beamforming, spatial filtering) rely heavily on the phase difference between the signals captured by the microphones. This phase difference is used to distinguish signals coming from desired and unwanted directions. Owing to the small size of these devices, the phase difference is small and practically buried in ambient and instrumental noises. This is why the differences in the magnitudes of the captured signals should play an important role in distinguishing desired from undesired signals.

5.8.1 Linear Beamforming Using the Directivity of the Microphones

Theoretically speaking, two omnidirectional microphones placed at a distance of 5–30 mm (Figure 5.39) can generate all the first-order directivity patterns discussed in Chapter 3. The difference of the two signals produces a figure-8 directivity pattern. Changing the proportions would give all other directivity patterns – hypercardiod, supercardiod, cardiod, subcardiod, and omnidirectional. These directivity patterns cannot be steered; that is, the cardioid directivity pattern will point either at $0°$ or at $180°$.

Another limitation is the frequency compensation as discussed in Chapter 3. In an acoustical first-order microphone, conversion to an electrical signal happens after acoustically mixing the signals from front and back. The microphone's self noise is added in the process of convertion to an electrical signal. In the case of two microphones, conversion to an electrical signal happens first and then the signals are

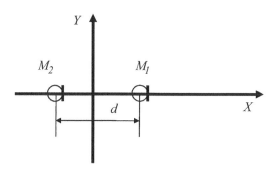

Figure 5.39 Two omnidirectional microphones in endfire configuration

mixed. This leads to higher noise levels and decreased performance in the lower part of the frequency band.

Using two or three microphones, closely positioned to each other, is attractive owing to the small size of the array and easy design. Elko and Pong [31] derive the directivity pattern of a small two-element microphone array. If we have two omnidirectional microphones, placed at a distance d, and our processing consists just of applying delay T to one of the microphones and summing the two signals, then the directivity pattern of the microphone array would be

$$U(\theta) = \alpha + (1-\alpha)\cos\theta,$$
$$\alpha = \frac{T}{T+d/c}. \tag{5.135}$$

With a fixed d we can vary T to achieve any of the first-order directivity patterns. These directivity patterns have a single null in the rear half-plane, given by

$$\theta_{\text{null}} = \text{arc}\cos\left(\frac{\alpha}{\alpha-1}\right). \tag{5.136}$$

In the same paper the authors propose a simple NLMS adaptive algorithm to vary the delay (i.e., the parameter α) to achieve the best noise suppression. In the case of ambient isotropic noise, the filter should converge to a hypercardiod directivity pattern with the maximum for the first-order directional microphone directivity index of 6 dB. If there is a point noise source, the adaptive algorithm would place the null towards its direction. In a later paper [32], the authors use three closely positioned omnidirectional microphones to generate steerable and variable directivity patterns.

The design approach so far is simply a reduced version of the delay-and-sum beamformer. With two microphones nothing more than achieving a first-order directivity pattern can be done. With two cardiod microphones, pointing in opposite directions, the same set of first-order directivity patterns can be generated as with two omniridirectional microphones. The benefit of using unidirectional microphones is that the resulting microphone would have less instrumental noise in the output for the majority of real cases.

In [33], Mihov *et al.* use the generalized beamformer (Equation 5.15) and optimization methods as described in Section 5.4.6 to design the time-invariant beamformer for a mobile phone device. The microphone array consists of two directional microphones, pointing in opposite directions, as shown in Figure 5.40. The goal for such devices can be to design a set of beamformers that capture the signals from:

- *Front only* – talking-on-the-phone mode – holding the headset close to the mouth for videophone mode and also when the device is held at arm's length.
- *Rear only* – video camera mode – pointing the device towards the subject of interest and the video camera is in the rear of the device. In this case the device screen is used as a viewfinder.
- *Both front and rear* – video camera mode – when the user wants to capture their own voice for adding comments or narration.

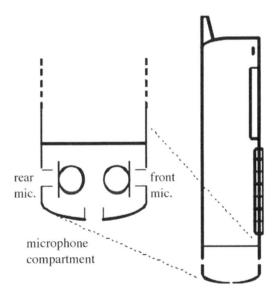

Figure 5.40 Small microphone array with two cardiod microphones facing in opposite directions

One of the problems in this much more realistic design is that the directivity patterns of the used microphones are not cardioid and cannot be described with a frequency-independent coefficient α as in Equation 5.135. Even if they were, placing them in the device enclosure will change the directivity pattern and the frequency response. This requires a measured directivity pattern $U_m(\theta, f)$, which can be done as described in Chapter 3. With these real patterns, ideal first-order directivity patterns cannot always be achieved. In some cases, achieving them will boost the instrumental noise to unacceptable levels. Then the directional microphones have a directivity index of 3.5 dB – below the expected 4.8 dB from a cardioid microphone. As the result of the optimization for the third case above, the authors achieve a good figure-8 directivity pattern, frequency response equalization, and 0.29 points improvement in the perceptual sound quality (MOS score measured with the PESQ algorithm). The designed beamformer can be implemented as a time-domain 64- or 128-taps filter and integrated directly into the ADC chip, as many of these chips have stereo input, digital filters for frequency correction of each channel, and mono mode where they sum the two signals. Thus the improvement in sound capture quality does not consume CPU time from the main processor. The downside of this simple and inexpensive design is that this beamformer cannot steer and switch modes.

5.8.2 Spatial Suppressor Using Microphone Directivity

With two closely positioned microphones, the maximum noise suppression that can be achieved is 6 dB – the hypercardioid directivity pattern. Using the unidirectional microphone from the design above with 3.5 dB directivity index, the improvement in SNR is 2.5 dB. If more noise suppression is necessary, a non-linear spatial suppression algorithm should be employed. Usually, mobile devices are constrained by both CPU power and available memory. While the approaches for microphone-array post-processors described earlier in this chapter may work well, usually they are CPU- and memory-expensive. In most cases they rely directly (IDOA) or indirectly (coherence matrix) on the phase difference, which in small devices is not useful. Such small microphone arrays require use of the magnitude as well.

A three-staged approach for spatial filtering for small microphone arrays is described in [34]. It is implemented for the same device described in the previous subsecton. The processing block diagram is shown in Figure 5.41. The signals from

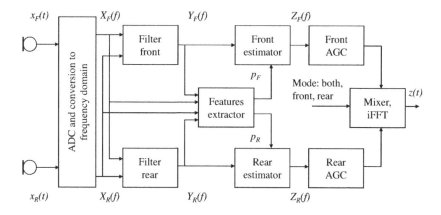

Figure 5.41   Block diagram of the processing algorithm for a small microphone array

the two microphones are processed by two beamformers optimized to provide maximum front–back difference. Then, a feature extractor, assisted by a binary voice activity detector (VAD), computes the differences in the outputs of the beamformers. Based on dynamically updated statistical models, the probability of the speech signal to be coming from the desired direction is computed, which is then applied as suppression gain.

5.8.2.1 Time-invariant Linear Beamformers

Using measured directivity patterns $U_F(\theta,f)$ and $U_R(\theta,f)$ of the front and rear microphones, as they are installed in the device, two beamformers are designed using the direct optimization approach. The optimization criteria are selected to provide the maximum difference between the captures from the front $[-\Delta\theta, +\Delta\theta]$ and rear $[-\pi+\Delta\theta, +\pi-\Delta\theta]$ sectors, as shown in Figure 5.42. The constrained optimization goals are

$$Q_{F\,const} = \max_{\mathbf{W}_{FF},\mathbf{W}_{FR}} \left(\frac{\int\limits_{-\Delta\theta}^{+\Delta\theta} (\mathbf{W}_{FF}\cdot\mathbf{X}_F(\theta) + \mathbf{W}_{FR}\cdot\mathbf{X}_R(\theta))d\theta}{\int\limits_{-\pi+\Delta\theta}^{\pi-\Delta\theta} (\mathbf{W}_{FF}\cdot\mathbf{X}_F(\theta) + \mathbf{W}_{FR}\cdot\mathbf{X}_R(\theta))d\theta} \right)$$

$$Q_{R\,const} = \max_{\mathbf{W}_{RF},\mathbf{W}_{RR}} \left(\frac{\int\limits_{-\pi+\Delta\theta}^{\pi-\Delta\theta} (\mathbf{W}_{RF}\cdot\mathbf{X}_F(\theta) + \mathbf{W}_{RR}\cdot\mathbf{X}_R(\theta))d\theta}{\int\limits_{-\Delta\theta}^{+\Delta\theta} (\mathbf{W}_{RF}\cdot\mathbf{X}_F(\theta) + \mathbf{W}_{RR}\cdot\mathbf{X}_R(\theta))d\theta} \right). \tag{5.137}$$

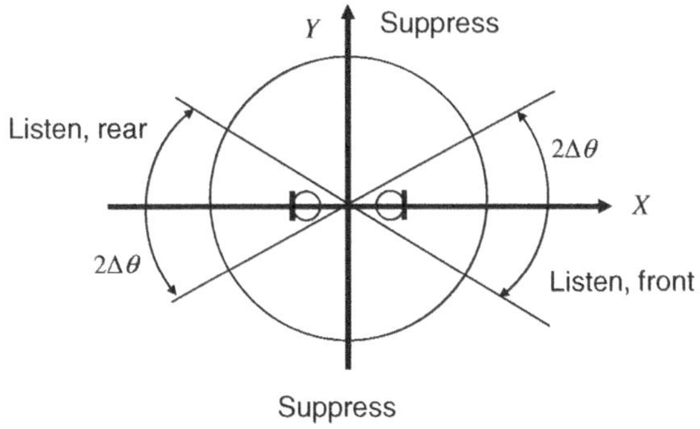

Figure 5.42 Front and rear listening sectors and suppression zones

Here, $\mathbf{W}_{FF}$, $\mathbf{W}_{FR}$, $\mathbf{W}_{RF}$, and $\mathbf{W}_{RR}$ are the weights vectors. We can assume the independence of the signals in the frequency bins and do the optimization for each frequency bin separately. To convert the constrained optimization goals in Equation 5.137 to unconstrained, we can add the constraints as penalty functions. For the front beam the unconstrained optimization goal looks like

$$Q_F = Q_{F\,const} + g_1|C_F - 1.0|^2 + g_2 \sum |W|^2. \tag{5.138}$$

Here, g_1 and g_2 are weight parameters. C_F is the array gain towards the front direction, and the second term enforces the unity gain and zero phase shift constraint. $\Sigma|W|^2$ is the sum of absolute values of the optimized gains and reduces the instrumental noise gains. The unconstrained optimization goal for the rear beam is defined in the same way. After completing the optimization, the designed directivity pattern has close to cardioid shape and the frequency response towards the main response axis is well equalized.

5.8.2.2 Feature Extraction and Statistical Models

Distinguishing between desired and unwanted signals is based on their spatial position. As the beamformers are optimized to increase the difference between front and rear signals, four features are selected: difference in signal level for the whole audio frame, difference in signal magnitude per frequency bin, delay for the whole frame, and delay per frequency bin.

Under ideal conditions (perfect microphone matching, identical directivity patterns, and isotropic ambient noise), the levels should be the same and the average level difference ΔL should be zero. Potential non-zero differences in the pauses can occur when the real microphones have different-than-modeled characteristics owing to channel mismatch. To compensate for this, both the mean and variation of the level differences are computed. The noise model update should be in the noise frames only:

$$
\begin{aligned}
L_C^{(n)} &= \left(1 - \frac{T}{\tau_W}\right) L_C^{(n-1)} + \frac{T}{\tau_W} \Delta L^{(n)} \\
\lambda_W^{(n)} &= \left(1 - \frac{T}{\tau_W}\right) \lambda_W^{(n-1)} + \frac{T}{\tau_W} (\Delta L^{(n)} - L_C^{(n)})^2.
\end{aligned}
\tag{5.139}
$$

Here, T is the frame duration and τ_W is the adaptation time constant. Then the level difference for the current frame is

$$L_W^{(n)} = \Delta L^{(n)} - L_C^{(n)} \tag{5.140}$$

and this is the first of the four features. The couple (L_C, λ_W) characterizes a Gaussian process for the level differences fluctuation during noise-only frames. The level difference during speech frames is modeled as a statistical process with asymmetric PDF: exponential for the positive differences and Gaussian-shaped for the negative differences:

$$
P_{FW}(\Delta L_W | \theta_{FW}, \sigma_W) = \left|
\begin{array}{ll}
\dfrac{1}{\theta_{FW}} \exp\left(-\dfrac{\Delta L_W}{\theta_{FW}}\right) & \Delta L_W > 0 \\[4mm]
\dfrac{1}{\theta_{FW}} \exp\left(-\dfrac{\Delta L_W^2}{2\sigma_w^2}\right) & \text{otherwise.}
\end{array}
\right.
\tag{5.141}
$$

Here the exponential distribution parameter is estimated during voiced frames and positive level differences as

$$
\theta_{FW}^{(n)} = \left(1 - \frac{T}{\tau_W}\right)\theta_{FW}^{(n-1)} + \frac{T}{\tau_W}\Delta L_W^{(n)}.
\tag{5.142}
$$

The statistical model parameter for the sound coming from the rear, $\theta_{RW}^{(n)}$, is estimated and updated in the same way when the level differences are negative.

The second feature is the magnitude difference per frequency bin. The same statistical models as above are created by estimating the parameters $L_{Cb}^{(n)}(k)$, $\theta_{FWb}^{(n)}(k)$, and $\theta_{RWb}^{(n)}(k)$ for each frequency bin. The adaptation time constant is τ_{Wb}.

The third feature is the time delay between the signals from the two microphones. The delay is estimated using PHAT weighting and the "generalized cross-correlation method" (explained in the next chapter):

$$
\mathbf{C}_{FR}(\tau) = \mathbf{iFFT}\left[\frac{\mathbf{X}_F \cdot \mathbf{X}_R^*}{|\mathbf{X}_F| \cdot |\mathbf{X}_R|}\right].
\tag{5.143}
$$

Quadratic interpolation can be used to find the maximum with sub-sampling period resolution. Based on the classification from the VAD and the delay sign (negative or positive), three statistical models are built: noise (updated during non-voiced frames), front (updated during voiced frames and positive delays), and rear (updated during voiced frames and negative delays). The models assume Gaussian distribution, same variances, and means computed from the geometrical positions of the microphones. This leaves only one parameter to estimate in real time – the variance λ_D. The adaptation time constant is τ_D.

The fourth feature is a delay per frequency bin, estimated from the phase differences of the microphone signals:

$$
D_b(k) = \frac{\text{norm}\left[\arg(X_F(k)) - \arg(X_R(k))\right]}{2\pi f}
\tag{5.144}
$$

normalized in the range $[-\pi, +\pi]$. The adaptation time constant is τ_{Db}.

5.8.2.3 Probability Estimation and Features Fusion

Given a frame level difference $L_W^{(n)}$ between the front and back beams, the probability that this frame is dominated by a signal coming from the front is

$$\hat{p}_{FW}^{(n)} = \frac{P_{FW}(\Delta L_W^{(n)})}{P_{FW}(\Delta L_W^{(n)}) + P_{RW}(\Delta L_W^{(n)}) + P_{NW}(\Delta L_W^{(n)})} \tag{5.145}$$

where $P_{FW}(\Delta L_W^{(n)})$, $P_{RW}(\Delta L_W^{(n)})$, and $P_{NW}(\Delta L_W^{(n)})$ are the values of the front, rear, and noise PDFs for this level difference. The probabilities for the other three features are estimated in the same manner.

Once we have the probability estimations for a speech signal coming from the desired direction, we can combine them:

$$p_k^{(n)} = \prod_{i=1...4} ((1-G_i)\hat{p}_i^{(n)}(k) + G_i) \tag{5.146}$$

where $p_k^{(n)}$ is the probability of a signal coming from desired direction in n-th frame and k-th bin, $p_i^{(n)}(k)$ is the probability for the i-th feature, and G_i is the feature gain. When the gain is 1 the feature is disabled; when it is 0 the feature is in its full weight. The overall probability can be used as a suppression gain, reducing the presence of sounds coming from unwanted directions. This is an MMSE solution for the time-domain waveform.

5.8.2.4 Estimation of Optimal Time-invariant Parameters

The algorithm above (besides means and variances which are estimated in real time from the input signals) has adaptation time constants and gains with values that cannot be estimated theoretically. To find the optimal values for them, mathematical optimization can be used with the values of the adaptation time constants and gains as parameters – a total of eight parameters: τ_W, τ_{Wb}, τ_D, τ_{Db}, G_W, G_{Wb}, G_D, and G_{Db}.

To estimate the desired suppression gains, a pre-recorded corpus is used. A source consisting of a recorded human voice played through a mouth simulator, placed in the desired position, can be recorded with the sound capture system in low noise and reverberation conditions – in an anechoic chamber, for example. This can be repeated for various speakers with different genders and ages. Ambient noises in various conditions – cafeteria, office, street, and so on – can be recorded using the same device. The sum of any combination of these two signal types is what would be recorded in the same conditions with a real human speaker. This allows the creation of a substantial number of test recordings. All three files (clean, noise-only, mixture) from each set are processed in parallel. Clean speech and noise recordings having been separated, a precise Wiener gain for each frame and frequency bin is possible:

$$H_W^{(n)}(k) = \frac{|X_k^{(n)}|^2}{|X_k^{(n)}|^2 + |N_k^{(n)}|^2} \tag{5.147}$$

which is an MMSE estimator when applied to the input signal. Here, $X_k^{(n)}$ and $N_k^{(n)}$ are the clean-speech and noise signals, respectively. As the probability estimator in Equation 5.146 is an MMSE estimator as well, then we should minimize

$$Q_{\text{contsr}} = \min_{\mathbf{R}} \left(\sum_{n=0}^{N-1} \sum_{k=0}^{K-1} (H_w^{(n)}(k) - \hat{P}_k^{(n)})^2 \right) \tag{5.148}$$

where $\mathbf{R}$ is the vector of the parameters for optimization. To keep the values of the adaptation time constants, especially the gains, in the allowed boundaries, we convert the constrained optimization goal (Equation 5.148) into non-constrained by adding penalty functions:

$$\begin{aligned} Q_{\text{non-contsr}} = Q_{\text{contsr}} \ &+ \sum_{i=0}^{R-1} (\max (0, r_i - r_{\max}(i))^2) \\ &+ \sum_{i=0}^{R-1} (\min (0, r_i - r_{\min}(i))^2). \end{aligned} \tag{5.149}$$

For estimation of the parameter optimal values, almost any algorithm for multidimensional mathematical optimization can be used.

In [34] the authors use a 16 kHz sampling rate and a frame size of 512 samples. For minimizing Equation 5.149 they used a set of 16 files with various voices and input SNRs. The first 80% of each file is used for optimization, the last 20% for testing. After each iteration of the gradient descent algorithm, the test parts should be evaluated using the optimization criterion. The optimization procedure should be stopped when there is no further improvement in the test set evaluation after five iterations in a row. This is done to prevent overtraining of the optimization parameters. The estimated optimal parameters are used to evaluate a second set of recordings, which did not participate in the optimization. The optimal gains per feature are shown in the first line of Table 5.7. It is obvious that the optimization procedure practically turned off the last two features: delay per frame and delay per frequency bin. This is due to the small distance between the microphones, equivalent to the delay of one quarter of the sampling period. The same table presents the estimated gains for various feature combinations. The combination level difference per frame and level difference per bin produces the best results and only these two features are used for generating the results in Table 5.8,

Table 5.7 Optimal gain for various features combinations

Feature combination	Gain				Av. SNR (dB)
	Lev/fr	Lev/bin	Del/fr	Del/bin	improv.
All four	0.00	0.00	0.89	0.99	11.06
Lev/bin&fr, Del/fr	0.02	0.00	0.68		10.82
Lev/bin&fr	0.00	0.19			10.43
Lev/bin, Del/fr		0.00	0.48		4.83
Lev/fr	0.00				5.25
Lev/bin		0.00			5.12
Del/fr			0.00		6.21
Del/bin				0.00	1.96

Table 5.8 Improvements in SNR and MOS for each processing stage

Processing stage	BF	SF	Total
Av. SNR improv. (dB)	5.12	5.31	10.43
Av. MOS improv.	0.15	0.24	0.39

which shows the results of the processing using two evaluation parameters: SNR and MOS, estimated using the PESQ algorithm. The noise-suppression and speech-enhancement chain is well balanced across the processing blocks, and shows very good noise suppression and an audible increase in perceptual sound quality.

5.9 Summary

This chapter was dedicated to the algorithms and approaches for capturing sound with multiple microphones – microphone arrays. The time-invariant beamformers are simpler to design and are computationally efficient in real time. Adaptive beamformers show their strength in non-isotropic sound fields and point noise sources, but require more computational power. Manufacturing tolerances of the microphones cause channel mismatch and reduction in the beamformers' performance. Either designs that are robust to the channel mismatch beamformer, or proper calibration or auto-callibration procedures, or a combination of both, should be used to mitigate the channel mismatch. Further improvement of noise-suppression capabilities of microphone arrays can be done by using non-linear post-processors and spatial filters. These algorithms are similar to the noise suppressors and, besides suppressing noise, introduce distortions and musical noise. While the classic beamformers and spatial filters rely mostly on phase differences in the captured signals, small microphone arrays should use mostly differences in magnitudes, owing to their small size.

Bibliography

[1] Frost, O. (1972) An algorithm for linearly constrained adaptive array processing. *Proceedings of IEEE*, **60**(8), 926–934.

[2] Van Trees, H.L. (2002) *Optimum Array Processing*, Part IV of Detection, Estimation and Modulation Theory, John Wiley & Sons, New York.

[3] Cook, R., Waterhouse, R., Berendt, R. *et al.* (1955) Measurement of correlation coefficients in reverberant sound fields. *Journal of the Acoustical Society of America*, **27**, 1072–1077.

[4] Elko, G.W. (2000) Superdirective microphone arrays, in *Acoustic Signal Processing for Telecommunications* (eds S. Gay and J. Benesty), Kluwer Academic, Norwell, MA, Chapter 10, pp. 181–237.

[5] Cox, H., Zeskind, R. and Koou, T. (1986) Practical supergain. *IEEE Transactions on Acoustics, Speech, and Signal Processing*, **ASSP-34**(3), 393–398.

[6] Shoup, T. (1979) *A Practical Guide to Computer Methods for Engineers*, Prentice-Hall, Englewood Cliffs, NJ.

[7] Tashev, I. and Malvar, H.S. (2005) A new beamformer design algorithm for microphone arrays. Proceedings of International Conference of Acoustic, Speech and Signal Processing ICASSP 2005, Philadelphia, PA.

[8] Tashev, I. (2005) Beamformer sensitivity to microphone manufacturing tolerances. Nineteenth International Conference Systems for Automation of Engineering and Research SAER 2005, St. Konstantin Resort, Bulgaria.

[9] Wong, G.S.K. and Embleton, T.F.W. (eds) (1995) *AIP Handbook of Condenser Microphones: Theory, Calibration, and Measurements*, American Institute of Physics, New York.

[10] Nordholm, S., Claesson, I. and Dahl, M. (1999) Adaptive microphone array employing calibration signals: an analytical evaluation. *IEEE Transactions on Speech and Audio Processing*, **7**(3), 241–252.

[11] Wu, H., Jia, Y. and Bao, Z. (1996) Direction finding and array calibration based on maximal set of nonredundant cumulants. Proceedings of ICASSP '96.

[12] Tashev, I. (2004) Gain self-calibration procedure for microphone arrays. Proceedings of International Conference for Multimedia and Expo ICME 2004, Taipei, Taiwan.

[13] Er, M.H. (1990) A robust formulation for an optimum beamformer subject to amplitude and phase perturbations. *Signal Processing*, **19**(1), 17–26.

[14] Doclo, S. and Moonen, M. (2007) Superdirective beamforming robust against microphone mismatch. *IEEE Transactions on Audio, Speech, and Language Processing*, **15**(2), 617–631.

[15] Doclo, S. and Moonen, M. (2003) Design of broadband beamformers robust against gain and phase errors in the microphone array characteristics. *IEEE Transactions on Signal Processing*, **51**(10), 2511–2526.

[16] Doclo, S. and Moonen, M. (2003) Design of broadband beamformers robust against microphone position errors. Proceedings of International Workshop on Conference on Acoustic, Echo, and Noise Control IWAENC, Kyoto, Japan.

[17] Doclo, S. and Moonen, M. (2003) Design of broadband speech beamformers robust against errors in the microphone characteristics. Proceedings of International Conference of Acoustic, Speech and Signal Processing ICASSP 2003, Hong Kong.

[18] Chen, H., Ser, W. and Yu, Z. (2007) Optimal design of nearfield wideband beamformers robust against errors in microphone array characteristics. *IEEE Transactions on Circuits and Systems*, **54**(9), 1950–1959.

[19] Widrow, B., Mantey, P.E., Griffiths, L.J. and Goode, B.B. (1967) Adaptive antenna systems. *Proceedings of the IEEE*, **55**(12), 2143–2159.

[20] Griffiths, L.J. and Jim, C. (1982) An alternative approach to linearly constrained adaptive beamforming. *IEEE Transactions on Antennas and Propagation*, **AP-30**(1), 27–34.

[21] Hoshuyama, O., Sugiyama, A. and Hirano, A. (1999) A robust adaptive beamformer for microphone arrays with a blocking matrix using constrained adaptive filters. *IEEE Transactions on Signal Processing*, **47**, 2677–2684.

[22] Herbordt, W. and Kellermann, W. (2001) Computationally efficient frequency-domain robust generalized sidelobe canceller. Proceedings of the International Workshop on Acoustic Echo and Noise Control (IWAENC), Darmstadt.

[23] Zelinski, R. (1988) A microphone array with adaptive post-filtering for noise reduction in reverberant rooms. Proceedings of ICASSP-88, vol. 5, pp. 2578–2581.

[24] McCowan, I. and Bourlard, H. (2003) Microphone array post-filter based on noise field coherence. *IEEE Transactions on Speech and Audio Processing*, **11**(6), 709–716.

[25] Tashev, I., Seltzer, M. and Acero, A. (2005) Microphone array for headset with spatial noise suppressor. Proceedings of Ninth International Workshop on Acoustic, Echo and Noise Control IWAENC 2005, Eindhoven, The Netherlands.

[26] Wolfe, P. and Godsil, S. (2003) A perceptually balanced loss function for short-time spectral amplitude estimator. Proceedings of IEEE ICASSP, Hong Kong, China.

[27] Seltzer, M., Tashev, I. and Acero, A. (2007) Microphone array post-filter using incremental bayes learning to track the spatial distribution of speech and noise. International Conference on Audio, Speech and Signal Processing ICASSP 2007, Honolulu, USA.

[28] Tashev, I. and Acero, A. (2006) Microphone array post-processor using instantaneous direction of arrival. International Workshop on Acoustic, Echo and Noise Control IWAENC 2006, Paris.

[29] Yoon, B., Tashev, I. and Acero, A. (2007) Robust adaptive beamforming algorithm using instantaneous direction of arrival with enhanced noise suppression capability. International Conference on Audio, Speech and Signal Processing ICASSP 2007, Honolulu, USA.

[30] Seltzer, M. and Tashev, I. (2008) A log-MMSE adaptive beamformer using a nonlinear spatial filter. International Workshop on Acoustic, Echo and Noise Control IWAENC 2008, Seattle, WA.

[31] Elko, G. and Pong, A. (1995) A simple adaptive first-order differential microphone. Proceedings of IEEE Workshop on Applications of Signal Processing to Audio and Acoustics (WASPAA), Mohonk Mountain Resort, New York.

[32] Elko, G. and Pong, A. (1997) A Steerable and Variable First-order Differential Microphone Array. Proceedings of IEEE International Conference on Audio, Speech and Signal Processing ICASSP 1997, Munich, Germany.

[33] Mihov, S., Gleghorn, T. and Tashev, I. (2008) Enhanced sound capture system for small devices. International Scientific Conference on Information, Communication, and Energy Systems and Technologies, Nis, Serbia.

[34] Tashev, I., Mihov, S., Gleghorn, T. and Acero, A. (2008) Sound capture and spatial filter for small devices. Interspeech 2008, Brisbane, Australia.

[35] Brandstein, M. and Ward, D. (eds) (2001) *Microphone Arrays*, Springer-Verlag, Berlin, Germany.

[36] Cox, H. (2004) Super-directivity revisited. Proceedings of Instrumentation and Measurement Technology Conference IMTC 2004, Como, Italy.

[37] Hänsler, E. and Schmidt, G. (eds) (2004) *Acoustic Echo and Noise Control: A Practical Approach*, John Wiley & Sons, New York.

[38] Huang, Y. and Benesty, J. (eds) (2004) *Audio Signal Processing for Next-Generation Multimedia Communication Systems*, Kluwer Academic, Norwell, MA.

[39] Seltzer, M. and Raj, B. (2001) Calibration of Microphone arrays for improved speech recognition. Mitsubishi Research Laboratories, TR-2002-43, December 2001.

[40] Tashev, I. and Seltzer, M. (2008) Data driven beamformer design for binaural headset. International Workshop on Acoustic, Echo and Noise Control IWAENC 2008, Seattle, WA.

[41] Vary, P. and Martin, R. (2006) *Digital Speech Transmission: Enhancement, Coding and Error Correction*, John Wiley & Sons, New York.

6

Sound Source Localization and Tracking with Microphone Arrays

The localization and tracking of sound sources has multiple applications. Among the most important is finding the correct direction in which to point the listening beam. In the previous chapter we always assumed that the direction to the desired sound source is known and the problem we solved was to find the best way to capture the sound originating from that direction. From this perspective, a sound source localizer and tracker is part of every sound capture system with a microphone array that uses beamsteering. The precision of localization is affected by the ambient noise and reverberation.

In this chapter we will continue to look at algorithms and scenarios for single and compact microphone arrays, locating – in real time – sound sources in room or office environments. The general problem of localization of sources with one or multiple arrays of transducers is beyond the scope of this book. The same is true for computationally heavy, iterative algorithms for off-line post-processing.

There are two distinct approaches to estimate the direction to the sound source: based on time-delay estimates, and based on steered response. They can both produce estimation using a single audio frame; but to improve the precision, and for tacking of multiple sound sources, post-processing algorithms are frequently used. The chapter ends with some practical considerations.

6.1 Sound Source Localization

6.1.1 Goal of Sound Source Localization

The goal of sound source localization algorithms is to detect and track the position of one or multiple sources. This is a passive method that does not emit any sounds; it uses the sounds generated by the object. The major property of the sound, related to sound source localization, is its limited and relatively low propagation speed. The speed of sound is approximately 343 m/s for room temperature and normal atmospheric pressure. With this speed, a sound wave will travel 10 cm in 291 μs, which is 4.66 sampling periods when sampling the sound with 16 000 samples per second. This means that even a relatively small two-element microphone array with distance between the microphones of 10 cm can detect where the sound comes from. Of course, as was discussed in the previous chapter, this small microphone array cannot distinguish signals coming from front or back.

Based on the number of microphones and the microphone-array geometry, the direction estimation can range from $\pm 90°$ (for linear arrays) to detection of the azimuth and the elevation, or full direction of arrival (DOA). Compact microphone arrays are not very good at estimation of the distance to the sound source unless it is comparable with the size of the array. To detect full 3D coordinates of a sound source in an office environment, several microphone arrays can be used – placed usually on the walls – and triangulation algorithms. Such algorithms are beyond the scope of this book, but the algorithms used in each of the microphone arrays are quite similar to what will be described further in this chapter.

6.1.2 Major Scenarios

One of the earliest devices for localization of sound sources is patented by Professor Mayer. The purpose of this device is to locate other ships on the river and to assist navigation in fog. It consists of two sound receptors and uses a human operator to detect the direction to the other ship. A sketch of this device, called "topophone," was published in *Scientific American* in 1880 and is shown in Figure 6.1.

The idea of extending human capabilities to detect the direction of sound sources was used between the two world wars for early detection of enemy aircraft. The Royal Observation Corps in Britain operated devices like the one shown in Figure 6.2. The picture was published in the issue of *Popular Mechanics* for December 1938. It contains four horns and is controlled by two operators; one detects the azimuth towards the aircraft, the second the elevation. Similar devices were used at that time in Germany, Japan, the Soviet Union and the United States. With the invention of radar these human-assisted devices for localization of aircrafts became history. Now, in the military context, sound source localization is mostly used under water to locate ships and submarines. Hydrophone arrays have been part of every submarine since World War II, and some countries have extended sets of stationary hydrophones on their coastlines for the same purpose. All these scenarios and environments have their specifics and are beyond the scope of this book.

Sound source localization is used in biology to track and study the behavior of certain animals. Perfect subjects for such studies are animals that use sound waves for navigation and orientation, such as dolphins and bats. They emit specific sounds that allow recognition and tracking of the animal. For such an experiment to localize and track a flying bat, see [1].

Professor Mayer's topophone [1880]

Figure 6.1   Topophone (*Scientific American*, 1880)

Figure 6.2   Sound source localization of aircraft (*Popular Mechanics*, 1938)

Here we will study compact microphone arrays, in most cases used in parallel for sound capture of human speech. The sound source localizer is part of the sound capture system with a microphone array. Whereas in the previous chapter we discussed beamforming algorithms under the assumption that the direction towards the desired sound source is known, the sound source localizer provides the listening direction.

For scenarios such as the recording of a meeting and speaker identification, the location of the speaker is an important clue to who is talking. People can, of course, move in the meeting room, so a different position does not necessarily mean a different person; but a sudden change of the current talker's position most probably means a change of speaker.

In many cases the microphone array works in conjunction with a video camera. The sound source localization can be combined with the output of a face detection algorithm, running on the video stream, for improved precision and reliable tracking. The combined results can be used for controlling the pan/tilt/zoom functions of the camera to best capture the face of the current talker.

6.1.3 Performance Limitations

The precision of localization estimates with a microphone array is limited by the ambient noise and the room reverberation. To a lesser degree, another limiting factor is the self noise of the microphones.

The *ambient noise* is not correlated to the sound emitted by the sound source. It causes smearing of the localization estimates, usually around the ground truth. This problem can be mitigated by collecting multiple localization estimates and averaging them.

Reverberation is the reflection of sound waves from walls and other objects in the room. The direct signal and multiple delayed and decayed copies of the same signal arrive at the receiving point (the microphone). Note that they are highly correlated and the mixture of the direct path and reflections can introduce a constant shift in the localization estimates. In addition, the combined room impulse response is highly volatile and change even after small movements of the sound source. As humans move their heads when speaking, localization of a speaker in a reverberant environment is not easy. For example, strongly reflective surfaces such as whiteboards and windows can create "ghost" sound sources, or substantially decrease the precision of localization estimates.

6.1.4 How Humans and Animals Localize Sounds

Humans have two ears and can locate sound sources in three dimensions. Two microphones in the air can detect the direction in one dimension and in the range of $\pm 90°$. In a three-dimensional space, the geometric positions of sources generating the same sound sensation in the two microphones are called the "cone of confusion."

To achieve 3D sound source localization, humans and other animals use multiple cues:

- Interaural time difference (ITD) is the difference between the times when the sound wave reaches the eardrums of the two ears. It depends on the sound source position and how the sound wave diffracts around the head.
- Interaural intensity difference (IID) is the difference between the sound intensity registered by the two ears. It is different for different sound source positions and frequencies.
- Head-related transfer function (HRTF).
- Reverberation.
- Prior knowledge of the sound.

ITD and IID as functions of the sound source position can be derived from the head-related transfer function. This function combines the effects of the sound wave diffraction around the head, reflections from the shoulders and from the torso, pinna effects, and the resonance characteristics of the ear canal. In general it is a time-invariant complex function of the azimuth φ, elevation θ, distance ρ, and frequency f. All radial coordinates are for a system with center at the middle of the line between the ears. HRTF can be measured directly or synthesized analytically (less precise). The measurement can happen in an anechoic chamber, with multiple speakers playing chirp signals from all directions. Small microphones, placed at the end of each ear canal of a human subject, record what the ears hear. Follows computing the transfer functions from the test points and finding proper interpolation for the points in between. HRTF varies from subject to subject, so using an averaged HRTF decreases the modeling precision. Using known HRTFs for the left and right ear, the signals from the two ears, and the large number of neurons between them, humans can estimate the direction towards the sound source. A sound from a given location is convolved with the HRTF function from this location for the left and the right ear.

If we have the inverted HRTFs, then we can form a hypothesis: If the sound is at location (φ, θ, ρ) then it should look like

$$\hat{S}_L(f) = X_L(f) \cdot H_L^{-1}(\varphi, \theta, \rho, f)$$
$$\hat{S}_R(f) = X_R(f) \cdot H_R^{-1}(\varphi, \theta, \rho, f)$$

(6.1)

when estimated from the signal at the left and the right ear, respectively. Then the potential sound source location would be the one that minimizes the differences between these two estimations:

$$(\varphi, \theta, \rho) = \underset{\varphi, \theta, \rho}{\arg\min} \left(\int_{f_{min}}^{f_{max}} (\hat{S}_L(f) - \hat{S}_R(f))^2 df \right).$$

(6.2)

This approach is used in some binaural sound source localizers in humanoid robots. It is computationally expensive; there are some complications around inverting the measured HRTFs as well. Of course it is unknown how exactly humans and animals perform three-dimensional sound source localization. What we know is that they do it pretty well in a horizontal plane where the localization precision has deviation less than 5 degrees [2]. The elevation is estimated with lower precision, most probably owing to the fact the humans live on the ground (practically in a "two-and-a-half"-dimensional space) and do not need so precise elevation estimation.

This is not the case for birds. The famous barn owl (Tyto Alba) is recognized as having the best sense of hearing of all the owls and can pinpoint prey in total darkness. The practically noiseless way these birds fly helps the owl's highly sensitive ears to detect if a rodent is moving around, even under snow. To achieve full 3D sound source localization these owls have asymmetrically set ear openings – the left ear is higher than the right. In addition, these species have a very pronounced facial disc, which acts like a "radar dish," guiding sounds into the ear openings. The shape of the disc can be altered at will, using facial muscles. Also, the owl's bill is pointed downward, increasing the surface area over which the sound waves are collected by the facial disc (Figure 6.3). These features create HRTFs that are unique, variable, and different for the two ears. The translation of left, right, up, and down signals are combined instantly in the owl's brain, and will create a mental image of the space where the sound

Figure 6.3   Barn owl Tyto Alba. (Credit: http://en.wikipedia.org/wiki/Barn_owl)

source is located. Studies of owl brains have revealed that the medulla (the area in the brain associated with hearing) is much more complex than in other birds. A barn owl's medulla is estimated to have at least 95 000 neurons – three times as many as a crow [3].

Humans and most animals are not very good at estimating the distance to a sound source. (Here we exclude active hearing by sending sound pulses and listening to the reflected sounds, as bats and dolphins do). The major distance cue is reverberation, and especially the arrival time of the first reflection, presumably from the ground (Figure 6.4a). If the first reflection arrives with larger delay after the direct path, the sound source is closer. Lower delay means greater distance. Figure 6.4b shows the delay of the first reflection as a function of the distance between two humans. It decreases from 10 ms for a very close position to 2 ms for a distance of 10 m. The

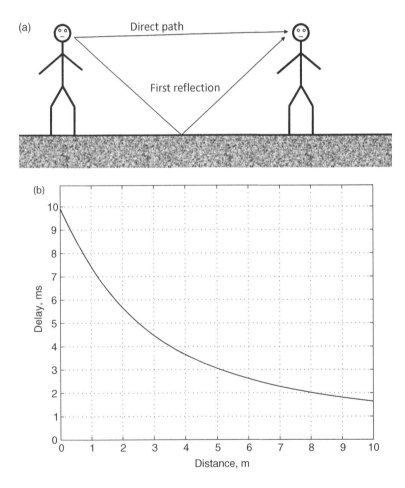

Figure 6.4   (a) Direct path and reflection from the ground; and (b) the arrival time difference as function of the distance

proportions of the direct sound and the overall reverberation plays an important role as well: in general, a higher portion of reverberation means a larger distance.

It is a fact that humans can perform sound source localization even with one ear. The way we do this is most probably by slightly moving the head up and down, right and left – something we do even with two functional ears to improve the precision of the direction estimate. By these movements we sense the changes in HRTFs; that is, we sample the first derivative. This derivative and some clues based on experience of how this sound *should* sound allow us to do the sound source localization (with lower precision, of course, especially in certain zones).

6.1.5 Anatomy of a Sound Source Localizer

A block diagram of a sound source localizer (SSL) is shown in Figure 6.5. Most sound source localizers work in the frequency domain. The first block is conversion to the frequency domain, shared by the entire microphone-array processing block.

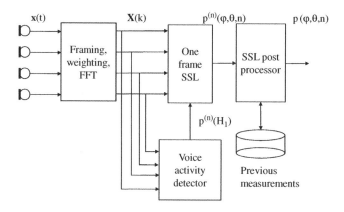

Figure 6.5 Block diagram of a sound source localizer (SSL)

The voice activity detector (VAD) provides the SSL with information about when to perform the localization. Obviously if there is no active sound source then processing is pointless and should be skipped. Even when the sound is continuous speech, the proportion of frames where SSL can do reliable location estimation is in the range of 30–50%; pauses are an integral part of human speech. The VAD can provide the probability of speech presence in each frequency bin, which can be used for weighting of localization estimates.

The one-frame SSL uses knowledge about the microphone-array geometry and the current audio frame to estimate the position of one or multiple sound sources. A good approach at this block is to terminate the estimation process if the confidence level is low. It is better to reject the results even when the VAD indicates that there is a voice activity, rather than provide a noisy or erroneous estimation. The output of this block can be either one position estimation, usually accompanied by a confidence level; or a list of multiple position estimations with their respective confidence levels; or the probability of sound source presence as a function of the direction. In all cases the estimation is accompanied by a time stamp. The position estimation can be just azimuth, or azimuth and elevation; or, in rare cases, azimuth, elevation, and distance. This is based on the microphone-array geometry and the purpose of sound source localization data. Position estimation at this stage is based solely on the information in the current audio frame. The precision is affected by noise and reverberation levels and limited by the array geometry and uncorrelated noise in the hardware.

The SSL post-processor uses all available sound source position estimations from a certain time interval. The length of this interval is based on the sound source dynamics, desired precision, and the algorithm used. Typical values are from one to four seconds. With an audio frame of 20 ms and 40% measurements in a 2 s time interval we will have around 40 measurements to work with. The post-processor can use these measurements to cluster around different hypotheses for sound source presence, track them, compute the speed and movement direction, and so on. The output of the SSL post-processor is quite similar to the output of the one-frame localizer. If we need the SSL to localize and track the sound source we want to capture, we need one result – the location to the desired sound source. Tracking multiple sound sources is necessary for conference room applications, integration with a video camera, and so on.

6.1.6 Evaluation of Sound Source Localizers

The evaluation parameters are very similar to those of any object detection and tracking system. The localization error varies based on ambient noise, other sound sources, the reverberation level, and the position of the sound source (depends on the microphone-array geometry). Usually typical noise and reverberation levels are used for testing and evaluation. The most common parameters are the localization error mean and deviation, defined as

$$\varepsilon = \frac{1}{N} \sum_{i=1}^{N} (\hat{\varphi}_i - \varphi_i)$$

$$\sigma_\varepsilon = \sqrt{\frac{\sum_{i=1}^{N} (\hat{\varphi}_i - \varphi_i)^2}{N^2}}. \tag{6.3}$$

Here, $\hat{\varphi}_i$ and φ_i are the sound source estimation and the ground truth in the i-th moment. This metric is valid for both moving and non-moving sound sources. If we do not have ground truth measurements for non-moving sound sources, it can be replaced by the mean of the measurements. In this case we assume that the distribution of the error is zero mean; that is, we do not have bias in our estimates.

All parameters from detection and estimation theory are applicable for evaluation of a sound source localizer: true positive rate, false positive rate, accuracy (of detection). In many cases the SSL provides a confidence level. This is the main decision for the presence of a sound source. ROC curves can be used to tune the threshold level for this decision (see Chapter 4).

The sound sources appear and disappear, humans start and stop talking. The third set of evaluation parameters are related to the SSL dynamics. A major parameter here is the *latency time*; that is, how soon after the sound source appears that the confidence level from the SSL will exceed a given threshold.

6.2 Sound Source Localization from a Single Frame

In this chapter we will use the sound capture model and the coordinate system described in Chapter 5 (see Section 5.2). As usual, we will assume processing in the frequency domain and will omit the frame indices where possible. A single sound source is captured by the microphones in the array as

$$X_m(f, p_m) = D_m(f, c)S(f) + N_m(f) \tag{6.4}$$

where the first term on the right-hand side

$$D_m(f, c) = \frac{e^{-j2\pi f \frac{\|c - p_m\|}{v}}}{\|c - p_m\|} A_m(f) U_m(f, c) \tag{6.5}$$

represents the phase rotation and the magnitude decay due to the distance to the microphone $\|c - p_m\|$, and v is the speed of sound. The $M \times K$ matrix $\mathbf{X}^{(n)}$ represents the n-th audio frame from the microphone array and is the input of the sound source localizer. M is the number of microphones and K is the number of the frequency bins – that is, the frame size. We assume knowledge of the microphone positions $\mathbf{p} = \{p_m, m = 0, 1, \ldots, M-1\}$, and the directivity pattern of the microphones $U_m(f, c)$. Based on this information, the SSL should tell us where the sound source position is.

6.2.1 Methods Based on Time Delay Estimation

6.2.1.1 Time Delay Estimation for One Pair of Microphones

Given two omnidirectional microphones positioned at distance d (Figure 6.6), the direction of arrival of the sound can be estimated as

$$\theta = \arcsin\left(\frac{\tau_D v}{d}\right) \tag{6.6}$$

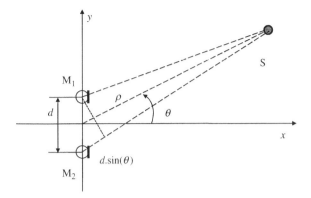

Figure 6.6 Time delay as an indication of the direction of arrival

where ν is the speed of sound, and τ_D is the difference between the times the sound from S reaches the two microphones M_1 and M_2 – that is, the time delay. In far-field sound propagation, $\rho \gg d$. The problem for sound source localization with one pair of microphones now is converted to estimation of the time delay between the two signals.

The most intuitive approach for time delay estimation is to find the maximum of the cross-correlation function of the two signals:

$$\mathbf{R}_{12} = \mathrm{iFFT}\left[\mathbf{X}_1\mathbf{X}_2^{\mathrm{H}}\right]. \tag{6.7}$$

Here we used the cross-power spectral density function

$$\mathbf{G}_{X_1X_2} = \mathbf{X}_1\mathbf{X}_2^{\mathrm{H}}$$

to estimate the cross-correlation function spectrum. The vector $\mathbf{R}_{12}$ is a function of time and peaks at the time shift between these two signals – that is, the delay. With this approach, frequency bins with higher energy will affect more the cross-correlation function than frequency bins with lower energy. Usually the distance between the microphones is small enough to prevent severe aliasing effects, which makes it much smaller than the wavelengths in the lower part of the speech frequency band. On the other hand this is where most of the speech energy is. The peak width of the cross-correlation function depends on the frequency components and is wider when more low-frequency components participate. This leads to a flatter and smeared peak and lower precision of the time delay estimation. To overcome this effect, Knapp and Carter [4] introduced the generalized cross-correlation (GCC) function, which applies two filters, $\mathbf{H}_1$ and $\mathbf{H}_2$, to the signals before computing the cross-power density function. This leads to frequency weighting in the generalized cross-correlation function:

$$\mathbf{R}_{12} = \mathrm{iFFT}[\boldsymbol{\psi} \cdot \mathbf{G}_{X_1X_2}] \tag{6.8}$$

where $\boldsymbol{\psi} = \mathbf{H}_1\mathbf{H}_2^H$ is the frequency weighting and the operator $(\cdot)$ denotes per-element multiplication. The same paper discussed several weightings that are optimal in one way or another.

The weighting proposed by Roth [5]:

$$\psi_{\text{Roth}}(f) = \frac{1}{G_{X_1X_1}(f)} \qquad (6.9)$$

is equivalent to applying a Wiener–Hopf filter to the first signal, which best approximates the mapping of $x_1(t)$ to $x_2(t)$. Considering the fact that $G_{X_1X_1}(f) = G_{SS}(f) + G_{N_1N_1}(f)$, where $G_{N_1N_1}(f)$ and $G_{SS}(f)$ are the noise power spectral density for the first channel and the sound source, respectively, the Roth weighting has the desirable effect of suppressing those frequency regions where $G_{N_1N_1}(f)$ is large and the estimation of $G_{X_1X_2}(f)$ is likely to be in error.

The *smoother coherence transform* (SCOT) weighting considers that the errors may be due to frequency bins where $G_{N_2N_2}(f)$ is high by combining the power spectral density from both channels:

$$\psi_{\text{SCOT}}(f) = \frac{1}{\sqrt{G_{X_1X_1}(f)G_{X_2X_2}(f)}} \qquad (6.10)$$

which, when combined with the cross-power spectral density of the two channels, gives the coherence function

$$\gamma_{12}(f) \triangleq \frac{G_{X_1X_2}(f)}{\sqrt{G_{X_1X_1}(f)G_{X_2X_2}(f)}}. \qquad (6.11)$$

One of the most commonly used weightings for time delay estimates is *phase transform* (PHAT) weighting:

$$\psi_{\text{PHAT}}(f) = \frac{1}{|G_{X_1X_2}(f)|} \qquad (6.12)$$

which simply eliminates the magnitudes and gives equal weight to the phases in each frequency bin. PHAT is developed purely as an ad-hoc technique. Theoretically after PHAT weighting we should see peaks as sharp as delta functions. This makes this approach most suitable for detecting multiple sources, or working in a reverberant environment. In practice the estimation is sensitive to the presence of noise as the errors are accentuated when the power is smallest (i.e., dominated by the noise power). In practical implementations, measures should be taken to prevent division by zero.

Eckhart [6] minimizes the deflection criterion – that is, the ratio of the change in mean correlator output due to the signal present to the standard deviation of correlator output due to noise alone:

$$\psi_{\text{Eckhart}}(f) = \frac{G_{S_1 S_2}(f)}{G_{N_1 N_1}(f) G_{N_2 N_2}(f)}. \tag{6.13}$$

An approximate but practical solution can be found by considering $G_{X_1 X_1}(f) = G_{SS}(f) + G_{N_1 N_1}(f)$, $G_{X_2 X_2}(f) = G_{SS}(f) + G_{N_2 N_2}(f)$, and $G_{SS}(f) \approx G_{X_2 X_2}(f)$ under a uncorrelated noise assumption:

$$\psi_{\text{Eckhart}}(f) \approx \frac{|G_{X_1 X_2}(f)|}{[G_{X_1 X_1}(f) - |G_{X_1 X_2}(f)|][G_{X_2 X_2}(f) - |G_{X_1 X_2}(f)|]}. \tag{6.14}$$

The Eckhart filter gives less weight to the frequency bins with high noise or low signal energy.

The processors so far can be justified on the basis of reasonable performance criteria, either heuristic or mathematical. Without providing the full derivation (see [4]), the *maximum-likelihood* (ML) estimator of the delay is derived to be equivalent to the Hannan and Thompson filter [7], called in some literature sources the "HT processor":

$$\psi_{\text{ML}}(f) = \frac{|\gamma_{12}(f)|^2}{|G_{X_1 X_2}(f)|[1 - |\gamma_{12}(f)|^2]}. \tag{6.15}$$

Practically, the ML processor modifies the PHAT processor with weighting for the signal coherence like the SCOT processor. In a different way, Jenkins and Watts [8] show that the variation of the direction of arrival estimation is

$$\text{var}[\theta(f)] \cong \frac{1 - |\gamma_{12}|^2}{|\gamma_{12}|^2} \frac{1}{L_1} \tag{6.16}$$

where L_1 is a proportionality constant dependent on how the data are processed. From this perspective, the ML processor gives more weight to frequency bins with lower variation, providing a more precise estimate. It can be shown that the ML processor achieves the Cramer–Rao lower bound of the estimation variance for the direction of arrival under conditions of precise estimation of the cross-power spectral density and coherence function. Theoretically it should provide the lowest variance of the estimates. In practice it is more important to be robust to estimation errors.

The derivation is under the assumptions of uncorrelated, stationary Gaussian signal and noise, and no multipath. The required coherence function is unavailable *a priori* and needs to be estimated from the data. This is typically done via a temporal averaging technique, such as the one in [9]. The coherence estimation can be problematic for non-stationary signals. This is why Brandstein *et al.* [10] propose an approximate, but more robust, ML weighting:

$$\psi_{\text{MLA}}(f) = \frac{|X_1(f)||X_2(f)|}{|N_1(f)|^2|X_2(f)|^2 + |N_2(f)|^2|X_1(f)|^2} \quad (6.17)$$

which is used in many real systems.

In general, the PHAT weighting is considered better in reverberant conditions, with the ML (or MLA) weighting more robust to ambient noise. Wang and Chu [11] proposed a weighting that is optimal in both a noisy and reverberant environment:

$$\psi_{\text{MLR}}(f) = \frac{1}{q|G_{X_1X_2}(f)| + (1-q)|N(f)|^2}. \quad (6.18)$$

Here, $0 < q < 1$ is the portion of the reverberant signal energy; that is, if we assume that the direct path energy is equal to the reverberation energy, $q = 0.5$. In the derivation it is assumed that $|N(f)| = |N_1(f)| = |N_2(f)|$, which is quite close to reality if we have omnidirectional microphones and a compact microphone array. It is valid if the array consists of unidirectional microphones pointing to the same direction. The noise energy can be computed during pauses using a simple VAD and the techniques discussed in Chapter 4. Further generalization for non-equivalent noise spectra and practical comparison of the weighting above is given in [12] and [13].

To compare the various weighting functions, we placed a human speaker 1.5 m in front of a two-element microphone array with distance between the microphones 195 mm in normal office noise and reverberation conditions. The generalized cross-correlation function between the two microphone signals for one of the voiced frames is shown in Figure 6.7. The functions are normalized to have maximum value of 1 for easier comparison. Better weighting functions should produce sharper peaks. The maximum-likelihood and Eckhart cases are worse, together with no weighting at all. SCOT, PHAT, and MLR weightings produce the sharpest peaks. Of course, a generalized cross-correlation computed from one frame is just an illustration.

To really compare various weightings, we estimated the direction of arrival by finding the maximum of the GCC for each voiced frame (a simple binary VAD was used). Around 20 s of speech produced 702 voiced frames at 20 ms frame size and 16 kHz sampling rate. To compare the robustness to noise and reverberation, we repeated the experiment by placing the speaker at direction 32° and 2.5 m distance. The results are shown in Table 6.1. The GCC from each frame was evaluated and the maximum found. The maximum point and its two neighbors were used for quadratic interpolation to increase the precision (see later in this chapter). Then the measurements within ±20° from the actual sound source direction were used to compute the means and deviations. The number of measurements used is shown in the first column.

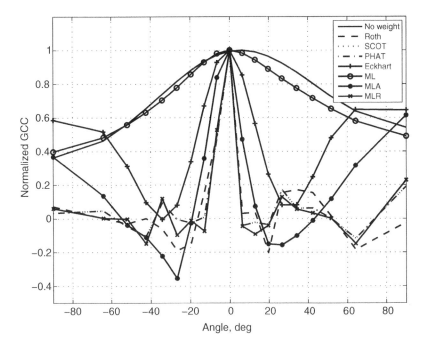

Figure 6.7 Generalized cross-correlation (GCC) functions with various weighting algorithms

Table 6.1 DOA precision for various weighting functions and two distances

Weighting	Equation	1.5 m			2.5 m		
		Number	Mean	Deviation	Number	Mean	Deviation
No weighting	$\psi(f) = 1$	502	−2.348	8.643	248	36.133	13.417
Roth	$\psi_{\text{Roth}}(f) = \frac{1}{G_{X_1X_1}(f)}$	656	−1.044	3.262	472	33.706	4.599
SCOT	$\psi_{\text{SCOT}}(f) = \frac{1}{\sqrt{G_{X_1X_1}(f)G_{X_2X_2}(f)}}$	694	−0.846	1.922	543	33.724	3.288
PHAT	$\psi_{\text{PHAT}}(f) = \frac{1}{|G_{X_1X_2}(f)|}$	694	−0.846	1.922	543	33.724	3.288
Eckhart	$\psi_{\text{Eckhart}}(f) = \frac{G_{S_1S_2}(f)}{G_{N_1N_1}(f)G_{N_2N_2}(f)}$	678	−2.305	4.807	452	30.147	7.479
ML	$\psi_{\text{ML}}(f) = \frac{|\gamma_{12}(f)|^2}{|G_{X_1X_2}(f)|\left[1-|\gamma_{12}(f)|^2\right]}$	683	−2.406	4.855	425	28.787	6.967
MLA	$\psi_{\text{MLA}}(f) = $ $\frac{|X_1(f)||X_2(f)|}{|N_1(f)|^2|X_2(f)|^2+|N_2(f)|^2|X_1(f)|^2}$	690	−1.529	3.128	532	32.205	5.197
MLR, 0.25/0.1	$\psi_{\text{MLR}}(f) = $ $\frac{1}{q|G_{X_1X_2}(f)|+(1-q)|N(f)|^2}$	699	−1.259	2.033	564	32.908	3.689

In general, the closer this number is to 702, the better is the weighting. The mean and deviation, computed from these points, are shown in the second and third columns. All numbers are in degrees. The table confirms that MLR, PHAT, and SCOT were definitely performing best for both distances. Note that the MLR weighting is computed with $q = 0.25$ for the 1.5 m distance and with $q = 0.1$ for 2.5 m, to reflect the worse noise and reverberation conditions.

EXERCISE

Copy in the MATLAB® directory file named *MicArrDesc.dat*, containing the description of a microphone array from Chapter 5. Look at the provided MATLAB script *MicArrSSL.m*. It is a skeleton for exploring the sound source localization algorithms in this chapter. The script reads a .WAV file, converts it to the frequency domain, and runs a simple binary VAD. The frames with voice activity are processed with various algorithms. The script writes the output into two text files, *PPP.dat* and *SSLTime.dat*. The first file contains a matrix with number of rows equal to the number of processed frames. The number of columns depends on the algorithm. For the GCC exploration it is `2*maxTaps + 1`; for the others which follow it is `numHypAngles`. This is the number of DOA hypotheses – see description further in this chapter.

Record your own .WAV file with a human speaking in front of the microphone array, or use the attached *Record1.WAV*. Process it with the script above to compute the generalized cross-correlation function for the first pair. Write a MATLAB script to load the text file and visualize the shape of the GCC. Repeat after changing the weighting function in `ComputeGCC()`. You should have shapes similar to Figure 6.7.

6.2.1.2 Combining the Pairs

The techniques above provide a way to estimate the direction of arrival using one pair of microphones. It is limited to the interval $[-90°, +90°]$. Adding more microphones allows estimation of the DOA for the entire sphere and to increase the precision of the localization estimates. An M-element microphone array has $M(M-1)/2$ unique pairs. This raises the question of how to combine the results from these pairs. It is not so trivial even in the case of a single sound source and linear microphone array where we have to estimate one directional angle. Averaging is not going to work reliably. Owing to reverberation, the generalized cross-correlation function can have several peaks, one of which is at the true delay. It can happen that, for certain frames and pairs, a reflection has a stronger peak. We will still have a peak at the right direction, but it will be lost if we take one measurement from each pair. In short, it is too early to discard the GCC function and convert it to a single delay at pair level. In [12] and [13] this is called the *least-commitment* approach.

A working algorithm for better combining the GCC functions from the microphone pairs is discussed by Birchfield and Gillmor [14]. The authors use a planar four-element array to localize sounds in the upper hemisphere. The GCC functions for each pair are computed using PHAT weighting and converted to functions of the azimuth and elevation $h_i(\varphi, \theta)$. Here i is the pair index, and the hemisphere is represented as a discrete set of points spaced at equal latitudes and longitudes. The resolution issues are mitigated using linear interpolation. The final direction is computed as the maximum of the combined cross-correlation functions from all pairs:

$$(\varphi, \theta) = \arg\max_{\varphi, \theta} \left(\sum_{i=1}^{\left(\frac{M(M-1)}{2}\right)} h_i(\varphi, \theta) \right). \tag{6.19}$$

The conversion of the generalized cross-correlation function to a function of the longitude and latitude in this paper is specific for the used geometry. There is no problem to generalize this function for arbitrary geometry. Two major concerns arise when looking at the details of this algorithm. The first is the required CPU performance. For a four-element microphone array we have six pairs. To compute the GCC function this will require six inverse Fourier transformations for each voiced frame. The second concern is related to the fact that the GCC function is discretized quite sparsely. For a distance of 20 cm between the microphones in the pair and a sampling rate of 16 kHz, there are only 17 points with valid delays. This means that computing $h_i(\varphi, \theta)$ with enough resolution will require interpolation, which introduces error and is another time-consuming operation.

This approach is further developed and presented by Brandstein and Ward [15] and called a *one-step time-delay direction of arrival* (1TDOA) estimator by Rui and Florencio [12,13]. Here we provide a slightly generalized version of this algorithm. The space where we expect the sound source to be (half circle, circle, hemisphere, sphere) is dicretized in L points selected in a way to provide sufficient spatial resolution. Then sound source location is where

$$p^*(l) = \arg\max_l \{E(l)\} \tag{6.20}$$

where E(l) is the estimated power, coming from direction l:

$$E(l) = \left| \sum_{m=1}^{M} \int_{-F_s/2}^{F_s/2} X_m(f) \exp(-j2\pi f \tau_m) df \right|^2. \tag{6.21}$$

Here, τ_m is the estimated travel time from position l to microphone m. Expanding the terms in Equation 6.21 leads to

$$E(l) = \sum_{m=1}^{M} \left| \int_{-F_s/2}^{F_s/2} X_m(f)df \right|^2 + \sum_{r=1}^{M}\sum_{s\neq r}^{M} \left| \int_{-F_s/2}^{-F_s/2} X_r(f)X_s^*(f)\exp(j2\pi f(\tau_r-\tau_s))df \right|^2 .$$

$$(6.22)$$

The first term is position-independent and constant for the current frame, while the second term is the cross-correlation for delay $\tau_r - \tau_s$. Owing to introducing it in this form the estimation is not discrete. The next logical step after removing the position-independent part and duplicated microphone pairs is to add some of the weightings, discussed above:

$$E(l) = \sum_{r=1}^{M-1}\sum_{s=r+1}^{M} \left| \int_{-F_s/2}^{F_s/2} \psi(f)X_r(f)X_s^*(f)\exp(j2\pi f(\tau_r-\tau_s))df \right|^2 . \qquad (6.23)$$

Note that here we assume omnidirectional microphones with spatially flat phase response. A closer look at Equation 6.21 reveals that this is nothing but the power on the output of a delay-and-sum beamformer. Looking at Equation 6.23 we see that this is the inverse Fourier transformation of weighted $X_r(f)X_s^*(f)$ in one point; that is, the GCC function for specific delay $\tau_r - \tau_s$. Practically this algorithm overcomes both problems of the previous approach: it computes the inverse Fourier transformation in one point (the one needed) and this point is at the exact delay $\tau_r - \tau_s$, which solves the sparse discretization issue.

This approach naturally leads us to the next group of methods for sound source localization. In some sources it is denoted as *steered response power PHAT* (SRP PHAT) when PHAT weighting is used to combine the bins.

EXERCISE

Implement the two algorithms above in the marked places inside *MicArrSSL.m*. Use the record from the previous exercise to store the shape of the spatial function. Write a MATLAB script to visualize it.

6.2.2 *Methods Based on Steered-response Power*

The steered-response power (SRP) group of methods forms a conventional beam, scans it over the appropriate region of the working space, and plots the magnitude squared of the output. The sound source is where the peak is (or peaks in case of multiple sound sources). In some literature sources this group of methods is called "beamsteering sound source localization" or "beamscan algorithms." Practically all SSL algorithms can be converted to finding a maximum (or maxima) of a certain function.

Another note about this group of algorithms is that, in what follows, all of them will be discussed for one frequency bin. The notation for the bin index is omitted for simplicity. At the end of the section we will discuss combining the steered power functions from all frequency bins into one SRP function.

6.2.2.1 Conventional Steered-response Power Algorithm

The power at the beamformer output as a function of the look-up direction c is

$$P_{\text{BF}}(c) = \mathbf{D}(c)^{\text{H}}\mathbf{S}\mathbf{D}(c). \tag{6.24}$$

Here, $\mathbf{S}$ is the cross-power matrix of the input sample, $\mathbf{S} \triangleq \mathbf{X}\mathbf{X}^{\text{H}}$. In many cases the cross-power matrix is averaged across the last several frames (N typicaly between 5 and 20) for increased stability of the estimation:

$$\mathbf{X} = \frac{1}{\sqrt{N}} \begin{bmatrix} X_1^{(n)} & X_1^{(n-1)} & \cdots & X_1^{(n-N+1)} \\ X_2^{(n)} & X_2^{(n-1)} & \cdots & X_2^{(n-N+1)} \\ \vdots & \vdots & \ddots & \vdots \\ X_M^{(n)} & X_M^{(n-1)} & \cdots & X_M^{(n-N+1)} \end{bmatrix}. \tag{6.25}$$

The vector $\mathbf{D}_c(f)$ is computed according to Equation 5.4. The locations $(c_1, c_2, \ldots, c_J)$ of the J sound sources are the J highest peaks in $P_{\text{BF}}(c)$. This estimator is frequently referred to as the "Bartlett beamformer." It scans the working space with a delay-and-sum beamformer, which is the most robust but least directive beamformer (see the previous chapter). This approach is the classic beamsteering or SRP algorithm for source localization.

6.2.2.2 Weighted Steered-response Power Algorithm

We can apply different weighing to each frequency bin as we can have several approaches for the beamformer weights design. Then:

$$P_{\text{BW}}(c) = \mathbf{D}(c)^{\text{H}}\mathbf{S}_w\mathbf{D}(c)$$
$$\mathbf{S}_w = w\mathbf{S}w^{\text{H}}. \tag{6.26}$$

Here, w is the weights vector for this frequency bin computed using some of the algorithms in the previous chapter.

6.2.2.3 Maximum-likelihood Algorithm

It is easy to derive the power steering function for the MVDR beamformer, which is both MMSE and ML solution for estimation of the source signal, by placing the MVDR weighting

$$w = \frac{\mathbf{D}^{\mathrm{H}}(c)\mathbf{S}^{-1}}{\mathbf{D}^{\mathrm{H}}(c)\mathbf{S}^{-1}\mathbf{D}(c)} \tag{6.27}$$

in Equation 6.26. After some simplifications we have

$$P_{\mathrm{ML}}(c) = \frac{1}{\mathbf{D}(c)^{\mathrm{H}}\mathbf{S}^{-1}\mathbf{D}(c)}. \tag{6.28}$$

For the full derivation see Chapter 9 in [16]. In some sources this algorithm is referred to as the "Capon algorithm," named after the author of the MVDR beamformer. Because the input signal is used for computation of $\mathbf{S}$, this is actually an MPDR beamformer, but we used the more common name for this algorithm.

6.2.2.4 MUSIC Algorithm

One well-known algorithm for sound source localization and signal estimation is the *multiple signal classification* (MUSIC) algorithm proposed by Schmidt [17]. We provide here the slightly generalized version from Chapter 9 in [18]. Given J sound sources F_j, the sound capturing equation is an extension of Equation 6.4:

$$\begin{bmatrix} X_1 \\ X_2 \\ \vdots \\ X_M \end{bmatrix} = \begin{bmatrix} \mathbf{D}(c_1) & \mathbf{D}(c_2) & \cdots & \mathbf{D}(c_J) \end{bmatrix} \begin{bmatrix} F_1 \\ F_2 \\ \vdots \\ F_J \end{bmatrix} + \begin{bmatrix} N_1 \\ N_2 \\ \vdots \\ N_M \end{bmatrix} \tag{6.29}$$

where $\mathbf{D}(c_j)$ is the capturing vector for all microphones and source location c_j – see Equation 6.5. In matrix form this can be written as

$$\mathbf{X} = \bar{\mathbf{D}}\mathbf{F} + \mathbf{N}. \tag{6.30}$$

Here, $\bar{\mathbf{D}}$ is the capturing matrix with size $M \times J$, for M microphones and J sound sources. Then the $M \times M$ spectral matrix of the $\mathbf{X}$ vector is

$$
\begin{aligned}
\mathbf{S}_x &\triangleq \mathbf{X}\mathbf{X}^H \\
&= \bar{\mathbf{D}}(\mathbf{F}\mathbf{F}^H)\bar{\mathbf{D}}^H + \mathbf{N}\mathbf{N}^H \\
&= \bar{\mathbf{D}}\mathbf{S}_f\bar{\mathbf{D}}^H + \sigma^2\mathbf{I}.
\end{aligned}
\tag{6.31}
$$

Assuming that J is known, the goal is to estimate $(c_1, c_2, \ldots, c_J)$. The spectral matrix can be written in terms of eigenvalues and eigenvectors as

$$
\mathbf{S}_x = \sum_{i=1}^{N} \lambda_i \mathbf{\Phi}_i \mathbf{\Phi}_i^H = \mathbf{\Phi}_i \Lambda \mathbf{\Phi}_i^H
\tag{6.32}
$$

where $\Lambda = \text{diag}\,[\lambda_1, \lambda_2, \ldots, \lambda_N]$. Assuming that the eigenvalues are in decreasing size – that is, $\lambda_1 \geq \lambda_2 \geq \cdots \geq \lambda_J \geq \lambda_{J+1} = \cdots = \sigma^2$ – then the first J eigenvalues are the signal subspace eigenvalues and the first J eigenvectors are the signal subspace eigenvectors:

$$
\mathbf{U}_S \triangleq [\mathbf{\Phi}_1 \mathbf{\Phi}_2 \cdots \mathbf{\Phi}_J]
\tag{6.33}
$$

which is an $N \times J$ matrix. The rest of the eigenvectors define the noise subspace:

$$
\mathbf{U}_N \triangleq [\mathbf{\Phi}_{J+1} \mathbf{\Phi}_{J+2} \cdots \mathbf{\Phi}_N]
\tag{6.34}
$$

and is an $N \times (N-J)$ matrix. Note that there is still a noise component in the signal subspace, but there is no signal component in the noise subspace. In real conditions we do not know the eigenvalues and eigenvectors and will have to estimate them from the data. The spectral matrix $\hat{\mathbf{S}}_x$ can be simply the sample spectral matrix or proper averaging can be applied (see above). Then we can estimate $\hat{\Lambda}$, sort it, and determine $\hat{\mathbf{U}}_N$ or $\hat{\mathbf{U}}_S$. The steering function is derived to be

$$
\begin{aligned}
P_{\text{MUS}}(c) &= \frac{1}{\mathbf{D}(c)^H \mathbf{U}_N \mathbf{U}_N^H \mathbf{D}(c)} \\
&= \frac{1}{\mathbf{D}(c)^H [\mathbf{I} - \mathbf{U}_S \mathbf{U}_S^H] \mathbf{D}(c)}.
\end{aligned}
\tag{6.35}
$$

The first J maxima of $P_{\text{MUS}}(c)$ are the coordinates of the J sound sources. The second form is more convenient for use in practical applications. For most cases we can assume a single sound source in the current bin and just use the first eigenvector $\mathbf{U}_S = \mathbf{\Phi}_1$. In MATLAB the function `[F,l]=eigs(S)` provides the first M eigenvalues and eigenvectors. Then `Us=squeeze(F(:,1))`. This algorithm is the most stable and precise, but requires the highest CPU power.

6.2.2.5 Combining the Bins

The steered-response power algorithms discussed so far work for one frequency bin. Some of them were derived for antenna arrays, which work in a very narrow band. For them, the derived SRP function – we can denote it by $P_{\text{SSL}}(c)$ – has to be processed and the source locations found. In contrast, in audio we deal with wideband signals (so a larger number of frequency bins) that are sparse in the sense that there is almost no speech energy in many of the bins. Also, the noise and reverberation (clutter in radio) are substantially higher in audio signals. All this requires combining the signals $P_{\text{SSL}}(c, k)$ from different frequency bins – the index k is added to denote the dependency on frequency.

The simplest approach is just to average the SRP functions, while hoping to reduce the noise and enhance the peaks if they coincide in many frequency bins:

$$P_{\text{SSL}}(c) = \frac{1}{K} \sum_{k=1}^{K} P_{\text{SSL}}(c, k). \tag{6.36}$$

Of course, the averaging in real systems will be in the working frequency band, between some beginning and ending bins.

Another approach is to use some of the weightings derived for the GCC function. PHAT weighting will look like this:

$$P_{\text{SSL}}(c) = \frac{1}{K} \sum_{k=1}^{K} \frac{M}{\mathbf{X}_k^{\text{H}} \mathbf{X}_k} P_{\text{SSL}}(c, k). \tag{6.37}$$

Here, $\mathbf{X}_k$ is the input vector with length M containing the input signals for this frequency bin from all microphones. We can benefit from the advantages of MLR weighting if we have the noise variance $\bar{N}_k$ averaged across all channels:

$$\bar{N}_k = \frac{1}{M} \sum_i N_i(k)$$

so that

$$P_{\text{SSL}}(c) = \frac{1}{K} \sum_{k=1}^{K} \frac{M}{q\mathbf{X}_k^{\text{H}} \mathbf{X}_k + (1-q)\bar{N}_k} P_{\text{SSL}}(c, k). \tag{6.38}$$

6.2.2.6 Comparison of Steered-response Power Algorithms

The SRP algorithms discussed so far were evaluated using the four-element linear microphone array shown in Figure 5.4. Actually, evaluation of the various weightings for GCC computation was done using the two outer microphones and the same set of two recordings: person speaking in front of the array at 1.5 m distance and person speaking at 2.5 m distance and direction $+32°$. The SRP functions were evaluated in the interval $[-90°, +90°]$ discretized with steps of 5 degrees. They were computed only for the 702 voiced frames. Only one location was used for each frame – the highest peak of the $P_{SSL}(c)$ function. The results within $\pm20°$ from the actual direction to the sound source were used to compute the means and deviations. The number of potential combinations of algorithms to compute the steered response per frequency bin and the weightings for combining all the bins in the current frame is large. That is why we reduced the comparison to the following combinations:

- **TDE interpolation:** time-delay estimates with PHAT-weighted GCC and interpolation as described in Equation 6.19. Spline interpolation was used to obtain the values of GCC between the sampling points.
- **TDE estimation:** time-delay estimates with PHAT-weighted GCC, estimated for the exact delay and squared according to Equation 6.23.
- **SRP BF:** classic steered-response power approach according to Equation 6.24 and bins combined by summation using Equation 6.36.
- **SRP PHAT:** classic steered-response power approach according to Equation 6.24 and bins combined by PHAT weighting using Equation 6.37.
- **SRP MLR:** classic steered-response power approach according to Equation 6.24 and bins combined with MLR weighting using Equation 6.38. In the 1.5 m distance case $q = 0.25$, and for 2.5 m distance $q = 0.1$.
- **SRP ML:** maximum-likelihood steered-response power approach according to Equation 6.28 and bins combined by PHAT weighting using Equation 6.37.
- **SRP MUSIC:** implementation of the MUSIC algorithm with steered-response power computed by summation using Equation 6.36.

The shapes of the SRP function for one frame are shown in Figure 6.8. It is obvious that the peak of the power steering function for SRP BF is wider than any other function. Definitely the SRP MUSIC provides the sharpest peak.

These observations are confirmed in Table 6.2. It shows the number of valid measurements ($\pm20°$ from the ground truth), the mean, and the deviation of these valid measurements. The first new observation is that most of the measurements lie in this $\pm20°$ and are valid. The deviations are much lower than measured with a single pair, and the number of measurements in the range is substantially higher – practically all of them. All algorithms actually perform quite well, the highest deviation being with SRP BF, as expected from the width of the peak. The best performing algorithm is SRP

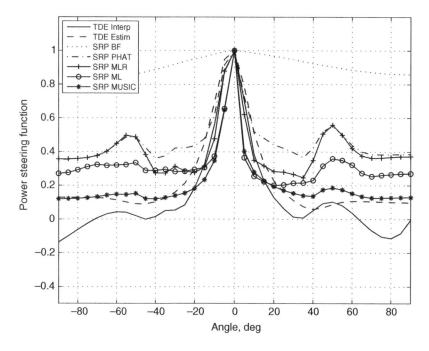

Figure 6.8   Steering-response power (SRP) functions from various algorithms

MUSIC, closely followed by SRP PHAT. The latter algorithm is most commonly used in microphone-array-based sound capture systems because of its lower computational requirements. The precision for sound source localization is quite good: at 2.5 m an angle of 0.97° (the precision of SRP MUSIC) covers 4.2 cm, roughly the width of a human mouth. We will see later in this chapter that this precision can be increased further.

EXERCISE

Implement the steered power sound source localization algorithms above in the marked places inside *MicArrSSL.m*. Note that the script computes the spectral matrix and the noise variances for you. The angles where we want the steered response computed are in an array named `hypAngles`. Store the steered-response power function from each processed frame into the vector `P`; later it will be placed in a `PPP` matrix and stored on the disk. Write a MATLAB script to read and visualize the steered response power functions. The SRP functions should look like these in Figure 6.8.

6.2.2.7 Particle Filters

All the steered-response power algorithms discussed so far find the sound source DOA (DOAs) by finding the maximum (maxima) of a certain function of the direction. If we

Table 6.2 Sound source localization for various steered-response power functions

Algorithm	Equation	1.5 m			2.5 m		
		Number	Mean	Deviation	Number	Mean	Deviation
TDE interpolation	$P_{\text{TDE-I}}(c) = \sum_{r=1}^{M-1}\sum_{s=r+1}^{M} GCC_{rs}(c)$	702	−0.447	1.836	537	33.619	2.558
TDE estimation	$P_{\text{TDE-E}}(c) = \sum_{r=1}^{M-1}\sum_{s=r+1}^{M} |GCC_{rs}(c)|^2$	698	−1.154	1.902	560	34.595	3.814
SRP BF	$P_{\text{BF}}(c) = \mathbf{D}(c)^{\text{H}}\mathbf{SD}(c)$	702	0.082	2.179	621	28.890	3.880
SRP PHAT	$P_{\text{BF-PHAT}}(c) = \dfrac{M}{\mathbf{X}^{\text{H}}\mathbf{X}}\mathbf{D}(c)^{\text{H}}\mathbf{SD}(c)$	701	−1.041	0.660	619	32.736	1.606
SRP MLR, $q = 0.25/0.1$	$P_{\text{BF-MLR}}(c) = \dfrac{M}{q\mathbf{X}^{\text{H}}\mathbf{X}+(1-q)\mathbf{N}^{\text{H}}\mathbf{N}}\mathbf{D}(c)^{\text{H}}\mathbf{SD}(c)$	702	−1.097	0.893	625	31.631	2.115
SRP ML	$P_{\text{ML}}(c) = \dfrac{1}{\mathbf{D}(c)^{\text{H}}\mathbf{S}^{-1}\mathbf{D}(c)}$	699	−1.640	2.012	617	34.537	2.875
SRP MUSIC	$P_{\text{MUS}}(c) = \dfrac{1}{\mathbf{D}(c)^{\text{H}}[\mathbf{I}-\mathbf{U}_S\mathbf{U}_S^{\text{H}}]\mathbf{D}(c)}$	702	−0.729	0.248	626	33.157	0.976

have a linear microphone array that works in the range of angles $[-60°, +60°]$, usually estimating this function every 5 (and even every 10) degrees is sufficient to find the DOA with enough resolution. The overall precision can be increased by quadratic interpolation (see next section) to acceptable levels. This means estimation of the steered power function $P_{SSL}(c)$ in 26 (or 13) points and does not require a substantial amount of computational resources. If we want to do full 3D sound source localization, however, the number of points increases substantially, to almost 500 for every 5 degrees (or 120 for every 10-degree case). This requires a substantial number of estimations. The question is: Can we estimate the location of the maximum with fewer estimations of the steered-response power function?

One potential approach is to use particle filters. The SRP function is estimated in a set of points, which move randomly from frame to frame. The weight of each point participates in estimation of the sound source location, and is updated in each frame using statistical methods. Under certain conditions, some particles die and are replaced by particles duplicated from particles with higher weight. The number of particles can be smaller than the number of points needed to cover the working field evenly.

Particle filters, also known as the "sequential Monte Carlo method," are sophisticated model estimation techniques. In the literature the same approach is variously called "bootstrap filtering," the "condensation algorithm," "interacting particle approximations," and even "survival of the fittest." We are not going to provide here the full theory of particle filtering and all of its methods, but will give an example of using them for sound source localization. A very good tutorial on particle filters with algorithm examples is provided in [18].

Particle filtering is a technique for implementing the recursive Bayesian filter by Monte Carlo simulations. The key idea is to represent the required posterior density function by a set of random samples with associated weights and to compute estimates based on these samples and weights. As the number of samples becomes very large, this Monte Carlo characterization becomes an equivalent representation of the posterior PDF, and the particle filter approaches the optimal Bayesian estimate. Fortunately, in many practical cases, we can use a lower number of particles and still have satisfactory precision.

Vermaak and Blake [19] used particle filtering to combine the time-delay estimates from multiple microphone-array pairs. All maxima of the generalized cross-correlation function are considered as potential directions to sound sources. The proposed particle filter quickly converges to the real position of the sound source. Subsequently, Ward and Williamson [20] applied a similar approach to a steered-response power (delay-and-sum beamformer) sound source localizer. In both cases satisfactory precision is achieved by using only 50 particles – a number much lower than the necessary sampling points to cover evenly the working field to achieve the same precision of localization estimates. Here we provide a generalization of the solution in [20], applicable for use with any of the power steering functions above.

The sound source localization problem can be formulated in a state–space estimation framework by associating the source location at discrete time nT with an unobserved state vector $\alpha^{(n)} = [x_s, y_s, z_s, \dot{x}_s, \dot{y}_s, \dot{z}_s]^{\mathrm{T}}$, where $c_s = [x_s, y_s, z_s]^{\mathrm{T}}$ is the source position and $\nu_s = [\dot{x}_s, \dot{y}_s, \dot{z}_s]^{\mathrm{T}}$ is the source velocity. In the same way, c_s and ν_s can be replaced with direction and elevation angles $c_s = [\varphi_s, \theta_s]^{\mathrm{T}}$ and the corresponding angular velocities $\nu_s = [\dot{\varphi}_s, \dot{\theta}_s]^{\mathrm{T}}$. If $\mathbf{X}^{(1:n)}$ is the concatenation of all inputs of the microphone-array frames up to frame n, estimating the state $\alpha^{(n)}$ would be easy if we can calculate the conditional density $p(\alpha^{(n)}|\mathbf{X}^{(1:n)})$ which is a direct measure of how likely a particular state is based on the microphone signals. It can be calculated indirectly by

$$p\left(\alpha^{(n)}|\mathbf{X}^{(1:n)}\right) \propto p\left(\mathbf{X}^{(n)}|\alpha^{(n)}\right) p\left(\alpha^{(n)}|\mathbf{X}^{(1:n-1)}\right) \tag{6.39}$$

where $p(\mathbf{X}^{(n)}|\alpha^{(n)})$ is the likelihood (or measurement density). The prediction density $p(\alpha^{(n)}|\mathbf{X}^{(1:n-1)})$ is given by

$$p\left(\alpha^{(n)}|\mathbf{X}^{(1:n-1)}\right) = \int p\left(\alpha^{(n)}|\alpha^{(n-1)}\right) p\left(\alpha^{(n-1)}|\mathbf{X}^{(1:n-1)}\right) d\alpha^{(n-1)} \tag{6.40}$$

where $p(\alpha^{(n)}|\alpha^{(n-1)})$ is the state transition density, and $p(\alpha^{(n-1)}|\mathbf{X}^{(1:n-1)})$ is the prior filtering density. Although closed-form solutions exist for Equations 6.39 and 6.40, these recursions can be approximated through Monte Carlo simulation of a set of particles (representing the source state) having associated discrete probability masses.

Assume that we have an initial set of particles $\{\alpha_i^{(0)}, i = 1 : N\}$ and give them initial weights $\omega_i^{(0)} = 1/N, i = 1 : N$. These initial particles are spread randomly in the working space and have randomly generated speeds with given distribution. Then the algorithm for each consequent frame goes as follows:

1. Resample the particles from the previous frame $\{\alpha_i^{(n-1)}\}$ according to their weights $\{\omega_i^{(n-1)}\}$ to form the resampled set of particles $\{\bar{\alpha}_i^{(n-1)}, i = 1 : N\}$. This is protection from the so-called *degeneracy problem*, when, after a few iterations, all but one particle have negligible weight. A suitable measure of degeneracy of the algorithm is the effective sample size, approximated by $\hat{N}_{\mathrm{eff}}^{(n)} = 1/\sum_{i=1}^{N}(\omega_i^{(n)})^2$. Note that $N_{\mathrm{eff}} \leq N$ and small N_{eff} indicates severe degeneracy. When N_{eff} goes below a certain threshold N_{T}, the degeneracy problem should be addressed properly and one of the possible approaches is resampling of the particles. Some algorithms do resampling at each iteration unconditionally. There are several algorithms to do the resampling, one of which is described in [18] and consists of the following steps:
   a. Construct a cumulative distortion function (CDF) as follows: $c_1 = 0$, $c_i = c_{i-1} + \omega_i^{(n)}$.
   b. Draw a starting point $u_1 \sim U[0, N^{-1}]$.
   c. Resample each element $\bar{\alpha}_i^{(n-1)} = \bar{\alpha}_j^{(n-1)}$, $\bar{\omega}_i^{(n-1)} = N^{-1}$, where j is the first index and $u_1 + N^{-1}(j-1) \leq c_i$.

Note that effectively we reset the weights to N^{-1} as in fact the resulting samples are i.i.d. samples from the discrete density. During this process some particles with lower weight will disappear, while particles with higher weight will be duplicated (hence the name "survival of the fittest"). Thus the particles will concentrate around the areas where the weights are higher, which means where the desired maxima are due to the way we recompute the weights in the next steps.

2. Predict the new set of particles $\{\alpha_i^{(n)}\}$ by propagating the resampled set $\{\bar{\alpha}_i^{(n-1)}\}$ according to the source propagation model for each coordinate:

$$x_i^{(n)} = x_i^{(n-1)} + T\dot{x}_i^{(n-1)}$$

$$\dot{x}_i^{(n)} = \frac{\dot{x}_i^{(n-1)} + TF_{xi}^{(n)}}{1 + \beta_x T}.$$

Here, β_x is the coefficient of friction, T is the discretization time step, and

$$F_{xi}^{(n)} = \begin{cases} \mathbb{N}(0, \sigma_x^2) & \text{if} \quad i^{(n)} = 1 \\ 0 & \text{if} \quad i^{(n)} = 0 \end{cases}$$

is the excitation force. When the particle is in motion ($i^{(n)} = 1$) the excitation force is assumed to be an i.i.d. zero-mean Gaussian random variable with variance σ_x^2. If $i^{(n)} = 0$, the excitation force is removed, but the particle continues to move with decreasing speed due to the friction. The transition probabilities for the indicator variable are specified by

$$p\left(i^{(n)}|i^{(n-1)}\right) = \Pr\left\{i^{(n)} = i|i^{(n-1)} = j\right\} = p_{ij}, \quad i,j = 0, 1.$$

If the average amount of time the model is expected to spend in state i before switching the states is given by τ_i, the self-transition probability for state i can be obtained as $p_{ii} = 1 - T/\tau_i$. In [19], the motion parameters are fixed to $\beta_x = 10\,\text{s}^{-1}$, $\sigma_x = 5\,\text{ms}^{-2}$, $\tau_i = 5\,\text{s}$, and $i = 0,1$. Note that the duplicated particles during the resampling process will start to diverge owing to the random elements (speed and start/stop) in the propagation.

3. For the location $c_i^{(n)}$ of each particle $\alpha_i^{(n)}$ calculate the steered response power function $P_{\text{SSL}i}^{(n)}(c_i^{(n)})$.

4. Weight the new particles according to the likelihood function

$$\omega_i^{(n)} = p\left(\mathbf{X}^{(n)}|\alpha_i^{(n)}\right) = \phi\left(P_{\text{SSL}i}^{(n)}(c_i^{(n)})\right)$$

and normalize so that

$$\sum_i \omega_i^{(n)} = 1.$$

A good candidate for the likelihood shaping function (LSF) is $\phi(x) = x^i$, where typically $i = 1, 2, 3$, or 4. In general, a higher value of i increases the peakedness of the function, but overdoing it may cause the loss of some peaks. The recommended value is 2 or 3, but for some of the functions derived above the best results may be achieved with $i = 1$.

5. Estimate the current source location as the weighted sum of the particle locations:

$$E\left\{c^{(n)}\right\} = \sum_{i=1}^{N} \omega_i^{(n)} c_i^{(n)}.$$

6. Store the particles and their weights for use with the next frame.

Both papers ([19] and [20]) claim sufficient precision of the localization estimates using two pairs of microphones placed on neighboring walls of a 3 m × 3 m room. The reported location estimation precision of ~0.1 m can be achieved if we place 900 points in the working area (every 10 cm). Both papers use only 50 particles, which means that particle filtering is computationally much more efficient.

6.3 Post-processing Algorithms

6.3.1 Purpose

The algorithms discussed in the previous section operate with a single set of frames. They are invoked when the VAD indicates that there is voice activity. At the end, the single-frame sound source localizer can decide to reject the current measurement if in doubt that there is no distinct sound source or sources. If the measurement is not rejected it is time-stamped and sent for post-processing, usually accompanied by a confidence level. The measurement can be a set of one or more discrete locations (i.e., where the maxima of $P_{SSL}(c)$ are) or the function $P_{SSL}(c)$ itself. The set of discrete measurements has the advantage of smaller size and improved precision by using quadratic interpolation. Sending the localization function requires more memory and processing power. Its advantage is that, after long averaging, weak sound sources can be detected.

Figure 6.9 shows the steered power function – output of the SRP MUSIC algorithm – as a function of time and direction. We have two distinct sound sources at around 0° and −50°. The power steering function is computed only when the VAD indicates speech activity. The sound source in front of the microphone array is a human speaker at 1.5 m distance, the second sound source is radio at 2.5 m distance.

In most practical cases the single-frame sound source localizer sends for post-processing one (or eventually more) measurements, each of which contains three elements: direction (one or two angles), confidence level, and time stamp (either frame number or actual time). Figure 6.10 shows the processed steered-response power

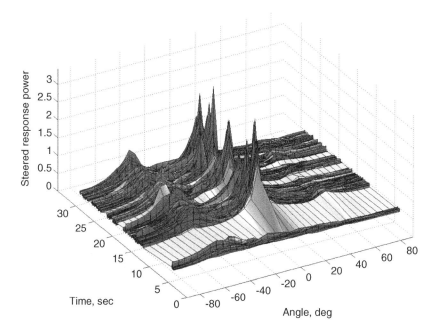

Figure 6.9   Steering-response power as a function of direction and time

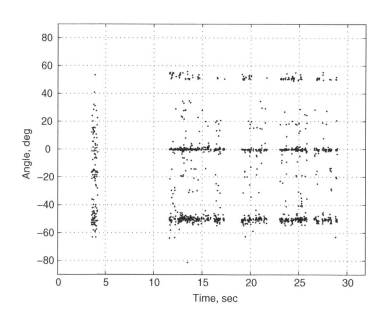

Figure 6.10   Single-frame sound source localizations for two sources

functions. Each dot is location of a maximum. The measurements from the human speaker and the radio are clear; there are some random measurements and a strong reflection at $+55°$, which is the direction to the whiteboard in the office where the recording was made. The histogram of these measurements is shown in Figure 6.11, which confirms the location of the two sound sources.

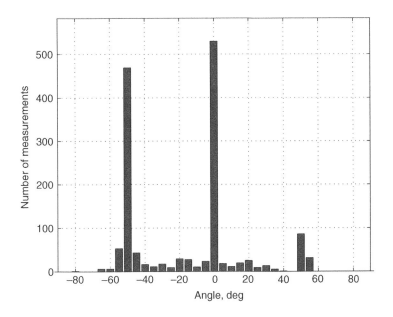

Figure 6.11   Histogram of the sound source localizations for two sources

The role of the post-processing block is to combine several measurements and increase the precision of the sound source localization, to remove noisy measurements, and to track one or more sound sources. Its output is quite similar to the output of the single-frame sound source localization: one or more data blocks with the same three elements of direction, confidence, and time stamp. In some rare cases it can include velocities as well.

In addition, the post-processor organizes storage of the per-frame measurements and removal of those older than a given time interval. More measurements means better precision, but larger lag in the case of a moving sound source. A good compromise here is a measurement lifetime between one and four seconds. The normal speech signal produces around 20 measurements per second assuming a frame rate of 50 per second (20 ms duration). This means that when somebody is actively speaking we will have in the buffer between 20 and 80 measurements for processing.

6.3.2 Simple Clustering

Improving the localization estimates by averaging several measurements can happen if we properly classify which measurement belongs to which sound source. There are many clustering algorithms described in the literature. They are capable of grouping a large amount of data into a variable number of clusters. For the relatively small number of measurements there are in the buffer, it is possible to do the same using much simpler algorithms. One algorithm that works well in this case is so-called "bucket clustering."

6.3.2.1 Grouping the Measurements

Assume that we have certain prior information about the precision of the single-frame sound source localizer for direction and elevation: standard deviations σ_φ and σ_θ, respectively. The microphone-array working volume (i.e., the place where the sound sources are) is defined with minimum and maximum direction angles φ_{min} and φ_{max} and minimum and maximum elevation angles θ_{min} and θ_{max}. The idea is to divide the working volume on overlapping areas with a size that is six times the standard deviation in each dimension. For 50% overlapping this leads to

$$M = 4 \frac{\varphi_{max} - \varphi_{min}}{6\sigma_\varphi} \cdot \frac{\theta_{max} - \theta_{min}}{6\sigma_\theta} \tag{6.41}$$

different sections (or buckets) in the space. During the first step each measurement is posted into the corresponding section. Owing to overlapping, one measurement can go to two or more neighboring sections. If the working area covers the entire circle, measures should be taken to ensure the overlapping around the area where the angle switches from $-\pi$ to π. More overlapping leads to better resolution for separating sound sources but increases the necessary computations.

6.3.2.2 Determining the Number of Cluster Candidates

To determine the number of sound sources, we apply a threshold defined as the average confidence of sections with more than one measurement:

$$C_{Th} = \frac{1}{L} \sum_{i=1}^{M} \sum_{j=1}^{N_i} C_{ij} \tag{6.42}$$

where C_{ij} is the confidence of the jth measurement in the i-th section, N_i is the number of measurements in the i-th section, M is the number of sections, and L is the number of sections with number of measurements larger than 1. The section confidence is the sum of the weights of all measurements in the section:

$$C_i = \sum_{j=1}^{N_i} C_{ij}.$$ (6.43)

If the section confidence is greater than the threshold and the number of measurements is greater than 1, we assume that there is a potential sound source in this section and mark it for further processing.

6.3.2.3 Averaging the Measurements in Each Cluster Candidate

To ensure that sound sources with close coordinates will be considered together and averaged properly, we project them on a sphere with radius 1. This converts the radial coordinate system to Cartesian:

$$p_{ij} = [x, y] = [\cos \varphi_{ij} \cos \theta_{ij}, \sin \varphi_{ij} \cos \theta_{ij}].$$ (6.44)

Obtaining the position of the sound source and the standard deviation in the sections with confidence above the threshold is done by simple weighted averaging:

$$p_i = \frac{1}{C_i} \sum_{j=1}^{N_i} C_{ij} p_{ij}$$ (6.45)

$$\sigma_i = \sqrt{\frac{\frac{1}{C_i} \sum_{j=1}^{N_i} C_{ij} \|p_i - p_{ij}\|^2}{N_i - 1}}.$$ (6.46)

Owing to the overlapping and the size of the sections, portion of the measurements belonging to the same sound source might be in two neighboring sections and we may have some noise measurements in the section as well. To eliminate these measurements and to improve the precision, a second pass of weighted averaging is performed only for the measurements in the range $p_i \pm 2\sigma_i$ to obtain the final position estimate $\tilde{p}_i$ for this section. Denote the set of these measurements by $\tilde{n}_i$. Then the new sound source position estimate $\tilde{p}_i$ and the weighted standard deviation $\tilde{\sigma}_i$ are recalculated for precision estimation as follows:

$$\tilde{C}_i = \sum_{j \in \tilde{n}} C_{ij}$$

$$\tilde{p}_i = \frac{1}{\tilde{C}_i} \sum_{j \in \tilde{n}} C_{ij} p_{ij}$$ (6.47)

$$\tilde{\sigma}_i = \sqrt{\frac{\frac{1}{\tilde{C}_i} \sum_{j \in \tilde{n}} C_{ij} \|p_i - p_{ij}\|^2}{\tilde{N}_i(\tilde{N}_i - 1)}}.$$

Here $\tilde{N}_i$ is the number of measurements left in the i-th cluster. Note that $\tilde{\sigma}_i$ is the precision estimate for the averaged result $\tilde{p}_i$, while σ_i is estimation of the measurements deviation.

6.3.2.4 Reduction of the Potential Sound Sources

At this point we have an array of hypothetical sound sources, each represented by position, standard deviation, and weight. The goal of this next step is to eliminate from this list sound source hypotheses that belong to the same sound source. This can happen owing to the overlapping of the sections. The criterion for elimination is that the distance between any two hypothetical sound sources k and l ($k \neq l$) is less than the overlapping in this direction. For 50% overlapping of sections, if

$$\|\tilde{p}_k - \tilde{p}_l\| \leq 3\sqrt{\sigma_\varphi^2 + \sigma_\theta^2} \tag{6.48}$$

then these two hypothetical sound sources are considered as one. The hypothetical sound source with lesser confidence is removed from the list. If there is a maximum number S of sound sources for tracking, only the first S (sorted by their confidence $\tilde{C}_i$) sound sources stay in the list.

The bucket clustering algorithm described above is straightforward and simple to implement and test. It provides good precision of the localization estimates. The major disadvantage is the limited resolution for adjacent sound sources. With the way bucketing and removal of duplicated sound sources is performed, the algorithm cannot distinguish sound sources closer than three times the standard deviation of a single measurement. This algorithm also has difficulties with moving sound sources.

EXERCISE

Set the SSL algorithm in *MicArrSSL.m* to SRP_PHAT or the one that performs better after your implementation. Process a .WAV file with a single sound source. Write a MATLAB script that reads the output files from *MicArrSSL.m*; these are *PPP.dat* and *SSLTime.dat*. For each steered power function, find all maxima and store them in a text file that contains three numbers on each line: time, direction, weight (see Equation 6.55).

Implement the bucket clustering algorithm above using as input data the text file generated in the previous step. Plot the estimated mean and variance as functions of the time. Compare them with the no-averaging and processing case.

6.3.3 Localization and Tracking of Multiple Sound Sources

To improve the localization and tracking capabilities of the sound source localization post-processor for multiple sound sources, some of the clustering algorithms well known in the literature can be employed.

6.3.3.1 *k*-Means Clustering

One of the most frequently used algorithms is "*k*-means clustering." The overall idea is simple, the algorithm tries to distribute N points in K clusters ($K < N$) in a way to minimize the total intra-cluster variance, or squared error function, or residual sum of squares (RSS):

$$E_K = \sum_{k=1}^{K} \sum_{x_j \in S_k} ||x_j - \mu_k||^2. \tag{6.49}$$

Here, μ_k are the cluster centroids, or in our case the estimated locations of the sound sources. The set x_j is the measurements we have in the buffer. One of the most commonly used algorithms for estimation of the cluster centroids is Lloyd's algorithm [21] (this paper is a published version of much earlier technical report from Bell Telephone Laboratories dated 1957). The algorithm is iterative and consists of the following steps:

- Given the set of K centroids μ_k, $k = 1:K$, for each point x_i, compute the distance to each centroid.
- Assign the point to the cluster with closest centroid.
- Re-compute the centroids for each cluster.
- Repeat until some stopping criterion is met.

There are several complications when using this algorithm to process sound source localization data. The first is that the *k*-means clustering algorithm does not make any sense for the case of one cluster plus noise, which is the most common. The second complication is that the original algorithm was designed to find the best quantization levels given the number of quanta. In sound source localization we do not know in advance how many sound sources we have. This means repeating the steps above for a hypothetical number of sound sources $K = 2$ to K_{max}, which adds computational complexity. A suitable criterion should be applied to stop at a certain number of sound sources. One of the most common approaches uses the *Bayesian information criterion* (BIC) [22]:

$$\mathrm{BIC}_K = N \ln\left(\frac{E_K}{N}\right) + K \ln(N). \tag{6.50}$$

The model with lower BIC is preferred.

EXERCISE

Record a .WAV file with two sound sources, or use the attached *Record9.WAV*. This was recorded with a four-element microphone array and has a radio playing at 2.5 m

distance and approximated direction of $-50°$. In front of the microphone array at
approximately $0°$ is a human speaker.

Set the SSL algorithm in *MicArrSSL.m* to SRP_PHAT or the one which performs
better after your implementation. Process the .WAV file with the two sound sources.
Use the MATLAB script from the previous exercise to process the output and generate
the sound source localization data. You can use the attached *SSLData.dat* file which
contains the output of processing *Record9.WAV*.

Implement a k-means clustering algorithm, at each moment using data only from
the previous 4 s. Average the localized sound sources and compute the deviation.
Evaluate the results. You may want to use the MATLAB implementation of the
algorithm in the function `kmeans()`.

6.3.3.2 Fuzzy C-Means Clustering

The third problem with using the k-means clustering algorithm is that it tries to
cluster all points in the set, while certain of them may not belong to any sound source;
that is, they can be sporadic reflections from walls and behave as random noise.
From this perspective, using some of the algorithms for fuzzy clustering makes much
more sense.

In fuzzy clustering, the data elements can belong to more than one cluster and
associated with each element is a set of membership levels. These indicate the strength
of association between each data element and a particular cluster. One of the most
commonly used algorithms in this class is the "fuzzy C-means algorithm" [23]. The
algorithm tries to minimize

$$E_K = \sum_{i=1}^{N}\sum_{k=1}^{K} u_{ik}^{\alpha}||x_i - \mu_k||^2, \quad \text{subject to} \quad \sum_{k=1}^{K} u_{ik} = 1. \quad (6.51)$$

Here, $\alpha \in [1, \infty]$ is the weighting exponent and $u_{ik} \in [0, 1]$ is the degree to which
the element x_i belongs to cluster k with centriod μ_k. The algorithm is iterative and for
each iteration has the following steps:

- Given data set x_i, $i = 1{:}N$, number of clusters K, and membership matrix $\mathbf{U}_{K\times N}$,
compute the cluster centroids as weighted sum

$$\mu_k = \sum_{i=1}^{N} u_{ik}^{\alpha} x_i \Big/ \sum_{i=1}^{N} u_{ik}^{\alpha}.$$

Before the first iteration the membership matrix can be initialized with equal or
random values, normalized according to Equation 6.51.

- Compute the new set of distances $||x_i - \mu_k||$, update the membership matrix

$$u_{ik} = ||x_i - \mu_k||^{-\frac{2}{\alpha-1}}$$

 and normalize it according to Equation 6.51.
- Compute the objective function E_K and compare its value with the previous iteration. Terminate if the improvement is below a certain small number.

The BIC can be used to determine the number of clusters. The nice property of this algorithm is that we can always apply a certain threshold and ignore the data points that are far from the cluster centriod, thus rejecting the noisy points.

EXERCISE

Implement the fuzzy C-means clustering algorithm, at each moment using data only from the previous 4 s. Average the localized sound sources and compute the deviation. Evaluate the results and compare them with the previous algorithm. You may want to use the MATLAB implementation of the algorithm in the function `fcm()`.

6.3.3.3 Tracking the Dynamics

The clustering algorithms above assume static clusters – that is, fixed centroids for the lifetime of the sound source localization measurements in the buffer. This is almost always true for sound source localization in a small office or conference room. In the worst case we will have a lag in the estimated position of a moving speaker, based on the current speed, distance to the microphone array, and lifetime of the measurements in the buffer. Once the speaker stops, the lag will quickly disappear with expiry of the older measurements in the buffer.

To better track moving sound sources we have to add to the estimated position estimation of the derivatives: velocity and, eventually but very rarely, the acceleration. The problem is complicated by the unknown number of sources, the presence of noise and ghost sound sources (reflections from the walls or other objects), and non-linearity in the object movements. In order to deal with the non-linear and/or non-Gaussian reality, two categories of techniques have been developed: parametric and non-parametric.

The best known algorithm to solve the problem of non-Gaussian, non-linear filtering is *extended Kalman filtering* (EKF) [24]. By linearizing non-linear functions around the predicted values, the extended Kalman filter is proposed to solve non-linear system problems. It was first introduced in control theory and later on applied in visual tracking. Because of its first-order approximation of a Taylor series expansion, EKF finds only limited success in tracking objects. Julier and Uhlmann [25] developed an *unscented Kalman filter* (UKF) that can accurately compute the mean and covariance under the assumption that it is easier to approximate a Gaussian distribution than it is to approximate arbitrary non-linear functions. While UKF is significantly better than EKF in density statistics estimation, it still assumes a Gaussian parametric form of the posterior, so it cannot handle general non-Gaussian distributions.

The non-parametric techniques are based on sequential Monte Carlo simulations, or particle filters. These have been discussed briefly earlier in this chapter, but see [26] and [27] for more details. They assume no functional form, but instead use a set of random samples (also called particles) to estimate the posteriors. When the particles are properly placed, weighted, and propagated, the posteriors can be estimated sequentially over time, so any statistical estimates such as the mean, modes, kurtosis, and variance can be easily computed.

These two approaches are combined in the unscented particle filter [28]. In this filter the authors deal with the sufficient statistics of the UKF for each particle, moving the particles towards the regions of high likelihood. This technique allows better tracking of moving objects, but is computationally expensive.

Another approach is the *joint probabilistic data association filter* (JPDAF), which in certain variants can be combined with the particle approach (sample-based JPDAF).

These approaches come from the general tracking field and have nothing specific for sound sources and localizations. As it is not very common to track the dynamics of moving sound sources, we stop here.

6.4 Practical Approaches and Tips

6.4.1 Increasing the Resolution of Time-delay Estimates

The generalized cross-correlation function after the inverse Fourier transformation in Equation 6.8 is sampled with a period equivalent to the sampling period of the input signals. Finding the time lag of the maximal sample gives estimation of the delay with a resolution of the sampling period. For a microphone-array pair with distance 20 cm between the microphones and sampling frequency 16 kHz, the range between $-90°$ and $+90°$ is unevenly discretized to 17 segments; that is, we have average resolution worse than $10°$. In most practical cases this resolution is not sufficient and is higher than the error caused by ambient noise and reverberation; that is, it is the dominant source of error.

The most straightforward approach to increase the resolution in time is to increase the sampling rate. Doubling the rate instantly increases the memory requirement by a factor of two and requires two to four times more computational power. A much less expensive approach is to use some form of interpolation. The maximum sample and the two neighbors can be used for quadratic interpolation. Given the maximum of the generalized cross-correlation function $G(k_m T)$ at point k_m and the two neighbors $G(k_m - 1)T)$ and $G(k_m + 1)T)$, the GCC function around its maximum can be approximated with a quadratic polynomial:

$$a[(k_m-1)T]^2 + b[(k_m-1)T] + c = G((k_m-1)T)$$
$$a[k_mT]^2 + b[k_mT] + c = G(k_mT) \tag{6.52}$$
$$a[(k_m+1)T]^2 + b[(k_m+1)T] + c = G((k_m+1)T).$$

This is a system of three equations with three unknowns: a, b, and c. Solving them and zeroing the first derivative gives us the interpolated value of the delay:

$$\tau_D = -\frac{b}{2a}. \tag{6.53}$$

In most practical systems this approach increases the TDE resolution four to six times, which is enough to take the estimation error due to resolution below the errors due to noise and reverberation.

6.4.2 Practical Alternatives for Finding the Peaks

For a given steered-beam target function $P_{SSL}(c)$ we have to find the maximum, or the maxima. One of the most commonly used approaches is to estimate the values of the target function for a given set of points, usually evenly spread in the working space. This is quite common when the sound source localization is in one dimension. Then we can estimate the value of the target function every 5 degrees, say, in the working interval of either $[-180°, +180°]$ or $[-90°, +90°]$. This means estimation of either 72 or 36 points. Then we can easily find the maxima by just scanning the values vector. The maximum point and the two neighbors can be used to increase the resolution by quadratic interpolation as described earlier in this chapter. If the sound source localization is performed in two dimensions it is trivial to convert the quadratic interpolation above to bi-quadratic.

If two-dimensional sound source localization is necessary, the number of evenly spread points increases rapidly. To cover the entire sphere with 5° resolution it is necessary to estimate ~500 points. In many cases this is practically impossible, especially in real time. Then we can use coarse-to-fine searching [29], or some of the gradient-based search algorithms. Unfortunately most of the later assume a single maximum in the working space and perform poorly in the presence of multiple maxima – they just find the closest maximum, and depend heavily on the starting point for the search.

6.4.3 Peak Selection and Weighting

Peak selection is another problem to overcome in practical sound source localization. There might be numerous fake peaks in the noise floor. To select only the significant peaks, they are compared with a threshold and if the value is below that the peak is rejected. A threshold commonly used is the average of the target function:

$$\eta = \frac{1}{N} \sum_{i=1}^{N} P_{\text{SSL}}(c_i). \qquad (6.54)$$

In many cases the sound source localization algorithms assume that one sound source is dominant in each audio frame. Even if there are two active speech sources in most of the frames, just one of them dominates. This is because the pauses are an integral part of human speech, and the speech signal is sparse not only in the frequency domain but in the time domain as well. Owing to noise and reverberation, the peaks are not so sharp and smear when multiple sound sources are presented. The normal constraints for the microphone-array size reduce its ability to distinguish closely positioned sound sources. Then it is better just to discard the frames with multiple sound sources instead of filling the post-processor buffer with noisy data. An ad-hoc measure of how good a peak we have is the confidence level

$$C^{(n)} = \frac{P_{SSL\,\text{max}} - P_{SSL\,\text{min}}}{\eta}. \qquad (6.55)$$

This is a good measure for the confidence level of the sound source localization for the current frame. The sound source localization measurement for the current frame is sent to the post-processor only if the confidence level $C^{(n)}$ is above a certain level (a good practical value is 0.7). The confidence level accompanies this measurement for further processing. Rejecting any measurements below a certain threshold helps to remove noisy measurements. Figure 6.12 shows the measurements from Figure 6.10 with those under the threshold removed. The picture is much cleaner and contains fewer measurements for removal.

6.4.4 Assigning Confidence Levels and Precision

Each sound source should be attributed with a confidence level – a number between 0 and 1. A simple and efficient way to assign confidence level is used with the simple clustering algorithm, but is applicable with practically all of the post-processing algorithms discussed above.

The factors that affect the confidence level are the number of measurements $\tilde{N}_i$ used to estimate the position, the standard deviation $\tilde{\sigma}_i$, and the most recent time-stamp among these $\tilde{N}_i$ measurements. When the number of measurements is less than a given value N_{cnt}, the confidence level goes below 1. For practical purposes this value is between 5 and 15. For obvious reasons the confidence level will be lower when the standard deviation is higher, or the most recent measurement is in the middle of its

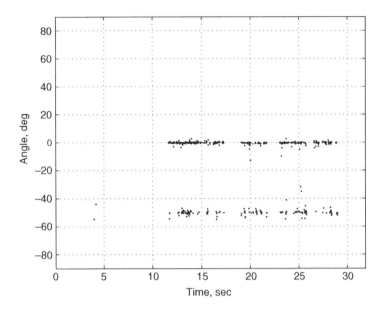

Figure 6.12 Single-frame sound source localizations after applying a confidence threshold

lifetime. The final confidence level consists of several sub-levels, calculated as follows:

$$c_{iN} = \tilde{N}_i/N_{cnt}$$
$$c_{i\sigma} = \sqrt{\sigma_\varphi^2 + \sigma_\theta^2}/\tilde{\sigma}_i \qquad (6.56)$$
$$c_{iT} = (t - T_{iLast})/(2T_L)$$

where c_{iN} is the confidence level based on the number of measurements; $c_{i\sigma}$ is the confidence level based on the standard deviation; c_{iT} is the confidence level based on the most recent measurement time-stamp T_{iLast}; where t is the current time and T_L is the measurement lifetime. After limiting the values of all sub-levels in the range between 0 and 1, the final confidence level is calculated as follows:

$$c_i = c_{iN} c_{i\sigma} c_{iT}. \qquad (6.57)$$

The final result from the post-processor is a list of sound sources, each presented with position, confidence level, and standard deviation. Note that the application should recognize which sound source belongs to which speaker, to do person tracking.

6.5 Summary

This chapter has discussed algorithms and approaches for sound source localization. A sound source localizer is an integral part of each sound capture system with microphone array and steered beam. The SSL usually consists of a per-frame localizer and post-processor. It is assisted by a voice activity detector, used to reject frames with no active sound source.

The per-frame SSL uses only the current frame to estimate the sound source position. There are two major groups of methods: using time-delay estimates and steered-response power. For microphone arrays with more than two elements, both approaches lead to finding maximum (or maxima) of a certain function. For easier estimates we can use various methods, such as quadratic interpolation, coarse-to-fine searches, gradient searches, and particle filtering.

The post-processor uses a set of measured results to provide the best estimate of the location of the sound source or sources. Various clustering algorithms are applicable here: bucket, k-means, fuzzy C-means, and so on. Tracking multiple sound sources and including their dynamics (speed and acceleration) increase the problem complexity. It can be solved using generic estimation and tracking algorithms, such as unscented particle filtering.

Bibliography

[1] Ghose, K., Zotkin, D., Duraiswami, R. and Moss, C.F. (2001) Multimodal localization of a flying bath. Proceedings of International Conference of Acoustic, Speech and Signal Processing ICASSP 2001, Salt Lake City, UT.

[2] Recanzone, G., Makhamra, S. and Guard, D. (1998) Comparison of relative and absolute sound localization ability in humans. *Journal of the Acoustical Society of America*, **103**(2), 1085–1097.

[3] König, C., Weick, F. and Becking, J. (1999) *Owls: A Guide to the Owls of the World*, Yale University Press.

[4] Knapp, C. and Carter, G. (1976) The generalized correlation method for estimation of time delay. *IEEE Transactions on Acoustic, Speech, and Signal Processing*, **ASSP-24**(4), 320–327.

[5] Roth, P. (1971) Effective measurements using digital signal analysis. *IEEE Spectrum*, **8**, 62–70.

[6] Eckhart, C. (1952) Optimal rectifier systems for the detection of steady signals, Univ. California, Scripps Inst. Oceanography, Marine Physical Lab. Rep SIO 12692, SIO Ref 52-11.

[7] Hannan, E. and Thomson, P. (1971) Estimation of coherence and group delay. *Biometrika*, **58**, 469–481.

[8] Jenkins, G. and Watts, D. (1968) *Spectral Analysis and its Applications*, Holden-Day, San Francisco, CA.

[9] Carter, G., Knapp, C. and Nuttall, A. (1973) Estimation of the magnitude-squared coherence function via overlapped fast fourier transform processing. *IEEE Transactions on Audio and Electroacoustics*, **AU-21**(4), 320–327.

[10] Brandstein, M., Adcock, J. and Silverman, H. (1995) A practical time-delay estimator for localizing speech sources with a microphone array. *Computer, Speech and Language*, **9**, 153–169.

[11] Wang, H. and Chu, P. (1997) Voice source localization for automatic camera pointing in videoconferencing. Proceedings of International Conference of Acoustic, Speech and Signal Processing ICASSP 1997.

[12] Rui, Y. and Florencio, D. (2003) Time delay estimation in the presence of correlated noise and reverberation. Proceedings of International Conference of Acoustic, Speech and Signal Processing ICASSP 2003.

[13] Rui, Y. and Florencio, D. (2003) New direct approaches to robust sound source localization. Proceedings of IEEE International Conference on Multimedia Expo (ICME), Baltimore, MD, pp. I: 737–740.

[14] Birchfield, S. and Gillmor, D. (2001) Acoustic source direction by hemisphere sampling. Proceedings of International Conference of Acoustic, Speech and Signal Processing ICASSP.

[15] Brandstein, M. and Ward, D. (ed.) (2001) *Microphone Arrays*, Springer-Verlag, Berlin, Germany.

[16] Van Trees, F. and Harry, M. (2002) Optimum array processing, in *Part IV of Detection, Estimation and Modulation Theory*, John Wiley & Sons, New York.

[17] Schmidt, R. (1986) Multiple emitter location and signal parameter estimation. *IEEE Transactions on Antennas and Propagation*, **AP-34**(3), 276–280.

[18] Arulampalam, M.S., Maskell, S., Gordon, N. and Clapp, T. (2002) A tutorial on particle filters for online nonlinear/non-gaussian bayesian tracking. *IEEE Transactions on Signal Processing*, **50**(2), 174–188.

[19] Vermaak, J. and Blake, A. (2001) Nonlinear filtering for speaker tracking in noisy and reverberant environments. Proceedings of International Conference of Acoustic, Speech and Signal Processing ICASSP, Salt Lake City, UT.

[20] Ward, D. and Williamson, R. (2002) Particle filter beamforming for acoustic source localization in reverberant environment. Proceedings of International Conference of Acoustic, Speech and Signal Processing ICASSP, Orlando, FL.

[21] Lloyd, S. (1982) Least squares quantization in PCM. *IEEE Transactions on Information Theory*, **IT-29**(2), 129–137.

[22] Schwarz, G. (1978) Estimating the dimension of a model. *Annals of Statistics*, **6**(2), 461–464.

[23] Bezdek, J. (1981) *Pattern Recognition with Fuzzy Objective Function Algorithms*, Kluwer Academic, Norwell, MA.

[24] Anderson, B. and Moore, J. (1979) *Optimal Filtering*, Prentice-Hall, Englewood Cliffs, NJ.

[25] Julier, S. and Uhlmann, J. (1997) A new extension of the Kalman filter to nonlinear systems. Proceedings of AeroSense: 11th International Symposium on Aerospace/Defense Sensing, Simulation and Controls. Vol. Multi-Sensor Fusion, Tracking and Resource Management II. Orlando, FL.

[26] Doucet, A., de Freitas, J. and Gordon, N. (eds) (2001) *Sequential Monte Carlo Methods in Practice*, Springer-Verlag, New York.

[27] Gordon, N., Salmond, D. and Smith, A. (1993) Novel approach to nonlinear/non-Gaussian Bayessian estimation. *IEE Proceedings-F*, **140**(2), 107–113.

[28] Van der Merwe, R., de Freitas, N., Doucet, A. and Wan, E. (2000) The unscented particle filter. Proceedings of 13th conference Advances in Neural Information Processing Systems (NIPS 2000).

[29] Duraiswami, R., Zotkin, D. and Davis, L. (2001) Active speech source localization by a dual coarse-to-fine search. Proceedings of International Conference of Acoustic, Speech and Signal Processing, ICASSP, Salt Lake City, UT.

[30] Benesty, J., Sondhi, M. and Huang, Y. (eds) (2008) *Speech Processing*, Springer-Verlag, Berlin, Germany.

[31] Huang, Y. and Benesty, J. (eds) (2004) *Audio Signal Processing for Next-generation Multimedia Communication Systems*, Kluwer Academic, Boston, MA.

7

Acoustic Echo-reduction Systems

Acoustic echo-reduction systems have been an integral part of telephones from very early, and the quality of the echo reduction is critical for the performance of every communication system that works in speakerphone mode. Acoustic echo cancellation (AEC) was one of the earliest applications of adaptive filters and one of the most studied. The purpose of the acoustic echo canceller is to remove from the microphone signal the sounds from the local loudspeaker. This is done by employing an adaptive filter to estimate the transfer function between the loudspeaker and the microphone. The adaptive filter processes the signal sent to the loudspeakers. Its output is subtracted from the microphone signal. This is why in many cases the entire acoustic echo-reduction system is called an acoustic echo canceller.

However, we shall see later in this chapter that AEC is just one of several processing stages. Owing to noise and reverberation, the transfer function estimation is not exact and the adaptive filter cannot remove completely the captured loudspeaker sound, called "echo." The second stage in acoustic echo-reduction systems is a suppression-based non-linear processor. It suppresses the residual energy similarly to a noise suppressor – by applying a time-varying real gain to each frequency bin.

Removing the captured sound from stereo and surround-sound audio systems is challenging owing to the non-uniqueness of the solution for the transfer function. There are several methods to mitigate this problem. In many practical echo-reduction systems additional blocks are used to quickly handle the case when feedback occurs, to track sampling rate drifts, and so on. They are described towards the end of the chapter.

7.1 General Principles and Terminology

7.1.1 Problem Description

The effect of sound reflection from walls and objects is called "reverberation." A human voice recorded in a studio with a closely positioned microphone has no reverberation. It sounds unnatural and usually is called "dry." With some reverberation the human voice sounds warmer and more natural. Echoes are distinct copies of the reflected sound. Humans can hear echoes when the difference between arrival times of the direct signal and the reflection is more than 0.1 s, but even with differences of 0.05 s the audio sounds echoic. In 0.1 s the sound travels 34.3 m, which means that if the object reflecting the sound is further away than 17 m (the sound travels to the object and back) the reflection will be heard as echo. Acoustic echo-reduction does not suppress the echoes in the room. It actually suppresses the *effect* when the local sound source is captured by the microphone, transmitted through the communication line, reproduced by the loudspeaker in the receiving room, captured by the microphone there, returned back through the communication line, reproduced from the local loudspeaker, and so on. This creates an echoic sound and in some cases causes feedback – that is, the entire system converts to a generator, reproducing an annoying constant tone. In the context of acoustic echo-cancellation, echo is the sound from the local loudspeaker captured by the local microphone.

These echo effects were a problem in telephones even before their official discovery. The first prototypes used four wires. They were practically two independent sets, each consisting of microphone, battery, two wires, and headphone. The patent application filed by Antonio Meucci on 28 December 1871 uses four wires in order to eliminate the "local effect," which is nothing but hearing your own voice in the headset. After failing to pay the patent application fee, two years later Meucci abandoned his patent application. This allowed Alexander Graham Bell to file, on 14 February 1876, his patent application for the invention we call today the "telephone." Later, the four wires were replaced by two wires; that is, the same two wires are used to carry the electrical signal from both sides. The first telephones actually used one wire and closed the circuit through the ground in a similar way that telegraphs were doing at that time. Mixing the incoming and outgoing signals caused problems with the "local effect," which was resolved by using a Wheatstone bridge to separate the two signals in telephone circuitry. As the telephone line impedance participates in the bridge, any change in the impedance impairs the suppression of the local effect. With increasing length of the telephone lines there appeared signal reflections when the impedances of the line and the telephone were not balanced. This required the presence of an echo suppressor in each telephone, implemented initially as passive circuitry that inserted signal losses (a pair of diodes which open when signal levels exceed a certain threshold). Later this circuitry was replaced by the "line echo canceller" (sometimes called a "network echo canceller"). Since the mid-1960s the line echo canceller has been implemented as an electronic adaptive filter. When speakerphones appeared they required suppression of the acoustic echo as well. This block is called the *acoustic echo canceller* and is the subject of this chapter.

The theory of acoustic echo cancellation was initially developed by AT&T Bell Labs [1] but was deployed only in the late 1970s owing to performance limitations of electronic blocks of that time. They become cost-effective in the 1990s, and currently adaptive echo cancellers are part of practically every mobile or stationary telephone with speakerphone capabilities.

7.1.2 Acoustic Echo Cancellation

In telecommunications, *near-end* denotes our end of the communication chain: microphone, loudspeaker, and sound. On the other side are the *far-end* microphone, loudspeaker, and sound. The schematic diagram is shown in Figure 7.1. The near-end speaker talks to the microphone and the audio signal is sent to the far-end room. There the near-end speaker's voice is reproduced by the far-end loudspeaker. The far-end microphone captures the sound from this loudspeaker and the voice of the far-end speaker. When both the local and the remote speakers talk simultaneously we have so-called *double talk*. The signal captured by the far-end microphone is transmitted through the communication line and reproduced by the near-end loudspeaker. Without acoustic echo cancellers the near-end speaker will reproduce the delayed and decayed copy of the near-end speech, which will be captured by the near-end microphone and the entire process will be repeated many times, causing annoying echoes. This system with feedback under certain conditions can become unstable and convert itself into a generator of a specific audio frequency, making communication impossible. Both near- and far-end stations have to have acoustic echo cancellers to remove the sound captured from the local loudspeakers. Then each station will transmit only the local voice. This breaks the feedback chain, the echoes are gone, and audio feedback is not possible.

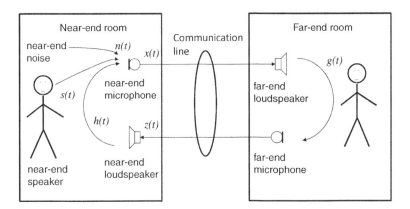

Figure 7.1   Speakerphone telecommunication system

The schematic diagram of the acoustic echo canceller is shown in Figure 7.2. The far-end signal $z(t)$ is sent to the loudspeaker. The microphone captures this signal convolved with the impulse response of the transfer path speaker-microphone $h(t)$. It captures the local voice $s(t)$ and noise $n(t)$. The transfer path speaker-microphone is

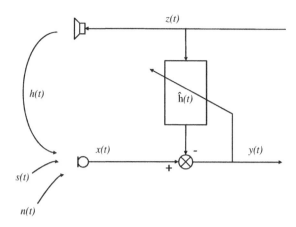

Figure 7.2 Schematic diagram of acoustic echo canceller

omitted for simplicity, as we will not deal with it in this chapter. The microphone signal is

$$x(t) = z(t)*h(t) + s(t) + n(t). \tag{7.1}$$

The acoustic echo canceller estimates the transfer path loudspeaker-microphone $\hat{h}(t)$ and subtracts the estimated portion of the loudspeaker signal from the microphone signal. At the acoustic echo canceller output we have

$$
\begin{aligned}
y(t) &= x(t) - z(t)*\hat{h}(t) \\
&= z(t)*h(t) - z(t)*\hat{h}(t) + s(t) + n(t).
\end{aligned}
\tag{7.2}
$$

As usual we will illustrate the algorithms with processing in the frequency domain. Then the convolution converts to multiplication and we have

$$
\begin{aligned}
Y_k^{(n)} &= Z_k^{(n)} H_k^{(n)} - Z_k^{(n)} \hat{H}_k^{(n)} + S_k^{(n)} + N_k^{(n)} \\
&= Z_k^{(n)} (H_k^{(n)} - \hat{H}_k^{(n)}) + S_k^{(n)} + N_k^{(n)}.
\end{aligned}
\tag{7.3}
$$

The modeling described so far assumes that the audio frame is longer than the reverberation process, which is incorporated in $h(t)$, and we model it with one tap filter for each frequency bin. The reverberation in a normal office or conference room lasts 200–400 ms to the moment that the reverberation energy goes below -60 dB. On the other hand, processing in the frequency domain and the overlap–add process, described in Chapter 2, adds one frame delay, increasing the latency of the entire system. A latency above 100 ms in communications is considered inconvenient for users, and a latency above 250 ms breaks the dialog. In most cases the overall latency of the communication channel should be kept below 50 ms. This leaves less than 20 ms for the audio frame; in many cases the audio frame duration is 10 ms. At 16 kHz sampling rate this is a 160-sample frame length. To accommodate the longer impulse response, the acoustic echo canceller uses a finite impulse response (FIR) filter with multiple taps for each frequency bin. This converts Equation 7.3 to

$$
\begin{aligned}
Y_k^{(n)} &= \sum_{i=0}^{L-1} Z_k^{(n-i)} H_k^{(n-i)} - \sum_{i=0}^{L-1} Z_k^{(n-i)} \hat{H}_k^{(n-i)} Z_k^{(n)} + S_k^{(n)} + N_k^{(n)} \\
&= \sum_{i=0}^{L-1} Z_k^{(n-i)} \left(H_k^{(n-1)} - \hat{H}_k^{(n-1)} \right) + S_k^{(n)} + N_k^{(n)}
\end{aligned}
\tag{7.4}
$$

where L is the number of taps in the FIR filter. Denoting

$$
\begin{aligned}
\mathbf{Z}_k^{(n)} &= [Z_k^{(n)}, Z_k^{(n-1)}, \ldots, Z_k^{(n-L+1)}]^T \\
\mathbf{H}_k^{(n)} &= [H_k^{(n)}, H_k^{(n-1)}, \ldots, H_k^{(n-L+1)}]^T \\
\mathbf{X}_k^{(n)} &= [X_k^{(n)}, X_k^{(n-1)}, \ldots, X_k^{(n-L+1)}]^T
\end{aligned}
\tag{7.5}
$$

the equation can be rewritten in vector form:

$$
Y_k^{(n)} = [\mathbf{H}_k^{(n)}]^T \mathbf{Z}_k^{(n)} - [\hat{\mathbf{H}}_k^{(n)}]^T \mathbf{Z}_k^{(n)} + S_k^{(n)} + N_k^{(n)}.
\tag{7.6}
$$

The total number of filter coefficients is $K \times L$, where K is the number of frequency bins and L is the number of taps for each frequency bin. The goal of the acoustic echo canceller is to estimate these coefficients as close as possible to the actual transfer function and to track eventual changes. If someone moves in the room, or a door opens, the transfer function between the loudspeaker and the microphone will change. This requires the use of adaptive filters in the acoustic echo cancellation.

7.1.3 Acoustic Echo Suppression

If we have perfect estimation of the transfer function, the signal captured from the loudspeaker will be completely suppressed. Owing to the near-end noise, shorter filters than the actual reverberation, and estimation errors, a portion of the captured loudspeaker signal will remain. This portion is called the *echo residual*:

$$
\mathfrak{R}_k^{(n)} = Z_k^{(n)} (H_k^{(n)} - \hat{H}_k^{(n)}).
\tag{7.7}
$$

Assuming that the adaptive filtering did its best, whatever phase information is left behind will be very difficult to track. Then the way to reduce the residual is to use

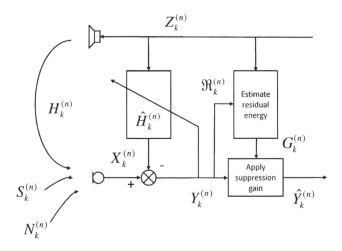

Figure 7.3 Acoustic echo canceller and acoustic echo suppressor

real-gain-based approaches similar to single-channel noise suppression (discussed in Chapter 4), as shown in Figure 7.3. If we can estimate the amount of residual energy $|\hat{\Re}_k^{(n)}|^2$ in each frequency bin, then we can estimate a real-valued suppression gain $G_k^{(n)}$ such as

$$\hat{Y}_k^{(n)} = G_k^{(n)} Y_k^{(n)} \approx S_k^{(n)} + N_k^{(n)} \tag{7.8}$$

that is optimal in one or another way. This technique is called *acoustic echo suppression*. The goal of the acoustic echo suppressor is to estimate the residual energy, to compute a suppression gain, and to apply it to the acoustic echo canceller output. It leads to improved suppression of the captured loudspeaker signal, but introduces distortion and artifacts typical of noise suppressors.

7.1.4 Evaluation Parameters

There is no way to optimize and tune the performance of any engineering system without specifying proper evaluation parameters. One commonly used parameter is the *system distance* [2]:

$$D^{(n)} = 10\log_{10}\left(\frac{||\mathbf{H}_k^{(n)} - \hat{\mathbf{H}}_k^{(n)}||^2}{||\mathbf{H}_k^{(n)}||^2}\right) \tag{7.9}$$

where $||\cdot||$ denotes the l^2-norm, $||\mathbf{H}_k^{(n)}||^2 = (\mathbf{H}_k^{(n)})^T \mathbf{H}_k^{(n)}$. This parameter measures the difference between the estimated and actual transfer functions. It is useful when the real transfer function is known – usually when the acoustic echo canceller is evaluated with simulated signals. A smaller difference means better estimation. In some sources the same parameter is called "normalized misalignment" [3]. This parameter is suitable only for evaluation of acoustic echo cancellers.

For evaluation of the residual energy, *echo return loss enhancement* (ERLE) is used. It is defined as

$$ERLE^{(n)} = 10\log_{10}\left(\frac{E\{(X_k^{(n)})^2\}}{E\{(\Re_k^{(n)})^2\}}\right) \tag{7.10}$$

and is the ratio of the signal and residual energies. The residual can be measured directly only if there is no local speech and the noise level is very low. Corrections for the noise energy should be made if this is not the case. A higher ERLE means better echo suppression.

One indirect parameter to measure the performance of an acoustic echo-reduction system is the perceptual sound quality. As discussed in Chapter 2, this is expressed with MOS points. Also, besides averaging the estimations from a large number of human listeners, the measurement can be done by using some of the algorithms for perceptual evaluation of sound quality. One of the standardized and frequently used algorithms is PESQ (perceptual evaluation of sound quality). It requires knowledge of the clean speech signal, which means that the voice of the local speaker should be recorded in parallel using a close-talk microphone. Another approach is a pre-recorded speech signal to be reproduced by either a loudspeaker of a head-and-torso simulator. The latter is better as it will create more realistic reverberation for the local speech.

An important part of the acoustic echo canceller evaluation is the *convergence time*. This is the time for estimation of the new filters $\hat{\mathbf{H}}_k^{(n)}$ after change in $\mathbf{H}_k^{(n)}$ or at the beginning of the process. As the convergence process may be considered complete when the residual reaches the noise level, then it will depend on the noise level and on the magnitude of the loudspeaker signal. To eliminate these dependencies, the converging process is often modeled with an exponential curve and a single parameter – the time constant.

7.2 LMS Solution for Acoustic Echo Cancellation

The block diagram in Figure 7.2 shows a classic application of an adaptive filter. At each audio frame the input signals are processed with the current filter and then the filter coefficients are updated. One of the potential goals is to minimize the least mean-square error – the classic LMS adaptive filter [4]. The gradient of the mean-square error is

$$\nabla^{(n)} = \frac{\partial E\{|\Re^{(n)}|\}}{\partial \mathbf{H}^{(n)}}$$

$$= 2E\left\{\Re^{(n)} \frac{\partial \Re^{(n)}}{\partial \mathbf{H}^{(n)}}\right\} \tag{7.11}$$

$$= -2E\{\Re^{(n)}\mathbf{Z}^{(n)}\}.$$

Note that the frequency-bin index k is omitted for simplicity. Here we replaced the error with the residual, which means that the gradient estimation is correct when there is no near-end speech and local noise. With estimation of the momentary gradient

$$\hat{\nabla}^{(n)} = \Re^{(n)}\mathbf{Z}^{(n)} \tag{7.12}$$

we can update the filter coefficients in the frames with no local talk:

$$\mathbf{H}^{(n+1)} = \mathbf{H}^{(n)} + \mu\Re^{(n)}\mathbf{Z}^{(n)}. \tag{7.13}$$

Here, μ is the step size and determines the adaptation speed. In most applications the step size is variable to provide faster adaptation. To guarantee convergence of the filter it should be smaller than [4]:

$$0 < \mu < \frac{2}{\lambda_{\max}} \tag{7.14}$$

where $\lambda_{\max}$ is the largest eigenvalue of the input correlation matrix. The convergence time constant is given by

$$\tau_{\mathrm{conv}} = \frac{1}{2\mu\lambda_{\mathrm{av}}}. \tag{7.15}$$

In summary, the LMS acoustic echo canceller should adapt only during frames without local speech and can use variable step size to converge faster. We will continue to use the notation μ in this chapter, but in most of the cases it is variable and dynamically computed; that is, it is actually $\mu^{(n)}$. In the literature there are many algorithms for step-size control, suitable for acoustic echo cancellation, so they are not discussed in detail here.

EXERCISE

Find the .WAV files *FarEndMono.WAV* and *AEC_Mono.WAV*. The first is the loud-speaker signal, the second is recorded in normal noise and reverberation conditions (a small office) without near-end speech. Modify the script *ProcessWAV.m* from Chapter 2 to have three input parameters: far-end file name, recorded file name, and output file name. Add reading of the second file, modify the conversion to the frequency domain to do conversion of the two files. Add the voice activity detector from Chapter 4 (*SimpleVAD.m*) to work on the far-end frames. Add the LMS adaptive filter to compensate the echo according to Equation 7.6, using $L = 10$. Do adaptation only when there is far-end speech (detected by the VAD) according to Equation 7.13. Evaluate the results by computing the ERLE (plot it as a function of time) and the convergence time. Adjust the adaptation speed by changing the value of μ. Save the script as *MonoAEC.m*.

Find a paper discussing dynamic step size and implement it. Compare the results.

7.3 NLMS and RLS Algorithms

The LMS adaptive filter has one known caveat. Assuming that the residual is proportional to the far-end signal, the adaptation rate will be proportional to the power of the far-end signal because it participates in the gradient estimation. To overcome this highly variable adaption speed, the *normalized least-mean-square* (NLMS) algorithm is preferred. It just adds normalization by the l^2 norm of the input vector:

$$\mathbf{H}^{(n+1)} = \mathbf{H}^{(n)} + \mu \frac{\mathfrak{R}^{(n)} \mathbf{Z}^{(n)}}{||\mathbf{Z}^{(n)}||^2}. \tag{7.16}$$

In real implementations a small number is added to the denominator to prevent division by zero. This is one of the most commonly used adaptive filters in acoustic echo cancellers.

NLMS adaptive filtering is well covered in the literature as well. Its convergence speed in the context of acoustic cancellers is critical for the quality of the cancellation and is well studied. Various modifications of the algorithm have been designed: proportionate NLMS (PNLMS) for better behavior with sparse impulse responses [5], and improved PNLMS (IPNLMS) for better convergence of the PNLMS algorithm [6].

Among other adaptive filter algorithms with application in acoustic echo cancellation should be mentioned the affine projection algorithm (APA), which is a further generalization of the NLMS [7]. The much faster adaptation comes at the cost of a substantial increase in computations, which led to creation of the fast affine projection (FAP) algoriothm [8].

The *recursive least-squares* (RLS) algorithm uses a recursive way to update the filter coefficients in each step and converges much faster than NLMS, but is computationally very expensive.

EXERCISE

Modify the script *MonoAEC.m* from the previous exercise to use the NLMS algorithm. Evaluate and compare the results with LMS.

Implement and compare with NLMS the performances of RLS, APA, and FAP algorithms.

7.4 Double-talk Detectors

The NLMS and other adaptive filters handle well the pauses in the far-end speech signal – the filter does not adapt when $\|\mathbf{Z}(n)\|^2 \to 0$. Still, in practical realizations of acoustic echo cancellers a voice activity detector (VAD) is used for the far-end signal and filters do not adapt when there is no far-end speech activity. The presence of near-end speech, however, can divert the adaptive filter to wrongly estimate the transfer function. A voice activity detector for the microphone signal can give an indication when there is far-end or near-end speech. To block the adaptation when there are both far-end and near-end speech signals we need a double-talk detector. The reaction of the adaptive acoustic echo canceller in both cases (no far-end speech or double talk) can be in a soft way; that is, instead of not adapting at all it can be implemented as reduction of the adaptation step size μ.

7.4.1 Principle and Evaluation

Double talk detectors (DTD) have the same evaluation parameters as those discussed in Chapter 4 for evaluation of voice activity detectors: true positive rate, false positive rate, and accuracy. The generic DTD computes a statistical parameter ξ, preferably data-independent, which is compared with a threshold η. If the value is higher than the threshold, double talk is detected; if it is below, there is no double talk. The threshold value can be adjusted using the ROC curves discussed in the same chapter. A good published paper about DTD evaluation criteria is [9].

Several improvements can be made to the classic comparison with a threshold:

- **Adding hysteresis**. Switch the state from "no double talk" to "double talk" when $\xi > \eta + \Delta\eta/2$, return to "no double talk" when $\xi < \eta - \Delta\eta/2$. Here, $\Delta\eta$ is the hysteresis and its value is adjusted together with the threshold η to be optimal in some way – best accuracy, minimal sum of the squares of false positives and false negatives, and so on. The hysteresis prevents frequent switching of the state when ξ is close to the threshold η.
- **Adding timing restrictions**. The speech signal has its own dynamics: probabilities to switch from pause to speech and from speech to pause, average duration of the speech segments, and so on. The simplest improvement is after switching to "double talk" state to stay there a certain minimal time.

Many DTD algorithms are developed for processing in the time domain. As we are here describing processing algorithms in the frequency domain, we will provide all equations using notation for that. Processing in the frequency domain uses L tap filters for each frequency bin, so the DTD algorithms designed for the time domain can be easily adapted to work for each frequency bin. The DTD output will be noisier owing to the shorter filter – in the time domain the number of taps is in the range of a couple of thousands, while in the frequency domain it is usually under ten. This is why the statistical parameter ξ, computed for each frequency bin (i.e., $\xi_k^{(n)}$), are combined to form the per-frame parameter $\xi^{(n)}$, which is compared with the threshold η. Combining is usually as a weighted sum:

$$\xi^{(n)} = \sum_{k=1}^{K} w_k \xi_k^{(n)}. \tag{7.17}$$

The weights are selected to be higher where there is more speech energy and lower where there is more noise. A typical shape of this frequency weighting is a band-pass filter in the range 200–3000 Hz. Some standardized weightings (C-message for example, see Chapter 3) can be used as well.

A good overview of various algorithms for double-talk detectors is given in Chapter 6 of [3]. When describing algorithms further we will omit the frequency bin indices whenever possible.

7.4.2 Geigel Algorithm

One of the earliest DTDs is the Geigel algorithm:

$$\xi^{(n)} = \frac{\max\{|\mathbf{X}^{(n)}|\}}{|Z^{(n)}|}. \tag{7.18}$$

This evaluates the ratio of the largest magnitude of the microphone signal $\mathbf{X}^{(n)}$ (see Equation 7.5) for the last L frames to the magnitude of the far-end speech $Z^{(n)}$. It assumes that the near-end speech is typically stronger in the microphone signal. The number of evaluated previous values is usually assumed the same as the length of the adaptive filter. This algorithm was designed for network echo cancellers where it works best. For acoustic echo cancellers the variability of the optimal threshold is higher and the Geigel algorithm works less reliably.

7.4.3 Cross-correlation Algorithms

Cross-correlation function based DTDs are considered more robust and reliable. The cross-correlation vector of $\mathbf{X}^{(n)}$ and $Z^{(n)}$ is

$$C_{XZ} = \frac{E\{X^{(n)}Y^{(n)}\}}{\sqrt{E\{|X^{(n)}|^2\}E\{|Y^{(n)}|^2\}}} = \frac{R_{XZ}}{\sigma_X\sigma_Z}. \qquad (7.19)$$

The statistical variable for comparing with a threshold can be either the l^{κ}-norm ($\kappa = 1, 2, \ldots$) of the correlation vector or the maximal value for the last L frames:

$$\begin{aligned} \xi_{CC}^{(n)} &= ||C_{XZ}||^{\kappa} \\ &= \max\{C_{XZ}\}. \end{aligned} \qquad (7.20)$$

The problem with this algorithm is that the cross-correlation function is not very well normalized. It is not quite robust when near-end noise is present. The *normalized cross-correlation* method is derived in [10]. In the absence of near-end speech and noise:

$$\sigma_X^2 = H^T R_{ZZ} H \qquad (7.21)$$

where $R_{ZZ} = E\{Z^{(n)}(Z^{(n)})^T\}$. Since $X^{(n)} = H^T Z^{(n)}$, then $R_{XZ} = R_{ZZ}H$ and Equation 7.21 can be rewritten in the form

$$\sigma_X^2 = R_{ZX}^T R_{ZZ}^{-1} R_{ZX}. \qquad (7.22)$$

When we have near-end noise and speech present, this converts to

$$\sigma_X^2 = R_{ZX}^T R_{ZZ}^{-1} R_{ZX} + \sigma_V^2, \qquad (7.23)$$

where $V = S + N$. The statistical parameter ξ for the DTD is the square-root of (7.22) divided by (7.23):

$$\begin{aligned} \xi_{NCC} &= \frac{\sqrt{R_{ZX}^T R_{ZZ}^{-1} R_{ZX}}}{\sqrt{R_{ZX}^T R_{ZZ}^{-1} R_{ZX} + \sigma_V^2}} \\ &= \sqrt{R_{ZX}^T (\sigma_X^2 R_{ZZ})^{-1} R_{ZX}} \\ &= ||C_{ZX}||^2. \end{aligned} \qquad (7.24)$$

Here, $C_{ZX} = (\sigma_X^2 R_{ZZ})^{-1/2} R_{ZX}$ is the normalized cross-correlation function.

The DTD as described can be computationally expensive. Later, a faster version of this algorithm was developed [11]. It is based on recursively updating $R_{ZZ}^{-1} R_{ZX}$ using the Kalman gain $R_{ZZ}^{-1}Z$. Then Equation 7.24 can be rewritten as

$$\xi_{FNCC}^2 = \frac{R_{ZX}^T R_{ZZ}^{-1} R_{ZX}}{\sigma_X^2(n)} = \frac{\chi^2(n)}{\sigma_X^2(n)} \qquad (7.25)$$

where the statistic parameter is squared for simplicity. Then for each frame the estimation of the correlation variables is as follows:

$$\sigma_X^2(n) = \lambda\sigma_X^2(n-1) + |X^{(n)}|^2$$
$$\chi^2(n) = \lambda\chi^2(n-1) + |X^{(n)}|^2 - \varphi(n)|\mathfrak{R}^{(n)}|^2$$
$$\mathfrak{R}^{(n)} = X^{(n)} - \mathbf{H}^\mathsf{T}\mathbf{Z}^{(n)}$$
$$\varphi(n) = \frac{\lambda}{\alpha(n)} \tag{7.26}$$
$$\alpha(n) = \lambda + (\mathbf{Z}^{(n)})^\mathsf{T}(\mathbf{R}_{ZZ}^{-1}(n-1))\mathbf{Z}^{(n)}$$
$$\mathbf{R}_{ZZ}^{(n)} = \lambda\mathbf{R}_{ZZ}^{(n-1)} + \mathbf{Z}^{(n)}(\mathbf{Z}^{(n)})^\mathsf{T}.$$

Here, λ is a forgetting factor. These two methods are among the most frequently used algorithms for double-talk detection.

7.4.4 Coherence Algorithms

Instead of using the cross-correlation function as a statistical variable we can use the squared magnitude of the coherence function [12]. If the coherence between $\mathbf{Z}^{(n)}$ and $\mathbf{X}^{(n)}$ is close to 1, then there is no double talk; if it goes below a certain threshold, then there is double talk. The squared magnitude of the coherence function for the frequency bin k is

$$\gamma_{ZX}^2(k) = \frac{|S_{ZX}(k)|^2}{S_{ZZ}(k)S_{XX}(k)}. \tag{7.27}$$

The statistics function then can be used per bin or as a weighted average of all frequency bins for a per-frame decision:

$$\xi_{\mathrm{COH}} = \sum_{k=1}^{K} w_k\gamma_{ZX}^2(k). \tag{7.28}$$

The weighting is usually by a band-pass filter in the range 200–2000 Hz with smooth slopes – see the beginning of this section.

EXERCISE

Copy the script *MonoAEC.m* from the previous exercise as *MonoAEC_DTDeval.m*. Add the double-talk detector. Implement all four algorithms from this subsection. Find and use for evaluation files *FarEndMono.WAV*, *AEC_Mono_wDoubleTalk.WAV*, and *NearEndMono.WAV*.

The first is the loudspeaker signal, the second is recorded in normal noise and reverberation conditions (small office) with near-end speech, and the third is a clean version of the near-end speech. The second file is recorded with near-end speech played by a head-and-torso simulator.

Add the near-end speech as a fourth parameter; add reading and conversion to the frequency domain for the near-end speech. Add a second simple VAD to work on the near-end speech. We have double talk when both VADs (far- and near-end) indicate speech activity. Compare this with the output of the DTD above. Build a table comparing the true positives, false negatives, and accuracies of the four algorithms. Use ROC curves to find the best thresholds for each algorithm.

Add the best DTD to the script *MonoAEC.m* for further use. Use the NLMS algorithm with a variable step size. Modify it to adapt only when there is far-end speech and no double talk. At this point you should have a decent MATLAB® implementation of a mono acoustic echo canceller. Evaluate the ERLE and convergence time.

7.5 Non-linear Acoustic Echo Cancellation

7.5.1 Non-linear Distortions

The adaptive filter assumes a linear transfer function between the far-end signal and the microphone. If the loudspeaker is not perfect (none is, but it is more valid for small loudspeakers) it will introduce non-linear distortions. Another potential source of non-linear distortions is clipping in the output amplifier. The effect of these distortions is that the reproduction tract adds harmonics of the far-end signal. The measure for harmonic distortions is called *total harmonic distortion* (THD) and is defined as the proportion of the power of all harmonics to the power of the first harmonic:

$$THD = \frac{\sum_{i=2}^{N} A_i^2}{A_1^2} = \frac{A_{RMS}^2 - A_1^2}{A_1^2} \qquad (7.29)$$

Here, A_1 is the amplitude of the sinusoidal signal sent to the loudspeaker, and A_i is the amplitude of the i-th harmonic that appears because of the non-linear distortions. This means that the microphone will capture signals (the harmonics) for which the acoustic echo canceller is not set up because they are not in the far-end signal. A small loudspeaker can have 10% THD at maximal power. This means that at least 10% of the echo energy, captured by the microphone, will not be cancelled, which is limiting factor to the performance of the acoustic echo canceller. ITU-T standards require reduction of the acoustic echo by at least 30 dB, which cannot be achieved even with a perfect echo canceller if the loudspeaker introduces more than 3% harmonic distortion. The quality requirements for loudspeakers used in systems employing acoustic echo cancellation are usually higher, but there are algorithmic ways to mitigate this problem as well. Birkett and Goubran [13] use a neural network and a second microphone (to provide an error signal) to achieve considerable improvement. With a very simple delay and saturation model, Stenger and Rabenstein [14] achieve ERLE improvement of 4 dB.

Practically all non-linear AEC algorithms proposed in the literature work in the time domain. Unfortunately their conversion to the frequency domain is not trivial owing to their complex structure. In the following subsections the notation is changed: $x[k]$ is the far-end signal and $y[k]$ is the microphone signal at moment kT.

7.5.2 Non-linear AEC with Adaptive Volterra Filters

Volterra series is a model for non-linear behavior, similar to the Taylor series. The difference is that with a Taylor series the output at any given moment depends only on the input at that moment, while in the Volterra series the output depends on the input at all times. This "memory" effect allows modeling of complex non-linear systems, containing capacitors and inductances. Initially defined with integrals, the Volterra series can be converted in discrete form and pruned to order M:

$$y[k] = \sum_{r=0}^{N} \sum_{k_1=0}^{M} \cdots \sum_{k_r=k_{r-1}}^{M} h_r[k_1, \cdots, k_r] x[k-k_1] \cdots x[k-k_r], \qquad (7.30)$$

where h_r are r-th-order Volterra kernels. As the Volterra kernels are symmetric, in Equation 7.30 only coefficients $k_r \geq k_{r-1}$ are used. An acoustic echo-cancellation algorithm using a second-order adaptive Volterra filter is proposed by Stenger et $al.$ [15]. Defining

$$\begin{aligned}
\mathbf{x}_1[k] &= (x[k], x[k-1], \cdots, x[k-M+1]) \\
\hat{\mathbf{h}}_1 &= (\hat{h}_1[0], \hat{h}_1[1], \cdots, \hat{h}_1[M-1])
\end{aligned} \qquad (7.31)$$

for the first-order and

$$\mathbf{x}_2[k] = (x^2[k], x[k]x[k-1], \cdots, x[k]x[k-M+1], x^2[k-1], x[k-1]x[k-2], \cdots,$$
$$x[k]x[k-M+1], \cdots x[k-M+1]x[k-M+1]) \qquad (7.32)$$
$$\hat{\mathbf{h}}_2 = (\hat{h}_2[0,0], \hat{h}_2[0,1], \cdots, \hat{h}_1[0, M-1], \hat{h}_2[1,1], \cdots, \hat{h}_1[M-1, M-1])$$

for the second-order Volterra kernel, the LMS adaptive Volterra filter is defined as

$$r[k] = y[k] - \hat{\mathbf{h}}_1[k]\mathbf{x}_1^T[k] - \hat{\mathbf{h}}_2[k]\mathbf{x}_2^T[k] \qquad (7.33)$$

$$\hat{\mathbf{h}}_1[k+1] = \hat{\mathbf{h}}_1[k] + \mu_1 r[k]\mathbf{x}_1^T[k] \qquad (7.34)$$

$$\hat{\mathbf{h}}_2[k+1] = \hat{\mathbf{h}}_2[k] + \mu_2 r[k]\mathbf{x}_2^T[k]. \qquad (7.35)$$

This can easily be converted to an NLMS algorithm by normalizing the step size:

$$\mu_1 = \frac{\alpha_1}{||\mathbf{x}_1[k]||^2} \qquad \mu_2 = \frac{\alpha_2}{||\mathbf{x}_2[k]||^2} \tag{7.36}$$

where α_1 and α_2 are the step-size parameters. To reduce computational complexity, the authors of the paper propose using different lengths of first- and second-order filters, $M_1 = 50$ taps and $M_1 = 25$ taps. With this second-order Volterra filter they report an ERLE improvement of 5.5 dB compared to a linear NLMS adaptive filter. This algorithm is applicable for systems working in the time domain. For systems using more processing steps (usually in the frequency domain) it does not fit well in the overall architecture.

7.5.3 Non-linear AEC Using Orthogonalized Power Filters

The non-linear transfer function can be modeled with power filters. These filters represent the output signal as a linear combination of a certain number of samples of the input signal (what linear filters do) and their square, their third power, and so on; that is, using the Taylor series. A power filter of P-th order is shown in Figure 7.4 and defined as follows:

$$y[k] = \sum_{p=1}^{P} \sum_{l=0}^{N-1} h[p,l]x^p[k-l] = \sum_{p=1}^{P} \mathbf{h}_p^{\mathrm{T}} \mathbf{x}_p[k] \tag{7.37}$$

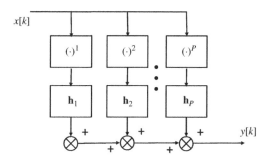

Figure 7.4   Power filter

where the vectors are defined as

$$\begin{aligned} \mathbf{x}_p[k] &= [x^p[k], x^p[k-1], \dots, x^p[k-N+1]]^T \\ \mathbf{h}_p &= [h_{p,0}, h_{p,1}, \dots, h_{p,N-1}]^T. \end{aligned} \tag{7.38}$$

Direct adaptation of the proposed power filter structure will be slow owing to the high correlation of the input signals (i.e., $x[k], x^2[k], \dots x^P[k]$) for these filters working in parallel. To improve the convergence speed, Kuech *et al.* [16] introduce a new set of mutually orthogonal input signals:

$$x_{o,1}[k] = x[k]$$

$$\cdots$$

$$x_{o,p}[k] = x^p[k] + \sum_{i=1}^{p-1} q_{p,i} x^i[k] \tag{7.39}$$

for $1 < p < P$. The orthogonalization coefficients $q_{p,i}$ are chosen such that

$$E\{x_{o,i}[k]x_{o,j}[k]\} = 0, \quad \text{for} \quad i \neq j \tag{7.40}$$

and determined using the Gram–Schmidt orthogonalization method [17]. After modifying the filter structure and using bias correction, the authors of [16] present experimental results showing improvement in ERLE. This method has the same problem as the previous one – it fits with difficulty in a large signal processing system operating in the frequency domain.

7.5.4 Non-linear AEC in the Frequency Domain

For acoustic echo cancellers working in the frequency domain the compensation for non-linearity of the loudspeaker is more complex. An interesting algorithm is proposed by Bendersky *et al.* [18]. After the linear AEC is placed a block that adaptively estimates the magnitudes of harmonics for each frequency bin and uses suppression methods to reduce the echo residual magnitude. This naturally leads us to the next section.

7.6 Acoustic Echo Suppression

The acoustic echo suppressor usually follows the acoustic echo canceller. Assuming that the adaptive filtering has already cancelled the trackable part of the echo signal, whatever phase information is left behind will be very difficult to estimate. This is why the next step is to remove the residual by using suppression techniques described in Chapter 4. The problem remains: we have a mixture of statistically independent signals (echo residual, local speech, and local noise). The goal is to estimate a real-valued suppression gain which, applied to the output of the acoustic echo canceller, suppresses the residual and lets the local speech pass undistorted.

7.6.1 Estimation of the Residual Energy

In noise suppressors, one of the most important components was building the noise model – that is, the noise variance for each frequency bin, $\lambda_d(k)$. It is estimated during the pauses of the speech signal, indicated by a voice activity detector (VAD). We assumed that the noise is almost stationary, so that the average noise power in each frequency bin changes much more slowly than the speech power. In acoustic echo suppressors, the equivalent of the noise model is estimation of the residual power. None of the assumptions for the noise above is valid. We can only assume that the residual power is a function of the far-end speech. We should estimate this function during frames when we have far-end speech (using a VAD on the far-end signal) and when there is no double talk.

Enzner et al. [19] propose using the coherence function for estimation of the residual energy. Given an acoustic echo-canceller output as a sum of the statistically independent near-end speech, noise, and AEC residual $Y_k^{(n)} = S_k^{(n)} + N_k^{(n)} + \Re_k^{(n)}$, then the squared coherence function is

$$C_{ZY}(k,n) = \frac{|\Phi_{ZY}(k,n)|^2}{\Phi_{ZZ}(k,n)\Phi_{YY}(k,n)} \tag{7.41}$$

and we can estimate the power spectral density of the residual energy as

$$\Phi_{\Re\Re}(k,n) = C_{ZY}(k,n)\Phi_{ZZ}(k,n). \tag{7.42}$$

This method uses a single tap filter in the frequency domain and underestimates the residual echo because the reverberation takes longer than the acceptable frame size. The authors generalize the residual energy estimation as

$$\Phi_{\Re\Re}(k,n) = \sum_{i=0}^{L-1} C_{ZY}(k,n-i)\Phi_{ZZ}(k,n-i). \tag{7.43}$$

The derivations in Equations 7.41 and 7.42 are valid under the assumption of uncorrelated echo and background noise signals, which is true in the long term. In the short term, in one frame, they are correlated and Equation 7.43 will overestimate the residual power. The authors propose a technique to compensate for the bias, which requires additional computational resources.

Instead of increasing the complexity of the model, Chhetri et al. [20] propose a direct regression model:

$$|\hat{\Re}_k^{(n)}| \approx \sum_{i=0}^{L-1} w_i |Z_k^{(n-i)}|. \tag{7.44}$$

On squaring Equation 7.44 we have

$$|\hat{\Re}_k^{(n)}|^2 \approx \left(\sum_{i=0}^{L-1} w_i |Z_k^{(n-i)}| \right)^2 \\ = \sum_{i=0}^{L-1}\sum_{j=0}^{L-1} w_i w_j |Z_k^{(n-i)}||Z_k^{(n-j)}| \tag{7.45}$$

which, according to the authors, is more powerful as it contains the cross-power terms that are missing in the power regression model

$$|\hat{\mathfrak{R}}_k^{(n)}|^2 \approx \sum_{i=0}^{L-1} w_i |Z_k^{(n-i)}|^2 \tag{7.46}$$

close to the estimation based on power spectral density in Equation 7.43. To compute the regression coefficients, the authors propose an adaptive algorithm that assumes knowledge of the noise magnitude. In each frame the residual magnitude is estimated according to Equation 7.44. Then the error signal and smoothed far-end power are computed:

$$E_k^{(n)} = \max(|Y_k^{(n)}| - \mathfrak{R}_k^{(n)}, N_k^{(n)}) \tag{7.47}$$

$$P_k^{(n)} = \alpha P_k^{(n-1)} + (1-\alpha)||Z_k^{(n)}||^2. \tag{7.48}$$

The adaptation happens after computing the normalized gradient:

$$\nabla_k^{(n)} = -\frac{2E_k^{(n)}|\mathbf{Z}_k^{(n)}|}{P_k^{(n)}}$$

$$\mathbf{w}_k^{(n+1)} = \mathbf{w}_k^{(n)} - \frac{\mu}{2}\nabla_k^{(n)}. \tag{7.49}$$

Here, $\mathbf{w}_k^{(n)}$ is the vector of regression coefficients from Equation 7.44. The adaptation is performed only in the absence of a near-end speech signal and the presence of a far-end speech signal. The authors use a variable step size μ, adjustable to ensure the positivity of $|Y_k^{(n)}| - \mathfrak{R}_k^{(n)}$ as much as possible, regardless of preventing $E_k^{(n)}$ to fall below the noise floor in Equation 7.47. The number of regression coefficients varies with the frequency and the room size.

7.6.2 Suppressing the Echo Residual

The suppression is based on estimation and applying a real-valued and time-varying suppression gain, usually between 0 and 1. The most straightforward approach is the Wiener suppression rule

$$V_k^{(n)} = \frac{\max(||Y_k^{(n)}||^2 - ||\hat{\mathfrak{R}}_k^{(n)}||^2, 0)}{||Y_k^{(n)}||^2} Y_k^{(n)} \tag{7.50}$$

with all the caveats discussed in Chapter 4. Many of the suppression rules from this chapter can be adapted and used for suppressing the echo residual.

An interesting approach is proposed by Madhu *et al.* [21]. They use an EM learning algorithm to estimate directly the probability of the presence of a residual signal. This probability is used as a suppression rule, in the same way as described in the sections on probability-based suppression rules. Besides the traditional ERLE as quality measurement, PESQ MOS is used. The paper reports better echo suppression and better perceptual sound quality than the regression AES.

The presence of near-end noise affects estimation of the echo residual and a noise model should be estimated, as described in the previous subsection. Most sound capture systems include a stationary noise suppressor immediately after the acoustic echo-reduction block. This justifies merging the noise and echo suppressors. Then the Wiener gain should be

$$H_k^{(n)} = \frac{||Y_k^{(n)}||^2 - ||\hat{\mathfrak{R}}_k^{(n)}||^2 - ||N_k^{(n)}||}{||Y_k^{(n)}||^2}. \tag{7.51}$$

Most of the a-priori and a-posteriori SNR estimators can be adapted and more sophisticated suppression rules used. Good results are achieved using probability as the suppression rule:

$$H_k^{(n)} = \frac{P_S^{(n)} p(S_k^{(n)}|\mu_Y, \sigma_Y^2)}{P_S^{(n)} p(S_S^{(n)}|\mu_Y, \sigma_Y^2) + P_N^{(n)} p(N_k^{(n)}|\mu_N, \sigma_N^2) + P_{\mathfrak{R}}^{(n)} p(\hat{\mathfrak{R}}_k^{(n)}|\mu_{\mathfrak{R}}, \sigma_{\mathfrak{R}}^2)} \tag{7.52}$$

where $P_X^{(n)}$ is the prior probability for the presence of X in the n-th frame, and $p(X|\mu, \sigma)$ is the PDF value for X given the mean and variance and know distribution. Practically this is a faster way to compute the probability than described in [21].

Overall, AES is an important part in acoustic-reduction systems. It provides 5–10 dB additional echo suppression. When it is well tuned, most of the artifacts, typical for every suppression algorithm, can be negligible. In such cases AES actually increases the perceptual sound quality. One more important thing – the AES is the safety net for the acoustic echo canceller. When the linear adaptive filter is not converged after rapid change in the transfer function, AES should be able to adapt faster and suppress the increased echo residual. The speech signal is quite sparse and the probability of having far- and near-end speech in the same frequency bin is relatively low. This means that, even under double-talk conditions, the quality of the output signal should be acceptable and in all cases better than no echo suppression at all.

EXERCISE

Copy the script *MonoAEC.m* from the previous exercise as *MonoAEC_AES.m*. Implement the two algorithms for estimation of the echo residual presented above.

Use the files with double talk. Compare the results using the same suppression rule. Select the better algorithm by comparing ERLE and performing listening tests. Not always does higher suppression mean better listening results. Adapt and implement some of the suppression rules from Chapter 4 and select the one that gives best results.

Experiment with an echo-reduction system with AES only by disabling the AEC part. Evaluate the results: echo suppression, and quality of the near-end sound.

Copy the script *MonoAEC.m* as *MonoAER.m* and add the best AES. At this point you should have a good working MATLAB implementation of a mono acoustic-reduction system.

7.7 Multichannel Acoustic Echo Reduction

7.7.1 The Non-uniquenes Problem

Most of communication systems operate with a mono far-end signal. Building a high-end telecommunication system with stereo sound would allow more comfortable communication and better perception of the positions of the different sound sources in the far-end room. Attempts to design a stereophonic acoustic echo canceller started in the early 1990s [22]. The first results were unsatisfactory and this raised the interest of the signal processing community. There followed many efforts to study the problem and offer solutions. At the beginning of the next decade there emerged scenarios such as controlling the stereo and surround-sound equipment with human voice and speech recognition, which requires stereo and a multichannel acoustic echo-reduction system.

When we refer to a stereo and multichannel acoustic echo-reduction system we mean highly correlated speaker channels. The most intuitive approach is to build the stereo acoustic echo canceller by chaining two mono units, as shown in Figure 7.5. This scheme will work flawlessly if the two loudspeaker channels are not correlated. Each of the filters will adapt independently from the other. Unfortunately this is not the case

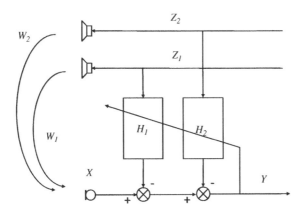

Figure 7.5   Stereo acoustic echo canceller with two mono acoustic echo cancellers

with stereo sound where the two channels are highly correlated. The problem is well described in [23].

Assume for a moment that we have a frame long enough to handle the reverberation; that is, we have one tap filter for each frequency bin (omitted in this section for simplicity). Then the sound source S in the far-end room is captured by the two microphones as follows:

$$Z_1 = G_1 S$$
$$Z_2 = G_2 S. \tag{7.53}$$

Here, G_1 and G_2 are the transfer functions from the sound source to each of the microphones in the far-end room. These two channels are reproduced by the speakers in the near-end room and captured by one of the microphones (no near-end speech and noise presented):

$$X = W_1 Z_1 + W_2 Z_2. \tag{7.54}$$

Here, W_1 and W_1 are the transfer functions from each of the loudspeakers to the microphone. We apply two acoustic echo cancellers with filters H_1 and H_2 and have on the output

$$Y = W_1 Z_1 + W_2 Z_2 - H_1 Z_1 - H_2 Z_2$$
$$= W_1 G_1 S + W_2 G_2 S - H_1 G_1 S - H_2 G_2 S \tag{7.55}$$

which, considering that $S \neq 0$ and completely converged canceller, leads to the equations

$$W_1 G_1 + W_2 G_2 - H_1 G_1 - H_2 G_2 = 0$$
$$G_1(W_1 - H_1) + G_2(W_2 - H_2) = 0. \tag{7.56}$$

The first thing to note is that we have two unknowns and one equation, which leads to an infinite number of solutions when $G_1 \neq 0$ and $G_2 \neq 0$, which is true if we have stereo sound capture. The two adaptive filters can converge to any of the infinite number of solutions. Unfortunately all of them depend on G_1 and G_2, except one: $W_1 = H_1$ and $W_2 = H_2$, which is the "true" solution – each adaptive filter converged to the corresponding transfer function. Any other solution is correct for the current situation, but changes in the far-end room – the speaker moves or another speaker starts to talk – will cause loss of convergence, appearing as echo on the output and the two adaptive filters will have to re-converge again. This is called the "non-uniqueness" problem.

In general, two adaptive filters, working in parallel, perform poorly if their input signals are highly correlated. Several approaches to create stereo and multichannel acoustic echo-reduction systems are discussed next.

7.7.2 Tracking the Changes

The general idea is to keep the structure in Figure 7.5 and to build an acoustic echo canceller that converges fast enough to track the changes in the far-end room, including changes of speaker position. This requires computationally complex adaptive filters such as RLS. Additional measures are taken to stabilize the convergence process. One example of such an approach is presented in [24].

7.7.3 Decorrelation of the Channels

The non-uniqueness problem occurs because the loudspeaker channels are highly correlated. If there is a way to de-correlate them, the structure with two adaptive filters in parallel will work. This can be achieved by adding non-linear elements, different for each channel, as shown in Figure 7.6. This idea is proposed in [25]. Unfortunately, to achieve stable working of the stereo acoustic echo canceller it is necessary to add a level of non-linear distortions, which are clearly audible and objectionable in high-end communication systems. Other attempts to decorrelate loudspeaker signals by adding uncorrelated noise to each channel, using comb filtering, using time-varying filters, and so on, either destroy the stereo picture or introduce unacceptable distortions and delays.

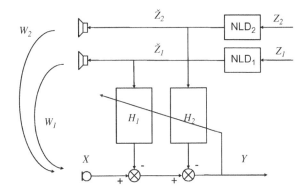

Figure 7.6 Stereo acoustic echo canceller with non-linear distortion of the far-end signal

The idea is developed further by Herre *et al.* [26], whereby the non-linear distortions are introduced accounting for psychoacoustics of human hearing. The authors achieve enough decorrelation to allow adaptive filters to converge in a surround-sound echo-cancellation system. On the other hand, their user studies show that human listeners cannot hear these distortions. The paper is a good overview of the approaches for multichannel acoustic echo cancellation and provides many literature references.

7.7.4 Multichannel Acoustic Echo Suppression

The energy transfer function is much less affected by correlation of the loudspeaker channels. Considering the fact that the speech signal is sparse in both time and frequency domains, building a stereo acoustic echo canceller based entirely on methods of suppression may not be such a bad idea as it seems at first look. Faller and Tournery [27] model the echo path as a delay and a single tap coloration filter for each frequency bin. The delay is computed and introduced before converting to the frequency domain. The coloration filter is estimated as

$$w_k^{(n)} = \frac{\left| E\{Z_k^{(n)*} X_k^{(n)}\} \right|}{\left| E\{X_k^{(n)*} X_k^{(n)}\} \right|}$$

$$E\{X_k^{(n)*} X_k^{(n)}\} = \frac{T}{\tau} |X_k^{(n)*} X_k^{(n)}| + \left(1 - \frac{T}{\tau}\right) E\{X_k^{(n-1)*} X_k^{(n-1)}\} \qquad (7.57)$$

$$E\{Z_k^{(n)*} X_k^{(n)}\} = \frac{T}{\tau} |Z_k^{(n)*} X_k^{(n)}| + \left(1 - \frac{T}{\tau}\right) E\{Z_k^{(n-1)*} X_k^{(n-1)}\}.$$

Here, T is the frame duration and τ is the adaptation time constant. The authors propose $\tau = 1.5$ s. Then the estimations of the echo residual and the suppression gain are

$$\hat{\Re}_k^{(n)} = w_k^{(n)} |Z_k^{(n)}|$$

$$H_k^{(n)} = \left[\frac{\max(|Z_k^{(n)}|^\alpha - \beta |\hat{\Re}_k^{(n)}|^\alpha, 0)}{|Z_k^{(n)}|^\alpha} \right]^{\frac{1}{\alpha}}. \qquad (7.58)$$

Here, α and β are design parameters. If $\alpha = 2$ the formula converts to a spectral subtraction suppression rule; $\beta < 1$ is used if the echo is underestimated, $\beta > 1$ otherwise. In the multichannel case (in the paper are discussed multiple reproduction and capture channels), the single-channel AES described above is used. The microphone and speaker energies are combined as follows:

$$|Z_k^{(n)}| = \left(\sum_{l=1}^{L} g_{Zl} |Z_k^{(n)}(l)|^\theta \right)^{\frac{1}{\theta}}$$

$$|X_k^{(n)}| = \left(\sum_{m=1}^{M} g_{Zm} |X_k^{(n)}(m)|^\lambda \right)^{\frac{1}{\lambda}}. \qquad (7.59)$$

The authors use weightings $g_{zl} = g_{Xm} = 1$ for all l and m, and $\theta = \lambda = 2$, which means that all loudspeaker and microphone channels are treated equally and Equation 7.59 combines the channel powers. Then the gain computed in Equation 7.58 is applied to all microphone channels.

7.7.5 Reducing the Degrees of Freedom

Looking at Equation 7.56 again, we can conclude that if we have one equation then we should use a single adaptive filter. A similar idea is proposed by Hirano and Sugiyama [28]. The stereophonic acoustic cancellation is for telecommunications and assumes that the sound source is captured by the two microphones according to Equation 7.53. The adaptive filters for both channels use the same input signal – the speaker channel which arrives earlier. One microphone channel of the proposed stereo echo-canceller structure is shown in Figure 7.7. Then Equation 7.55 changes to

$$Y = W_1 Z_1 + W_2 Z_2 - H Z_1$$
$$= W_1 G_1 S + W_2 G_2 S - H G_1 S. \tag{7.60}$$

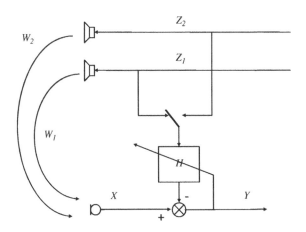

Figure 7.7 Stereo acoustic echo canceller with one adaptive filter

Assuming a converged filter and $S \neq 0$, Equation 7.56 in this case will look like

$$W_1 G_1 + W_2 G_2 - H G_1 = 0$$
$$H = (W_1 G_1 + W_2 G_2) G_1^{-1}. \tag{7.61}$$

The filter will have a solution and will converge if the far-end room impulse response is invertible. While the assumption $S \neq 0$ is safe, as we can adapt the filter only when it is met (i.e., we have speech activity at the far end) and the situation is in our control, this is not the case with the far-end room impulse response, which may not be invertible. Another caveat of this structure is that if the far-end speaker changes (rapid change of G_1 and G_2) the stereo echo canceller will have to reconverge.

To resolve these issues, the adaptive filter architecture shown in Figure 7.8 is proposed. Assume for a moment that we have good initial estimations of W_1 and $W_2 - H_{01}$ and H_{02}, respectively. Equation 7.55 with this structure looks like

$$
\begin{aligned}
Y &= W_1 Z_1 + W_2 Z_2 - H(H_{01} Z_1 + H_{02} Z_2) \\
&= W_1 G_1 S + W_2 G_2 S - H(H_{01} G_1 S + H_{02} G_2 S)
\end{aligned}
\tag{7.62}
$$

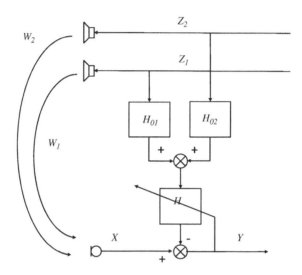

Figure 7.8   Stereo acoustic echo canceller with two fixed and one adaptive filters

and initially the adaptive filter is converged at $H = 1$; that is

$$
W_1 G_1 + W_2 G_2 - H(H_{01} G_1 + H_{02} G_2) = 0.
\tag{7.63}
$$

Now let something change in the far-end room (speaker move or change) and the impulse responses there change to $G_1 + g_1$ and $G_2 + g_2$. Then the echo canceller will remain converged because

$$
\begin{aligned}
&W_1(G_1 + g_1) + W_2(G_2 + g_2) - H(H_{01}(G_1 + g_1) + H_{02}(G_2 + g_2)) \\
&= W_1 G_1 + W_2 G_2 - H(H_{01} G_1 + H_{02} G_2) \\
&\quad + W_1 g_1 + W_2 g_2 - H(H_{01} g_1 + H_{02} g_2) \\
&= 0.
\end{aligned}
\tag{7.64}
$$

This filter structure is robust to changes in impulse responses in the far-end room. If the transfer functions in the near-end room change to $W_1 + w_1$ and $W_2 + w_2$, the adaptive filter will have to converge to $H + h$ to compensate for the changes. Is this compensation possible at all? We will have

$$(W_1 + w_1)G_1 + (W_2 + w_2)G_2 - (H+h)(H_{01}G_1 + H_{02}G_2)$$
$$= [W_1G_1 + W_2G_2 - H(H_{01}G_1 + H_{02}G_2)] + [w_1G_1 + w_2G_2 - h(H_{01}G_1 + H_{02}G_2)] \quad (7.65)$$
$$= w_1G_1 + w_2G_2 - h(H_{01}G_1 + H_{02}G_2)$$

as the first part of the second equation is equal to zero. Then the adaptive filter change is

$$h = (w_1G_1 + w_2G_2)(H_{01}G_1 + H_{02}G_2)^{-1}. \quad (7.66)$$

This solution exists because from Equation 7.63 we know that $H_{01}G_1 + H_{02}G_2$ is not zero. The adaptive filter can converge and compensate for the changes in the echo path. The two fixed filters thus reduce the degrees of freedom of the entire system; they act as constraints and hold the adaptive filter in a position where it can find the true solution. It is trivial to prove that, if there are changes in echo paths in both far- and near-end rooms, the adaptive filter can converge and the solution is Equation 7.66.

This structure will work well if we have good initial estimations of the echo paths in the near-end room. One easy way to do this is to play a short chirp signal from each loudspeaker consecutively at the beginning of each telecommunication session. Of course the chirp signals can be converted to something more melodic. It is important to have decorrelated wideband signals emitted from all loudspeakers for a short time. The entire initial estimation can take less than a second, including pauses before and after the calibration signal.

Another advantage of the proposed multichannel acoustic echo canceller is that the adaptive filter deals only with the small changes in the echo path. Most of the suppression comes from the fixed filters – the direct path and a substantial portion of the reverberation. People moving around in the near-end room cause relatively small increases in the residual, which the adaptive filter compensates for.

It is not a problem to extend this structure to surround-sound systems (five or seven loudspeaker channels) and for use with microphone arrays. Note that we have one adaptive filter per microphone channel, which means that this approach scales well for use with microphone arrays. With the traditional approach shown in Figure 7.5, a sound capture system with an eight-element microphone array and seven-channel surround-sound system will have to run 56 adaptive filters; while the approach in Figure 7.8 requires only eight. In addition, these 56 adaptive filters (if such a system can be made at all) should be computationally expensive RLS filters, while the eight can be regular NLMS filters with variable adaptation step.

7.8 Practical Aspects of the Acoustic Echo-reduction Systems

Building a robust acoustic echo-reduction system is a complex combination of research and engineering solutions. One of the most critical issues is guaranteeing that the adaptive filters will converge in all usage scenarios. Acoustic echo-reduction systems are no longer just for high-end professional equipment. Now personal computers are used for audio (and video) communication, with acoustic echo cancellation and suppression being part of the audio stack. People move their speakers during a session (which drastically changes the transfer function), they adjust the volume up and down using the knob on their loudspeakers or sound system (something AEC and AES are not aware of), and they place the microphones close to the loudspeakers, so the echo signal exceeds the local speaker voice by 20 dB or more. Laptop designers do the same (perhaps because both the microphone and the loudspeaker are part of the audio system and they should be together). All these factors require the addition of more processing blocks, many of which act as safety nets and engage only in critical situations – feedback, lost of convergence, and so on.

7.8.1 Shadow Filters

The basic idea is to have one fixed and one adaptive filter. The fixed filter is used to process the microphone signals. The adaptive filter works in parallel and adapts to changes in the transfer paths. When the adaptive filter starts to produce systematically better output its coefficients are copied to the fixed filter. The advantage here is that we can use more aggressive step sizes for faster convergence as we do not have to worry about the intermediate results during the convergence process. More details about shadow filtering can be found in [29].

7.8.2 Center Clipper

Humans can hear well even low-level signals with some organization. The echo residual is audible even when it is 20–30 dB below the level of the speech signal. For processing in the time domain the center clipper tracks the residual level and sets to zero all samples that are below this level:

$$\tilde{y}[k] = \begin{vmatrix} y[k] & |y[k]| \geq \mathbb{R} \\ 0 & \text{otherwise.} \end{vmatrix} \tag{7.67}$$

The clipping value $\mathbb{R}$ should be as small as possible and can be adaptive to track the residual.

The center-clipping equivalent in the frequency domain is the zeroing of all frequency bins with magnitude below the estimated residual level for that bin. In

Chapter 4 it was explained that zeroing a frequency bin is never a good idea. This very old type of processing has been superseded by more sophisticated acoustic echo-suppression systems. Its advantage, however, is simplicity and ease of implementation, especially when the acoustic echo canceller runs in the time domain on low-power processors.

7.8.3 Feedback Prevention

Feedback is not only annoying, it prevents the adaptive filter readapting to the changed transfer functions. The problem occurs when for a certain frequency the phase shift is close to $(2n + 1)$ times $180°$ and the gain is larger than 1 for the entire loop: far end + near end + far end.

One of the simplest ways to prevent feedback is to apply a variable gain to the signal that goes to the local loudspeakers when feedback is detected. Some professional echo-cancellation devices use a set of adjustable notch filters. Once feedback is detected they engage and suppress the feedback frequency. The missing narrow frequency band cannot be detected by humans and this approach is applicable even for high-end systems. Another potential solution is to put in the input processing chain a constant tone suppressor like the one described in Chapter 4. It will detect and suppress the feedback signal, which will allow the adaptive filter to converge. Another frequently used approach is so-called "frequency shift." The general idea is to translate the spectrum of the input signal 5–10 Hz. The technique was designed in the mid 1960s for public address systems; for more details see [30].

7.8.4 Tracking the Clock Drifts

This problem is typical for personal computers. In devices such as speakerphones or mobile telephones, the sampling frequencies of the analog-to-digital and digital-to-analog converters are synchronized by using the same clock generator. In personal computers we can have a loudspeaker connected to the output of the sound card and external USB microphone (usually together with the web camera), or we can have USB speakers and USB microphone, and so on. In general, even when the sampling frequency is set to be the same for both devices (say 16 kHz), the sampling does not happen synchronously and the sampling rates are different (within certain limits) owing to different clock generators. The adaptive filter should constantly readapt to track the sampling rate drift, which reduces its suppression abilities. This is why acoustic echo-reduction systems for personal computers have an integrated block that estimates the delay between the loudspeaker and microphone signals and constantly adjusts the delay of the loudspeaker signal. Usually this happens by shifting the weighting window for frame extraction, before conversion in the frequency domain.

7.8.5 Putting Them All Together

In various chapters of this book we have discussed multiple audio processing techniques. Some of them are linear (AEC, beamforming), some of them are not (noise suppressor, AES). Building an end-to-end audio processing stack requires a proper sequencing of these processing blocks. The general rules are linear processing first, slower blocks first.

An example of a multichannel sound capture system for an advanced telecommunication system or for voice control of multimedia equipment is shown in Figure 7.9. The system has stereo or surround-sound playback and uses a microphone array for sound capture. The multichannel acoustic echo cancellation is first and works on each microphone channel. The microphone-array beamformer follows and combines the signals from all microphones into one, filtering the noise and reducing the reverberation from the local room. In addition it does some suppression of the echo residuals. The adaptive filters in the AEC converge independently, under slightly different reverberation and noise conditions (both near and far end). We can say that the echo residuals have low correlation in the short term. From this perspective, the echo residuals behave as uncorrelated noise for the microphone array. This type of noise is suppressed by the instrumental gain. The increased level of uncorrelated noise should be accounted for during the microphone array design if it uses a time-invariant beamformer. The AEC is placed before the beamformer because the microphone array acts as a highly directional microphone. Changing the beam direction causes rapid change of the transfer function between the loudspeakers and the beamformer output, which the AEC cannot follow if placed after the beamformer. It is possible to have the AEC after the beamformer only if the microphone array works with a fixed (non-steerable) beam.

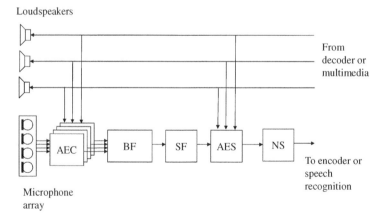

Figure 7.9 Block diagram of enhanced sound capture system with multichannel acoustic echo cancellation and microphone array

The beamformer output contains the enhanced local speech signal, decreased local noise levels, and still some echo residual. To increase the suppression from the microphone array, in Chapter 5 we proposed using a spatial filter, which in general is a sound source localizer per bin and suppressor for the bins where we have signals coming from an undesired direction. The echo residual should be processed by the acoustic echo suppressor discussed in this chapter. The stationary part of the local noise should be processed by one of the speech-enhancement algorithms discussed in Chapter 4. All of these algorithms compute and apply a time-varying real gain to the signal spectrum. They can be applied in the sequence above, or the gain computed jointly in a similar manner to Equation 7.52.

The output of this system goes either to the speech-recognition engine (if we want to do voice control of multimedia equipment) or to the encoder, which compresses the audio signal and sends it to the far-end room. The signal to the loudspeakers is either from the multimedia equipment (cable TV, DVD player, CD player, TV set, VCR, etc.) or from the decoder, converting the compressed audio from the far-end room to a waveform.

EXERCISE

Figure 7.8 is not complete. Identify the best places for the processing blocks discussed in this section. Where should the stationary tones compensation, or frequency shift, be placed? What about the sound source localizer from Chapter 6?

7.9 Summary

This chapter has discussed acoustic echo-reduction systems. They remove the sound from the loudspeakers which is captured by the microphone or microphones and is called echo. Such systems are part of all communication equipment and personal computers with speakerphone mode of operation. There are two major approaches for removing the echo: by cancellation and by suppression. Both are used in sequence to ensure the echo removal and maximal quality of the captured local speech signal.

Acoustic echo cancellers use adaptive filters to estimate the transfer path between the loudspeaker and the microphone. Then the signal sent to the loudspeakers is filtered and subtracted from the microphone signal. They should adapt to eventual changes in the transfer path. The adaptation process should happen when there is a loudspeaker signal (which can be detected by a voice activity detector) and there is no local speech signal, which is detected by a block called a double-talk detector. This block is an important part of each acoustic echo canceller.

Echo suppressors deal with the echo residual left after the acoustic echo canceller. They work in a similar to noise suppressors manner and use real-valued gain to reduce the residual echo. Estimating this residual is critical for acoustic echo suppressors and they use various adaptive algorithms to track and predict its power.

Multichannel echo reduction is part of high-end telecommunication systems and speech-controlled multimedia equipment. The main problem here is the non-uniqueness of the adaptive filter solution. This means that at any moment there are an infinite number of solutions, but they are different when some of the transfer paths change, except one – the true solution. This problem is resolved by faster adaptation, or advanced suppression algorithms, or by reducing the degrees of freedom of the adaptive system.

Building a robust system for echo reduction is a challenging research and engineering problem. Besides AEC and AES blocks, such a system may contain additional blocks such as algorithms for preventing feedback, and so on. An example of an end-to-end sound capture system has been described in this chapter.

Bibliography

[1] Sondhi, M. (1967) An adaptive echo canceller. *Bell System Technical Journal*, **46**, pp. 497–511.

[2] Vary, P. and Martin, R. (2006) *Digital Speech Transmission*, John Wiley & Sons, New York.

[3] Huang, Y. and Benesty, J. (eds) (2004) *Audio Signal Processing for Next-generation Multimedia Communication Systems*, Kluwer Academic, Norwell, MA.

[4] Haykin, S. (2002) *Adaptive Filter Theory*, 4th edn, Prentice-Hall, Upper Saddle River, NJ.

[5] Duttweiler, D. (2000) Proportionate normalized least-mean-squares adaptation in echo cancellers. *Transactions on Speech and Audio Processing*, **8**, 508–518.

[6] Benesty, J. and Gay, S. (2002) An improved PNLMS algorithm. Proceedings of International Conference on Acoustics, Speech, and Signal Processing, ICASSP '02. Orlando, FL.

[7] Ozeki, K. and Umeda, T. (1984) An adaptive filtering algorithm using an orthogonal projection to an affine subspace and its properties. *Electronic Communications of Japan*, **67-A**, 19–27.

[8] Gay, S. (2000) The fast affine projection algorithm, in *Acoustic Signal Processing for Telecommunication* (eds S. Gay and J. Benesty), Kluwer Academic, Boston, MA.

[9] Cho, J., Morgan, D. and Benesty, J. (1999) An objective technique for evaluating doubletalk detectors in acoustic echo cancellers. *IEEE Transactions on Speech and Audio Processing*, **7**, 6.

[10] Benesty, J., Morgan, D. and Cho, J. (2000) A new class of doubletalk detectors based on cross-correlation. *IEEE Transactions on Speech and Audio Processing*, **8**(2), pp. 168–172.

[11] Benesty, J. and Gänsler, T. (2006) The fast cross-correlation double-talk detector. *Signal Processing*, **86**, 1124–1139, Elsevier.

[12] Gänsler, T., Hansson, M., Invarsson, C.-J. and Salomonsson, G. (1996) A double-talk detector based on coherence. *IEEE Transactions on Communications*, **44**(11), 1241–1247.

[13] Birkett, A. and Goubran, R. (1995) Acoustic echo cancellation using NLMS-neural network structures. Proceedings of International Conference on Audio, Speech and Signal Processing ICASSP'95, Detroit, MI, pp. 2035–3038.

[14] Stenger, A. and Rabenstein, R. (1998) An acoustic echo canceller with compensation of nonlinearities. Proceedings of EUSIPCO 98, Isle of Rhodes, Greece.

[15] Stenger, A., Trautmann, L. and Rabenstein, R. (1999) Nonlinear acoustic cancellation with 2nd order adaptive Volterra filters. Proceedings of International Conference on Audio, Speech and Signal Processing ICASSP'99, Phoenix, AZ.

[16] Kuech, F., Mitnacht, A. and Kellermann, W. (2005) Nonlinear acoustic echo cancellation using adaptive orthogonalized power filters. Proceedings of International Conference on Audio, Speech and Signal Processing (ICASSP), Philadelphia, PA.

[17] Moon, T. and Stirling, W. (2000) *Mathematical Methods and Algorithms for Signal Processing*, Prentice-Hall, Englewood Cliffs, NJ.

[18] Bendersky, D., Stokes, J. and Malvar, H. (2008) Nonlinear residual acoustic echo suppression for high levels of harmonic distortion. Proceedings of International Conference on Audio, Speech and Signal Processing (ICASSP), Las Vegas, NV.

[19] Enzner, G., Martin, R. and Vary, P. (2002) Unbiased residual echo power estimation for hands-free telephony. Proceedings of International Conference on Acoustics, Speech, and Signal Processing (ICASSP), Orlando, FL.

[20] Chhetri, A., Surendran, A., Stokes, J. and Platt, J. (2005) Regression based acoustic echo suppression. Proceedings of International Workshop on Acoustic Echo and Noise Control (IWAENC), Eindhoven, The Netherlands.

[21] Madhu, N., Tashev, I. and Acero, A. (2008) An EM-based probabilistic approach for acoustic echo suppression. Proceedings of International Conference on Acoustics, Speech, and Signal Processing (ICASSP), Las Vegas, NV.

[22] Sondhi, M. and Morgan, D. (1991) Acoustic echo cancellation for stereophonic teleconferencing. Proceedings of Workshop on Applications of Signal Processing to Audio and Acoustics (WASPAA), Mohonk Mountain House, New Paltz, NY.

[23] Sondhi, M., Morgan, D. and Hall, J. (1995) Stereophonic acoustic echo cancellation: an overview of the fundamental problem. *IEEE Signal Processing Letters*, **2**(8), 148–151.

[24] Stokes, J. and Platt, J. (2006) Robust RLS with round robin regularization including application to stereo acoustic echo cancellation. Proceedings of International Conference on Multimedia and Expo (ICME), Toronto, Canada.

[25] Benesty, J., Morgan, D. and Sondhi, M. (1998) A better understanding and an improved solution to the specific problems of stereophonic acoustic echo cancellation. *IEEE Transactions on Speech and Signal Processing*, **6**(2), pp. 156–165.

[26] Herre, J., Buchner, H. and Kellermann, W. (2007) Acoustic echo cancellation for surround sound using perceptually motivated convergence enhancement. Proceedings of International Conference on Audio, Speech and Signal Processing (ICASSP), Honolulu, HI.

[27] Faller, C. and Tournery, C. (2006) Stereo acoustic echo control using a simplified echo path model. Proceedings of International Workshop on Acoustic Echo and Noise Control (IWAENC), Paris.

[28] Hirano, A. and Sugiyama, A. (1992) A compact multi-channel echo canceller with a single adaptive filter per channel. Proceedings of the IEEE International Symposium on Circuits and Systems, San Diego, CA, pp. 1922–1925.

[29] Hänsler, E. and Schmidt, G. (2004) *Acoustic Echo and Noise Control: A Practical Approach*, John Wiley & Sons, New York.

[30] Schroeder, M. (1964) Improvement of acoustic-feedback stability by frequency shifting. *Journal of Acoustic Society of America*, **36**, 1718–1724.

[31] Benesty, J., Amand, F., Gilloire, A. and Grenier, Y. (1995) Adaptive filter for stereophonic acoustic echo cancellation. Proceedings of International Conference on Audio, Speech and Signal Processing (ICASSP), Detroit, MI, pp. 3099–3102.

[32] Benesty J., Sondhi M. and Huang Y. (eds) (2008) *Speech Processing*, Springer-Verlag, Berlin, Germany.

[33] Buchner, H., Herbordt, W. and Kellermann, W. (2001) An efficient combination of multi-channel acoustic echo cancellation with a beamforming microphone array. Proceedings of International Workshop on Hands-free Speech Communication, Kyoto, Japan, pp. 55–58.

[34] Carini, A. (2001) The road of an acoustic echo controller for mobile telephony from product definition till production. Proceedings of International Workshop on Acoustic Echo and Noise Control (IWAENC), Darmstadt, Germany.

[35] Iqbal, M., Stokes, J., Platt, J. *et al.* (2006) Double-talk detection using real time recurrent learning. Proceedings of International Workshop on Acoustic Echo and Noise Control (IWAENC), Paris.

[36] Jiang, G. and Hsieh, S. (2004) Nonlinear acoustic echo cancellation using orthogonal polynomial. Proceedings of International Conference on Audio, Speech and Signal Processing (ICASSP), Montreal, Canada.

[37] Kim, S., Kim, J. and Yoo, C. (2003) Accuracy improved double-talk detector based on state transition diagram. Proceedings of Eurospeech 2003, Geneva, Switzerland, pp. 1421–1424.

[38] Kuech, F. and Kellermann, W. (2006) Ortogonalized power filters for nonlinear acoustic echo cancellation. *Signal Processing*, **86**, 1168–1181, Elsevier.

[39] McLachlan, G. and Krishnan, T. (1997) *The EM Algorithm and Extensions*, John Wiley & Sons, New York.

[40] Stenger, A. and Rabenstein, R. (1999) Adaptive Volterra filters for nonlinear acoustic echo cancellation. Proceedings of Nonlinear Signal and Image Processing (NSIP), Antalya, Turkey.

[41] Ye, H. and Wu, B. (1991) A new double-talk detection algorithm based on the orthogonality theorem. *IEEE Transactions on Communications*, **39**(11), 1542–1545.

8

De-reverberation

This chapter will deal with removing reverberation and its effects from recorded sound. Many times in this book it has been pointed out that sound reflects from walls and objects in a room and these reflected waves reach the microphone after the signal traveling a direct path. The microphone signal thus has superimposed multiple distorted copies of the desired signal. De-reverberation algorithms estimate the original speech from the microphone signal. As with acoustic echo-reduction systems, this problem can be approached from two sides: by estimation of the inverse room impulse response and de-convolving the microphone signal, and energetically by using estimation of the reverberation variance and suppression algorithms. A separate group of algorithms try to achieve de-reverberation using microphone arrays.

8.1 Reverberation and Modeling

8.1.1 Reverberation Effect

If we have a sound source $s(t)$ in a room and a capturing microphone placed at a certain distance, the sound wave from the source reflects from the walls and reaches the microphone shortly after the direct sound wave (Figure 8.1). Reflection of the already reflected waves repeats multiple times. After each reflection the wave energy decreases owing to energy absorption by the reflective surface. With time, the overall wave energy in the room decreases, the sound decays and goes below the noise level. This process is called *reverberation*. The sound captured directly from the source is called the *direct-path signal*. The ratio of the direct-path signal to the reverberated energy is called the *signal-to-reverberation ratio* (SRR).

If the microphone is positioned close to the sound source it captures mostly the direct path and a very small amount of reverberation. For a close-talk microphone, positioned 2–3 cm from the mouth, the effects of reverberation in the captured signal are negligible. Increasing the distance between the sound source and the microphone

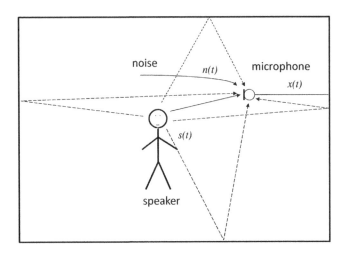

Figure 8.1 Direct path and reflections

causes the direct path sound to decrease, while the captured reverberated signal remains relatively the same. The distance at which the sound pressure level from the source is equal to the sound pressure level of the reverberated signal is called the *critical distance*. Note that it depends on room acoustics and on the microphone used. An omnidirectional microphone will have a shorter critical distance than a hypercardioid microphone because the second, pointed to the sound source, will capture less reverberation energy owing to its directivity. This is why the critical distance in a room is specified for omnidirectional microphones and is corrected for the specific microphone, usually with the *directivity index*. This means that if in a given room we have a critical distance of 1 m for omnidirectional microphone, for a hypercardioid microphone we will have a critical distance of 2 m owing to its directivity index of 6 dB.

The transfer function between the sound source and the microphone is called the *room impulse response* (RIR), denoted by $h(t)$. Regardless of the "room" in the name, it is different for every two points in the room. The measured room impulse response of a real room is shown in Figure 8.2a. In this there are three distinct parts, as shown in Figure 8.2b. The first is the direct path, followed by several distinct reflections. Based on the room size they arrive 5–10 ms after the direct path. For large rooms (concert halls, churches) the first reflections can arrive after 100 ms and then humans can hear them as separate, distinct sounds, called echoes. This second phase lasts around 20–30 ms and then follows the third part – the reverberation tail. Owing to exponentially increase of the number of reflected waves and volatility to smallest changes in the room, the reverberation tail is more of a stochastic than a deterministic process.

The microphone captures a signal that is the convolution of the room impulse response and the source signal plus some room noise:

$$x(t) = h(t)*s(t) + n(t). \tag{8.1}$$

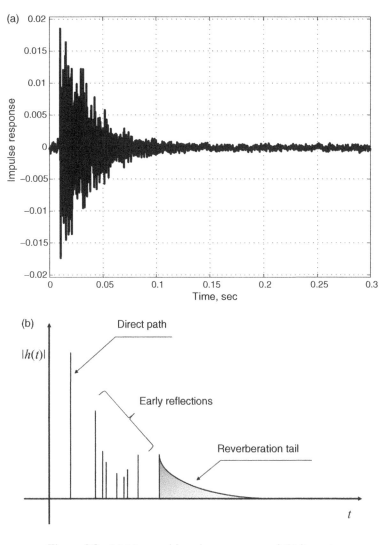

Figure 8.2 (a) Measured impulse response, and (b) its parts

In the frequency domain the convolution becomes multiplication and we have single tap filter for each frequency bin:

$$X_k^{(n)} = S_k^{(n)} H_k + N_k^{(n)}. \tag{8.2}$$

Due to the fact that the frame length is shorter than the reverberation process (explained in the previous chapter), the process is modeled with an L-tap filter:

$$X_k^{(n)} = \mathbf{H}_k^{\mathrm{T}} \mathbf{S}_k^{(n)} + N_k^{(n)} \tag{8.3}$$

where $\mathbf{S}_k^{(n)} = (S_k^{(n)}, S_k^{(n-1)}, \dots, S_k^{(n-L+1)})^{\mathrm{T}}$ and $\mathbf{H}_k = (H_k^{(0)}, H_k^{(1)}, \dots, H_k^{(L-1)})^{\mathrm{T}}$. This is the general model we will follow in this chapter.

The RIR decreases in magnitude with time. We consider the reverberation process over when the reverberation energy goes 60 dB below the direct path energy. The time for this is called T_{60} and is virtually constant for the room. It depends mostly on the room volume and was empirically derived by Sabine in late 1890s as

$$T_{60} = \frac{55.25V}{cS\bar{\alpha}} \tag{8.4}$$

where V is the room volume, c is the speed of sound, S is the total surface area of the room, and $\bar{\alpha}$ is the average absorption ratio, which can be estimated separately for each surface in the room:

$$\bar{\alpha} = \frac{1}{S} \sum_i \alpha_i S_i. \tag{8.5}$$

The absorption ratio varies with the material and the frequency, which means that we can have different reverberation times for different frequencies. The absorption ratio is usually between 0.01 and 0.9 for commonly used materials. A perfect sound-reflection surface would have an absorption ratio of 0, a surface that absorbs all the sound and reflects none of it has an absorption ratio of 1. Most materials increase their absorption ratio with increasing frequency (carpet, concrete), but some decrease with it (glass, gypsum). Typical values of T_{60} for a normal size room (office, home, conference) are in the range 200–400 ms. A large hall or church can have a T_{60} of several seconds. Recording studios usually have a reverberation time below 100 ms, in an anechoic chamber it is below 10 ms. The critical distance in a room with known T_{60} is given by

$$d_C = \sqrt{\frac{V}{100\pi T_{60}}}. \tag{8.6}$$

The reverberation time T_{60} as a parameter implicitly assumes that the reverberation energy decreases exponentially. For a practical measurement of T_{60}, the time is usually measured when the reverberation energy reaches -30 dB and this time is multiplied by 2. The reason for this is to avoid the noise floor. Some literature sources directly cite this time as T_{30}.

EXERCISE

Find the stereo file *RoomImpulseResponce.WAV*. The left channel contains a wideband chirp signal played from a mouth simulator in a normal office room. The right channel contains the microphone signal. Write a MATLAB® script to read the file and to compute the impulse response. Plot it and test by filtering the left channel – you should receive an almost exact copy of the right channel. Compute the MMSE.

8.1.2 How Reverberation Affects Humans

Humans have two ears and a brain between. The brain can use the head-related transfer functions of the two ears to localize and focus the hearing towards the sound source based on the sounds captured by the ears and prior knowledge of the sound. The brain constantly localizes the sound source and removes the effects of reverberation and noise. As a result, humans can hear and understand speech signals well for distances several times the critical distance. For sound sources further than that the understand-ability decreases, listening is less comfortable and requires a higher level of attention. If we use a single microphone to capture the sound and record or transmit it to the far-end room, the situation changes completely. All spatial cues are gone. Listening to this signal, the brain tries to remove the reverberation and noise, but there is not enough information to do that. This increases fatigue and reduces attention, and people tire faster, which reduces the efficiency of long remote meetings.

On the other hand, human voices, music, and sounds in general, recorded in a studio are considered by humans as "dry" and not very pleasant. A certain amount of reverberation makes them sound "warmer" and more pleasant. This is why the audio processing blocks for adding reverberation are part of the equipment in every radio station and recording studio. This is not a matter of adding echo effects, a small amount of reverberation is added to the voice of radio speakers as well. For telecommunication purposes, however, the goal is to remove as much reverberation as possible – the receiving room will add enough reverberation when in speakerphone mode.

To evaluate which part of the reverberation humans are more sensitive to, we conducted a controlled experiment. Ten subjects were asked to listen to pairs of recordings and to compare their quality on the scale of much worse, worse, somewhat worse, same, somewhat better, better, and much better (−3, −2, −1, 0, 1, 2, and 3 points, respectively). The subjects listened to close-talk microphone recordings convolved with an actual room impulse response, measured at a distance of 2 m in a room with $T_{60} = 300$ ms. Various files were generated by cutting the impulse response to durations of 50, 75, 100, 150, 200, and 300 ms. The averaged and normalized results of this differential MOS study are shown in Figure 8.3a. The perceptual sound quality goes down in the presence of the first reflections. Humans are more sensitive to the second part of the reverberation process and the tail contributes less to the negative effects of reverberation. Note that this is valid for moderate reverberation times. Increasing the reverberation time smears the speech signal and decreases understand-ability. A second set of differential MOS tests were conducted with simulated room impulse responses (see the image method later in this chapter) varying the room T_{60} and the distance between the speaker and the microphone. The results are shown in Figure 8.3b. In a room with a reverberation time of 100 ms, subjective quality is

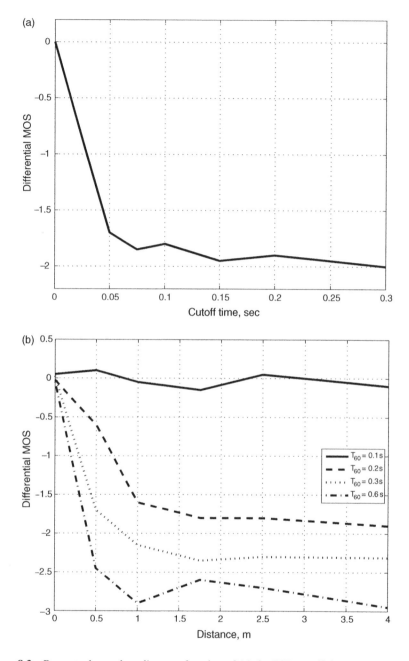

Figure 8.3 Perceptual sound quality as a function of (a) the RIR cutoff time, and (b) distance

practically the same for all distances, while increasing the reverberation time causes it to go down quicker with the distance.

8.1.3 Reverberation and Speech Recognition

A classic scenario for speech recognition is the dictation task, when the user is wearing a close-talk microphone and supplies the speech recognizer with very good sound quality. Advanced scenarios are speech recognition for voice commands (media room, office) and human–computer dialog systems (kiosk, car). In these scenarios users do not wear a microphone and there is far-field hands-free sound capture, which contains reverberation.

To evaluate the effect of different parts of the reverberation process on the speech recognizer, we convolved a clean speech corpus with the room impulse response from the previous subsection, cutting the impulse response in increments of 20 ms. For these tests we used a speech recognizer trained on clean speech. The *word error rate* (WER) as a function of the impulse response length is shown in Figure 8.4. It seems that the direct reflections have little effect on the recognizer capabilities, while the reverberation tail after 30 ms gradually increases the WER until 130 ms, when it reaches its stable state as the reverberation energy is drowned into the room noise floor.

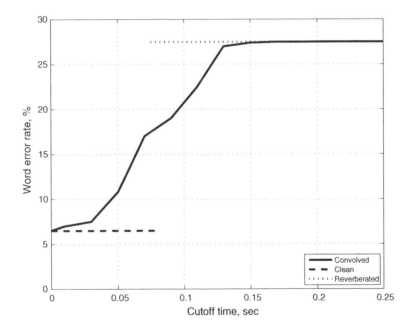

Figure 8.4 Word error rate as a function of the RIR cutoff time

The reason for this effect is the *cepstral mean normalization* (CMN), which is part of the front end of each modern speech recognizer. The front end converts the captured speech signal, which is a convolution of the clean speech signal $s(t)$ and the impulse response $h(t)$ of the whole capturing path, including the room impulse response:

$$x(t) = h(t)*s(t) \tag{8.7}$$

first to the frequency domain:

$$X = H \cdot S \tag{8.8}$$

and then to the cepstral domain:

$$X_{CC} = \mathrm{iFFT}[\log(S) + \log(H)] = S_{CC} + H_{CC} \tag{8.9}$$

which converts the multiplicative effect of the impulse response to additive. Assuming the long-term mean of the speech cepstrum is zero and a relatively constant impulse response, then subtracting the mean of the cepstrum from the cepstrum will remove the channel effects:

$$X_{CC} = S_{CC} + H_{CC} - \int X_{CV} dt \approx S_{CC}. \tag{8.10}$$

This processing is called cepstral mean normalization and compensates for the capturing-channel impulse response. The compensation is for impulse responses shorter than one audio frame, which is typically 25 ms. This explains the low effect of the early reflections in the speech recognition results. In our practical example the last 25% of the room impulse energy is accountable for 90% of the WER increase. Experiments with increasing the distance between the sound source and the capturing microphone show a WER curve similar in shape to Figure 8.4.

8.1.4 Measuring the Reverberation

Practical measuring of T_{60} can be done by playing rectangular bursts of a wideband signal (white noise). The sound is captured by a microphone, placed at a certain distance from the loudspeaker. The reverberation time T_{60} can be measured from the decay speed of the falling slope. It is abrupt for the source signal and exponentially decaying in the signal captured by the microphone. More stable and precise estimations can be obtained from the measured impulse response. Given knowledge of the sound sent to the loudspeakers, measuring the impulse response $h(t)$ is relatively easy. One of the most frequently used approaches is given by the Schroeder formula [1], which is a reverse-time integrated impulse response:

$$D(t) = \frac{1}{D_0} \int_t^\infty h^2(\tau) d\tau$$

$$D_0 = \int_0^\infty h^2(\tau) d\tau.$$

(8.11)

A plot of the Schroeder function of the impulse response from Figure 8.2 is shown in Figure 8.5. The T_{30} time is measured by linear extrapolation of the curve from a level of -5 dB to -15 dB. The room has a reverberation time T_{60} of 220 ms.

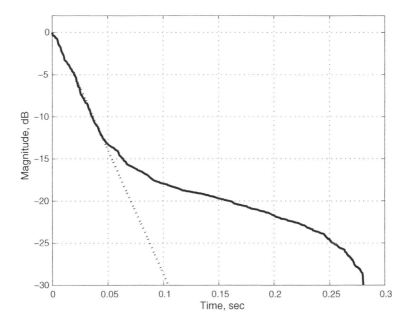

Figure 8.5   Schroeder function for the RIR from Figure 8.2a

While the Schroeder formula can give some information about the reverberation time, it is not relevant to human perception of the reverberated sound. We can measure the psychoacoustic quality of the sound by using some of the intrusive (i.e., with knowledge of the source signal) algorithms such as PESQ, or non-intrusive, such as ITU–T Recommendation P.835. They were both designed to evaluate decoder/encoder chains, and show good results for evaluation of sound quality after noise suppression and other non-linear processing algorithms. They are suitable for evaluation of the output of de-reverberation algorithms, but the correlation with the actual MOS is lower. Higher correlation with actual MOS is shown by the measure in [2], according to the authors.

EXERCISE

Compute and plot the Schroeder function for the impulse response, computed during the previous exercise. Find the T_{60} for this room.

8.1.5 Modeling

One of the most popular methods for synthesizing reverberated signals is the so-called "image method" [3]. The reflected sound from a wall can be modeled as an additional sound source, situated on the other side of the wall as an image of the original. Then given a room to model, the absorption coefficient of the walls, and the number of reflections, it is easy to represent the sound captured by the microphone in a given position as the sum of the sound sources (Figure 8.6). Each time a wall is crossed the sound magnitude is decreased by the absorption ratio and the phase inverted. (For non-rigid walls the reflected image will not be a point source.) The image method uses an angle-independent pressure wall reflection coefficient β, which leads to the energy absorption coefficient $\alpha = 1 - \beta^2$. Improvement of the image method can be found in [4]; the algorithm has been implemented in MATLAB and is available to use.

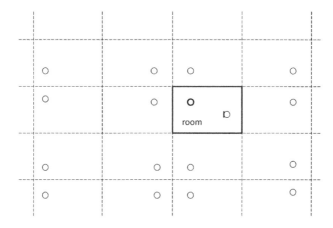

Figure 8.6   Image method for modeling of reverberation

The image method models the initial set of reflections well, but the reverberation tail is sparse and quite different from the actual measurements. This is due to a not very precise reflection model, which is frequency-independent and does not include energy diffusion, and to the fact that the room is modeled as empty with only eight reflective surfaces – the walls. Regardless of this, the image method is commonly used for generation of synthetic room impulse responses.

Using the idea of the image method we can do some interesting computations. The sound sources when a sound wave has already reached the microphone are inside a sphere of radius $R = ct$, where c is the speed of sound. In this sphere we have

$$N = \frac{\frac{4}{3}\pi R^3}{V} = \frac{4\pi c^3 t^3}{3V} \tag{8.12}$$

sound sources where V is the volume of the room. The rate at which reflected waves arrive at the microphone is the first derivative:

$$\frac{dN}{dt} = \frac{4\pi c^3 t^2}{V} \text{ [reflections/sec]}. \tag{8.13}$$

For a room of size $4\,\text{m} \times 3\,\text{m} \times 3\,\text{m}$ ($V = 36\,\text{m}^3$), the moment when the reflections will arrive with a rate equal to the sampling rate of $16\,\text{kHz}$ is $33.7\,\text{ms}$. This is roughly the moment where we can say that for that room the reverberation transforms from a discrete to a stochastic process.

8.2 De-reverberation via De-convolution

The general idea is to estimate an L-tap filter $\mathbf{G}_k$ applied to the captured signal to give us an estimation of the source:

$$\tilde{S}_k^{(n)} = \mathbf{G}_k^{\mathrm{T}} \mathbf{X}_k^{(n)} = \mathbf{G}_k^{\mathrm{T}} (\mathbf{H}_k^{\mathrm{T}} \mathbf{S}_k^{(n)}) + \mathbf{G}_k^{\mathrm{T}} \mathbf{N}_k^{(n)} \approx S_k^{(n)}. \tag{8.14}$$

In the ideal case $\mathbf{G}_k^{\mathrm{T}} \mathbf{H}_k^{\mathrm{T}} = 1$ and we achieve full restoration of the speech signal. There are two potential problems visible at first glance. The first is that the room impulse response $h(t)$ may not be invertible and the algorithm will have to find an approximate solution. A stable and causal system such as the transfer function $h(t)$ has a stable and causal inverse $g(t)$ only if it is a minimum-phase system. Unfortunately the room impulse response is almost never a minimum-phase system [12]. The second problem is that, even if it is invertible, we do not have access to the original speech, which was the case with acoustic echo-cancellation. Indirect criteria or properties of the speech signal have to be used as a criterion for updating the adaptive filter. This makes de-reverberation via direct estimation of the inverse filter a much more complex problem than acoustic echo cancellation.

EXERCISE

Use the .WAV file from the previous exercise. Even with complete knowledge of the sound source, can you find a good de-convolution filter? Evaluate the solution by filtering the microphone signal and comparing it to the speech signal.

8.2.1 De-reverberation Using Cepstrum

The main idea is to use the mean cepstrum as an estimate of the transfer function [5]. Once we have an estimation of H_{CC} from Equation 8.10, then it is converted back to the time domain, truncated to an appropriate length, and the de-convolution filter $g(t)$ is designed. Other algorithms try to use filtering in the cepstral domain based on empirical data. Most of the algorithms of this group are not seriously evaluated for improving speech-recognition results or human perception.

8.2.2 De-reverberation with LP Residual

Human speech production is modeled as an all-pole filter, while the reverberation adds only zeros. This means that the *linear prediction* (LP) coefficients will remain intact and only the LP residual will be affected. An entire group of algorithms use this fact. The general idea is to compute and subtract the LP and to work only with the residual. Once the reverberation effects are removed from the residual we can combine it back with the LP and resynthesize the speech signal. Such an algorithm is proposed by Yegnanarayana and Satyanarayana Murthy [6]. The LP residual is processed using the fact that the entropy is higher in the reverberant segments. Proper weighting is estimated and applied to regions with a high level of reverberation. The speech synthesis from the LP coefficients and processed residual inevitably introduces some distortions.

Similar LP properties are used by Gillespie *et al.* [7]. The authors use the fact that the reverberated LP residual has more of a Gaussian-shaped distribution, while the clean-speech LP residual is peakier; that is, it has higher kurtosis. To reduce the LP reconstruction artifacts, the authors propose a parallel structure (Figure 8.7). The adaptive filter works on the LP residual and maximizes the kurtosis. A copy of the filter works in parallel on the input signal. The algorithm works well in conditions of strong reverberation, where the difference is larger, but is weaker when the reverberation is

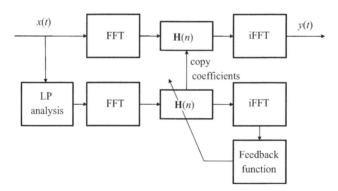

Figure 8.7   Block diagram of de-reverberator via maximizing kurtosis

low. The last requires longer adaption times, which harms the ability to track changes in the impulse response in the volatile tail.

8.2.3 De-reverberation Using Speech Signal Properties

Speech signals roughly consist of harmonic parts and noise-like segments. Nakatani *et al.* [8] claim that the second group is in general less affected by the reverberation, and a perceptual improvement can be obtained if the harmonic segments of the speech signal can be cleaned. The idea is to recognize these segments, to estimate the pitch, and to resynthesize the harmonic structure. Then the speech signal is assembled and evaluated. The algorithm is called HERB; it is iterative and computationally expensive which makes it more suitable for offline processing.

EXERCISE

Read one of the papers cited above and implement the algorithm. Record your own reverberant files and evaluate the algorithm. MATLAB provides good support for LP analysis. Process the .WAV file from the previous exercise and plot the Schroeder function. Compare it with the same function before the reverberation. Where are the strongest and weakest areas of the implemented algorithm?

8.3 De-reverberation via Suppression

The general idea is similar to the way noise suppressors and acoustic echo suppressors work. If we can estimate the reverberation energy in the signal we can use suppression techniques to estimate a real-valued gain for each frequency bin and reduce the bin magnitude proportionally to the portion of the reverberation energy. It is obvious that this approach will work better on the reverberation tail and will be weaker on the first reflections. On the other hand, most of the methods in the previous section can better estimate the first part of the correction filter, but usually fail to deal well with the volatile reverberation tail. From this perspective, the suppression group of methods provides good post-processors for the de-convolution methods. For speech recognition, reverberation suppressors can be the only de-reverberation algorithms used, as the cepstral mean normalization (CMN) in the front end deals well with the first reflections.

Essential for good reverberation suppression is a good estimate of the reverberation power. For the reverberation tail, a common model is the exponential energy decay both in the time domain and per frequency bin:

$$E_{\text{Rev}} = E_0 \exp\left(-\frac{t}{\tau}\right) \tag{8.15}$$

where E_0 is the initial signal power and τ is the time constant directly related to T_{60}. This simple model with only one parameter is very convenient to estimate, especially when the estimation is done per frequency bin and signals are quite noisy. Tashev and Allred [9] divide the frequency band into four or eight sub-bands and the decay time constant is estimated for each sub-band at the falling slope of the utterance, indicated by a voice activity detector. Via linear interpolation, this parameter is propagated for each frequency bin. It is trivial to convert the time constant to a single tap IIR (infinite impulse response) filter, which is used to process the input signal power and to receive the reverberation estimate. After this, an enhanced Wiener filter is used to suppress the reverberation. The algorithm's purpose is to improve the hands-free speech-recognition results, and it does not try to work with the first part of the impulse response.

More sophisticated methods for reverberation spectral variance estimation can be found in [10].

EXERCISE

Use the estimated room impulse response to compute the decaying time constant in several frequency bands. Model the reverberation spectral variance as a function of the input signal. Implement a simple Wiener suppression algorithm based on this estimation. Process the .WAV file from the previous exercise and plot the Schroeder function. Compare it with the same function before the reverberation. Where are the strongest and weakest areas of the implemented algorithm? Evaluate the results by listening to the output. Are there artifacts or musical noises?

8.4 De-reverberation with Multiple Microphones

8.4.1 Beamforming

In general, the more directional a microphone is the less reverberation it captures and the greater is the critical distance. From this perspective, microphone arrays that via beamforming can achieve a high directivity index are performing de-reverberation implicitly. Refer to Chapter 5 for the beamforming algorithms and microphone-array processing. This group of algorithms to date actually remains the most efficient, reliable, and commonly used method for de-reverberation.

8.4.2 MINT Algorithm

The authors of Chapter 12 in [11] propose this algorithm, based on the Bezout theorem. This says that, if the impulse responses from the sound source to each of the microphones have no common zeros, it is possible to perfectly compensate for the reverberation. This is a much lighter constraint than the one in single-channel de-convolution for minimum phase. In essence, this approach combines the signals from all the microphone channels using the assumption that the information lost in one channel (i.e., zero for this frequency) is available in another channel (no common zeros requirement). The authors derive the estimation of the compensation filter. The approach should theoretically give perfect compensation of the reverberation, but it is sensitive to noise.

EXERCISE

Use some of the multichannel .WAV file from Chapter 5 or Chapter 6 to process with some of the beamforming algorithms from Chapter 5. Compute and compare the Schroeder function for one of the microphone channels and for the beamformer output. Where is the beamformer more efficient, and what is left from the reverberation?

8.5 Practical Recommendations

De-reverberation is a complex problem that is not yet completely solved. Before employing some of the algorithms from the literature, a good evaluation should be conducted on why de-reverberation is necessary. Typical applications are far-field sound capture for communications, and voice control of multimedia equipment.

In the first case the target is human ears. The evaluation criteria should be the perceptual sound quality achieved by the end-to-end system. The PESQ (perceptual evaluation of sound quality) algorithm or the criterion from [2] are good for evaluation, separately or in combination.

If the target is speech recognition, there is no better criterion than the recognition rate. Build a speech corpus, recorded in the target conditions, and tune the de-reverberation algorithm to minimize the word error rate.

One of the most robust and efficient solutions for decreasing reverberation remains the combination of a microphone array with a beamformer and reverberation suppressor after that. As was mentioned in the chapter on echo reduction (and in many other places in this book), chaining suppressors is not a good idea. The suppression blocks for noise, echo residual, and reverberation should be combined for joint suppression. Each type of suppression should have what was called in Chapter 4 "minimal gain." This is a number that can adjust the amount of suppression from each algorithm: a minimal gain of 0 means full suppression, while going up to 1 turns off this type of suppression. The system should be tuned end-to-end to achieve the best results according to the design goals.

EXERCISE

Where should the reverberation suppression be placed on Figure 7.8? Add a reverberation suppressor to the microphone array processor and compare the results with those from the previous exercise.

8.6 Summary

This chapter has covered algorithms for removing the effects of reverberation, captured by a microphone or array during far-field sound capture. The reverberation smears the spectral and time properties of the speech signal. This decreases the understandability and increases fatigue in humans, and harms the automatic speech recognizer results. There are three major groups of approaches to fight the effects of reverberation: deconvolution filters, suppression algorithms, and microphone arrays.

The first group applies a filter to the microphone signal in an attempt to equalize the channel and remove the reverberation. Direct inversion of the room impulse response is not always possible. Several methods are used to find a criterion for adapting the compensation filter: working in the cepstral domain, using an LP residual, or using speech signal properties such as harmonic structure.

The second group works similarly to noise suppressors and echo suppressors. The reverberation power is modeled as an exponentially decaying signal, which leads to a single parameter for estimation – the time constant. Then most of the suppression techniques are applicable.

Multimicrophone approaches include the traditional beamforming, which is one of the most reliable and efficient ways to remove reverberation. Other methods and algorithms have been explored too.

Above all, using de-reverberation algorithms requires careful evaluation and specification of the design evaluation criterion. Based on this the system is designed and evaluated.

Bibliography

[1] Schroeder, M. (1965) New method for measuring reverberation time. *Journal of the Acoustical Society of America*, **37**, 409–412.

[2] Falk, T. and Chan, W. (2008) A non-intrusive quality measure of dereverberated speech. Proceedings of International Workshop on Acoustic Echo and Noise Control (IWAENC), Seattle, WA.

[3] Allen, J. and Berkley, D. (1979) Image method for efficiently simulating small-room acoustics. *Journal of the Acoustical Society of America*, **65**, 943–950.

[4] Lehmann, E. and Johansson, M. (2008) Prediction of energy decay in room impulse responses simulated with an image-source model. *Journal of the Acoustical Society of America*, **124**, 269–277.

[5] Bees, D., Blostein, M. and Kabal, P. (1991) Reverberant speech enhancement using cepstral processing. Proceedings of International Conference on Acoustics, Speech, and Signal Processing (ICASSP), vol. 1, pp. 977–980.

[6] Yegnanarayana, B. and Satyanarayana Murthy, P. (2000) Enhancement of reverberant speech using LP residual. *IEEE Transactions on Speech and Audio Processing*, **8**, 267–281.

[7] Gillespie, B., Malvar, H. and Florêncio, D. (2001) Speech dereverberation via maximum-kurtosis subband adaptive filtering. Proceedings of International Conference on Acoustics, Speech, and Signal Processing (ICASSP), Salt Lake City, UT.

[8] Nakatani, T., Miyoshi, M. and Kinoshita, K. (2004) One microphone blind dereverberation based on quasi-periodicity of speech signals, in *Advances in Neural Information Processing Systems,* vol. **16** (eds S. Thrun, L. Lawrence Saul and B. Schäolkopf), MIT Press, Cambridge, MA.

[9] Tashev, I. and Allred, D. (2005) Reverberation reduction for improved speech recognition. Proceedings of Hands-Free Communication and Microphone Arrays, Piscataway, USA.

[10] Habets, E. (2007) Single- and Multi-Microphone Speech Dereverberation using Spectral Enhancement. Ph.D. thesis, Eindhoven University, Eindhoven, The Netherlands.

[11] Benesty, J., Makino, S. and Chen, J. (2005) *Speech Enhancement*, Springer-Verlag, Berlin, Germany.

[12] Liu, Q., Champagne, B. and Kabal, P. (1995) Room speech dereverberation via minimum-phase and all-pass component processing of multi-microphone signals. Proceedings of IEEE Pacific Rim Conference on Communication, Computers, and Signal Processing, pp. 571–574.

[13] Attias, H., Platt, J., Acero, A. and Deng, L. (2001) Speech denoising and dereverberation using probabilistic models, in *Advances in Neural Information Processing Systems*, vol. **13** (eds S. Thrun, L. Lawrence Saul and B. Schäolkopf), MIT Press, Cambridge, MA.

[14] Benesty, J., Sondhi, M. and Huang, Y., (2008) *Speech Processing*, Springer-Verlag, Berlin, Germany.

[15] Gelbart, D. and Morgan, N. (2002) Double the trouble: handling noise and reverberation in far-field automatic speech recognition. Proceedings of ICSLP, Denver, CO.

[16] Gillespie, B. and Atlas, L. (2003) Strategies for improving audible quality and speech recognition accuracy of reverberant speech. International Conference on Acoustics, Speech, and Signal Processing (ICASSP) 2003.

[17] Griebel, S. and Brandstein, M. (2001) Microphone array speech dereverberation using coarse channel modeling. Proceedings of International Conference on Acoustics, Speech, and Signal Processing (ICASSP), vol. 1, pp. 201–204.

[18] Habets, E., Gaubitch, N. and Naylor, P. (2008) Temporal selective dereverberation of noisy speech using one microphone. Proceedings of International Conference on Acoustics, Speech, and Signal Processing (ICASSP), Las Vegas, NV, pp. 4577–4580.

[19] Hofbauer, H. and Loeliger, H. (2003) Limitations of FIR multi-microphone speech dereverberation in the low delay case. Proceedings of International Workshop on Acoustic Echo and Noise Control (IWAENC), Kyoto, Japan.

[20] Liu, J. and Malvar, S. (2001) Blind deconvolution of reverberated speech signals. Proceedings of International Conference on Acoustics, Speech, and Signal Processing (ICASSP), Salt Lake City, UT, vol. 5, pp. 3037–3040.

[21] Liu, Q., Champagne, B. and Kabal, P. (1996) A microphone array processing technique for speech enhancement in a reverberant space. *Speech Communication*, **18**, 317–334.

[22] Miyoshi, M. and Kaneda, Y. (1988) Inverse filtering of room acoustics. *IEEE Transactions on Acoustics, Speech, and Signal Processing*, **36**, 145–152.

[23] Nakatani, T., Juang, B., Yoshioka, T.*et al.* (2007) Importance of energy and spectral features in Gaussian source model for speech dereverberation. Proceedings of Workshop on Applications of Signal Prcoessing to Audio and Acoustics (WASPAA), New Paltz, New York, pp. 299–302.

[24] Nakatani, T., Yoshioka, T., Kinoshita, K.*et al.* (2008) Blind speech dereverberation with multi-channel linear prediction based on short time Fourier transform representation. Proceedings of International Conference on Acoustics, Speech, and Signal Processing (ICASSP), Las Vegas, NV, pp. 85–88.

[25] Subramaniam, S., Petropulu, A. and Wendt, C. (1996) Cepstrum-based deconvolution for speech dereverberation. *IEEE Transactions on Speech and Audio Processing*, **4**, 392–396.

[26] Unoki, M., Furukawa, M., Sakata, K. and Akagi, M. (2003) A method based on the MTF concept for dereverberating the power envelope from the reverberant signal. Proceedings of International Conference on Acoustics, Speech, and Signal Processing (ICASSP), vol. 1, pp. 840–843.

[27] Wu, M. and Wang, D. (2003) A one-microphone algorithm for reverberant speech enhancement. Proceedings of International Conference on Acoustics, Speech, and Signal Processing (ICASSP), vol. 1, pp. 844–847.

Index

Printed and bound by CPI Group (UK) Ltd, Croydon, CR0 4YY

16/04/2025

14658395-0003